FAST TRACK TO A  5

Preparing for the AP*
Chemistry
Examination

To Accompany
Chemistry
7th and 8th Editions
by Steven S. Zumdahl and Susan A. Zumdahl

Sheldon Knoespel
Michigan State University and Jackson Community College, Michigan

Tina Ohn-Sabatello
Maine East High School, Park Ridge, Illinois

Gordon Morlan
Wayne State University, Michigan

BROOKS/COLE
CENGAGE Learning™

Australia • Brazil • Japan • Korea • Mexico • Singapore • Spain • United Kingdom • United States

*AP and Advanced Placement Program are registered trademarks of the College Entrance Examination Board, which was not involved in the production of and does not endorse this product.

ISBN-13: 978-0-547-16861-6
ISBN-10: 0-547-16861-6

Brooks/Cole
10 Davis Drive
Belmont, CA 94002-3098
USA

Cengage Learning is a leading provider of customized learning solutions with office locations around the globe, including Singapore, the United Kingdom, Australia, Mexico, Brazil, and Japan. Locate your local office at: **international.cengage.com/region**

Cengage Learning products are represented in Canada by Nelson Education, Ltd.

For your course and learning solutions, visit **academic.cengage.com**

Purchase any of our products at your local college store or at our preferred online store **www.ichapters.com**

Printed in the United States of America
2 3 4 5 14 13 12 11

CONTENTS

ABOUT THE AUTHORS

SHELDON KNOESPEL currently teaches chemistry part time at Jackson Community College in Jackson, Michigan. He also serves as a science education instructor at Michigan State University and site coordinator for the Michigan Science Olympiad state tournament. Sheldon has also served as one of the committee members for the Laboratory Practical section of the American Chemical Society's United States National Chemistry Olympiad Exam. Previously Sheldon was employed with the MSU Department of Chemistry where he was Chemical Demonstrator for fourteen years. Prior to that he taught at Olivet High School, Olivet, Michigan, where he taught all of the physical sciences, including AP Chemistry, for more than twenty years.

TINA OHN-SABATELLO has been teaching honors and AP Chemistry for sixteen years at Maine East High School in Park Ridge, Illinois. Each year, her students achieve mostly 4's and 5's on the AP Chemistry exam. She also has taught for five summer terms at Triton Community College. Tina has B.S. and M.S. degrees from the University of Illinois at Chicago. She was one of five members of the SAT II Chemistry Test Development Committee. During her five-year term working for ETS, she wrote and revised chemistry questions for the subject matter test with one other high school teacher, three college professors, and two ETS consultants.

GORDON MORLAN taught chemistry for forty years; the last twenty years included AP Chemistry. He served as Science Department Chairman at Grosse Pointe North High School in Michigan for twenty-nine years. Gordon was also a College Board consultant and was a reader for the AP Chemistry Examination twelve times. He is an active competitive sailor and also enjoys reading, travel, and jazz.

PREFACE

Sheldon Knoespel revised the *Fast Track to a 5* for the changes in 8th edition of *Chemistry* and wrote new questions for this guide. He benefited from the reviews by Jeannie Meriwether of the Lovett School in Atlanta, Georgia. Jeannie is also a College Board Consultant for the Southern Region and leads summer workshops based on *Chemistry*.

January 2009

To learn chemistry well, you need to have a good text and a teacher to help you. No doubt, you already have a good text. I can't be there to help you review for the AP exam, but I can bring you the teaching techniques and ideas that have arisen from my classroom discussions. The chapter summaries, questions, and practice tests in this book will help lead you to success in the AP Chemistry exam.

I am very grateful to Cengage Learning for giving me the opportunity to reach students beyond my classroom. I couldn't have completed this task without the help and advice of reviewers Annis Hapkiewicz and Sheldon Knoespel. Most of all, I would like to thank my coauthor, Gordon Morlan, who worked diligently on the questions for this guide. I am very thankful for the guidance of Margaret Lannamann, my Cengage Learning liaison.

Lastly, I dedicate this book to my patient and loving family, Rick, Rina, and Antonio, who allowed me the time to write this book, and to my students for inspiring me to be creative in my teaching and to look forward to each year of teaching chemistry as if it were the first.

Tina Ohn-Sabatello
September 2005

Writing a book of this sort is never the result of only one or two people. The frank (sometimes brutally frank!) and carefully considered reviews of Annis Hapkiewicz are especially appreciated. So also is the continual upbeat attitude and encouragement of editor Margaret Lannamann. Credit for the entire final sample test belongs to Sheldon Knoespel.

My wife, Judy, has brought her understanding of how students learn to this project, as well as an understanding of the magnitude of this undertaking. Her encouragement and support were essential.

Gordon Morlan
September 2005

Part I

Strategies for the AP Exam

PREPARING FOR THE AP* CHEMISTRY EXAMINATION

Advanced Placement can be exhilarating. Whether you are taking an AP course at your school or you are working on AP independently, the stage is set for a great intellectual experience.

But sometime after New Year's Day, when the examination begins to loom on a very real horizon, Advanced Placement can seem downright intimidating—in fact, offered the opportunity to take the examination for a lark, even adults long out of high school refuse. If you dread taking the test, you are in good company.

The best way to deal with an AP examination is to master it, not let it master you. If you can think of these examinations as a way to show off how much chemistry you know, you have a leg up. Attitude *does* help. If you are not one of those students, there is still a lot you can do to sideline your anxiety. This book is designed to put you on a fast track. Focused review and practice time will help you master the examination so that you can walk in with confidence and score a 5.

WHAT'S IN THIS BOOK

This book is keyed to *Chemistry* by Steven and Susan Zumdahl, 7th and 8th editions, but because it follows the College Board Topic Outline, it is compatible with all textbooks. It is divided into three sections. Part I offers suggestions for getting yourself ready, from signing up to take the test and sharpening your pencils to organizing a free-response essay. At the end of Part I, you will find a Diagnostic Test. This test has all of the elements of the AP Chemistry Examination, but the 75 multiple choice questions are organized according to the College Board Topic Outline. When you go through the answers at the end of the Diagnostic Test, you will see how the examination is weighted for each content area.

Part II is made up of 16 chapters—again following the College Board Topic Outline. These chapters are not a substitute for your textbook and class discussion; they simply review the AP chemistry course. At the end of each chapter, you will find 15 multiple-choice questions and two free-response questions based on the material in that chapter. Again, you will find page references at the end of each answer directing you to the discussion on that particular point in *Chemistry*.

Part III has two complete AP Chemistry examinations. At the end of each test, you will find the answers, explanations, and references to *Chemistry* for the multiple choice and the free-response questions.

---

* AP and Advanced Placement Program are registered trademarks of the College Board, which was not involved in the production of and does not endorse this product.

WHAT'S IN THE *CHEMISTRY* TEXTBOOK THAT WILL HELP YOUR PREPARATION

As you work your way through the textbook there are some features that will assist you in getting the most out of the textbook:

■ Make use of the Conceptual Problem Solving Method introduced in Chapter 3 in which you break the problem down into three parts:(1)Where are we going? (2)How do we get there? and (3)Reality Check. This method is introduced in the 8th edition.

■ Make certain you carefully study the Sample Exercises with solutions in each chapter. These will often give you a guide to help develop your problem solving skills.

■ Read the For Review section that highlights the material presented in the chapter as a means to double check your understanding before you proceed.

■ Answer the Review Questions at the end of each chapter as a measure of your understanding. This will give you instant feedback as to how much of the chapter you may need to study again.

■ Working in small groups, answer the Active Learning Questions at the end of each chapter. This is an excellent opportunity to gauge your level of understanding and to receive assistance from your classmates.

■ Do as many Challenge and Marathon Problems as you can. These problems are intended to incorporate many different concepts into the solution. Problems like this are particularly good practice for the free-response portion of the AP exam.

Being successful in chemistry usually involves careful reading of the textbook and working as many different types of problems as possible. The successful student is not one who just memorizes a bunch of facts but one who is able to begin to tie the various facts and concepts together in the understanding of a new problem. Chemistry is like a lot of subjects in that an understanding of new material often depends on an understanding of previous concepts.

SETTING UP A REVIEW SCHEDULE

If you have been doing your homework steadily and keeping up with the course work, you are in good shape. Organize your notes, homework, and handouts from class by topic. Reference these materials as well as your textbook and this study guide when you have difficulty in a specific section. But even if you have done all that—or if it is too late to do all that—there are some more ways to get it all together.

To begin, read Part I of this book. You will be much more comfortable going into the exam if you understand how the exam questions are designed and how best to approach them. Then take the Diagnostic Test and see where you are right now.

Take out a calendar and set up a schedule for yourself. If you begin studying early, you can chip away at the review chapters in Part II. You will be surprised—and pleased—by how much material you can cover in half an hour a day of study for a month or so before the test.

Look carefully at the sections of the Diagnostic Test; if you missed a number of questions in one particular area, allow more time for the chapters that cover that area of the course. The practice tests in Part III will give you more experience with different kinds of multiple-choice questions and the wide range of free-response questions.

If time is short, skip reading the review chapters (although you might read through the chapter subheadings) and work on the multiple-choice and free-response questions at the end of each review. This will give you a good idea of your understanding of that particular topic. Then take the tests in Part III.

If time is *really* short, go straight from Part I to Part III. Taking practice tests over and over again is the fastest, most practical way to prepare. You cannot study chemistry by reading it like a novel. You must actively do problems to gain understanding and excel in your performance. Athletes do not perform well just by reading books about their sport or by watching others. They must get up and practice. So, you too, just like athletes, must practice, practice, practice if you want to do your best!

BEFORE THE EXAMINATION

By February, long before the exam, you need to make sure that you are registered to take the test. Many schools take care of the paperwork and handle the fees for their AP students, but check with your teacher or the AP coordinator to make sure that you are on the list. This is especially important if you have a documented disability and need test accommodations. If you are studying AP independently, call AP Services at the College Board for the name of the local AP coordinator, who will help you through the registration process.

The evening before the exam is not a great time for partying. Nor is it a great time for cramming. If you like, look over class notes or drift through your textbook, concentrating on the broad outlines, not the small details, of the course. You might also want to skim through this book and read the AP tips.

The evening before the exam *is* a great time to get your things together for the next day. Sharpen a fistful of no. 2 pencils with good erasers; bring a scientific calculator with fresh batteries. You may bring a programmable calculator; the memory will not be erased or cleared by the test administrator. It must not have a typewriter-style keyboard. Wind your watch and turn off the alarm if it has one; get a piece of fruit or a power bar and a bottle of water for the break. Make sure you have your Social Security number and whatever photo identification and admission ticket are required. Then relax. And get a good night's sleep.

On the day of the examination, it is wise not to skip breakfast—studies show that students who eat a hot breakfast before testing get higher grades. Be careful not to drink a lot of liquids, necessitating a trip to the bathroom during the exam. Breakfast will give you the energy you need to power you through the exam—and more. You will spend some time waiting while everyone is seated in the right room for the right exam before the test has even begun. With a short break between Section I and Section II, the AP Chemistry exam lasts for more than two and a half hours. So be prepared for a long morning.

You do not want to be distracted by a growling stomach or hunger pangs.

Be sure to wear comfortable clothes, taking along a sweater in case the heating or air-conditioning is erratic. Be sure, too, to wear clothes you like—everyone performs better when they think they look better—and by all means wear your lucky socks.

You have been on the fast track. Now go get a 5!

TAKING THE AP CHEMISTRY EXAMINATION

The AP Chemistry examination consists of two sections: Section I has 75 multiple-choice questions; Section II has six free-response questions. You will have 90 minutes for the multiple-choice portion. You will not be allowed to use a calculator for the multiple-choice questions. The questions are collected, and you will be given a short break. You then have 95 minutes for the free-response section which is broken into two portions.

You will be allowed to use a calculator in Part A, with a time limit of 55 minutes. All three questions in Part A are mandatory, with the first question involving an equilibrium problem and the next two questions based on a variety of topics, including a laboratory-based question. Each of these questions will be weighted equally at 20% of the overall Section II score. Time will be called at 55 minutes, at which time calculators must be put away for the remaining 40 minutes.

In Part B, you will be asked to answer three questions. Question 4, which is worth 10% of Section II, will ask you to write the balanced chemical equation for three separate chemical reactions and then answer one related question about each reaction. You will then answer two mandatory essays, one of which may be lab-based, if a lab-based question was not asked in Part A. Each of the two questions in Part B are weighted equally, counting 15% of the Section II score.

The table on the next page summarizes each part of the AP Chemistry examination.

STRATEGIES FOR THE MULTIPLE-CHOICE SECTION

Here are some rules of thumb to help you work your way through the multiple-choice questions:

■ **Guessing penalty** There are five possible answers for each question. Each correct answer is worth 1 point, and there is a 1/4-point guessing penalty for each incorrect answer. If you cannot narrow down the answers at all, it is against the odds to guess, so leave the answer sheet blank. However, if you can narrow down the answers even by eliminating one response, it is advantageous to guess. If you skip a question, be very careful to skip down that line on the answer sheet.

AP Chemistry Exam – Distribution of Questions		
Weighting	Possible Topics	Time Allowed
Section I 50%	Multiple choice	90 minutes
Section II 50%		
Part A (all mandatory)		55 minutes
Question 1	Equilibrium	
Questions 2 and 3	Thermochemistry, Thermodynamics, Kinetics, Stoichiometry, Redox, or Laboratory-Based	
Part B (all mandatory)		40 minutes
Question 4	Reactions (Three mandatory balanced equations with related second question for each one)	
Question 5	Essay—gases, states of matter, kinetics, redox, thermochemistry, electrochemistry, periodicity, or lab-based	
Question 6	Essay/Minor Calculations—topics same as those listed for Question 5	

- **Read the question carefully** Pressured for time, many students make the mistake of reading the questions too quickly or merely skimming them. By reading a question carefully, you may already have some idea about the correct answer. You can then look for it in the responses. Careful reading is especially important in EXCEPT questions.
- **Eliminate any answer you know is wrong** You can write on the multiple-choice questions in the test book. As you read through the responses, draw a line through any answer you know is wrong.
- **Read all of the possible answers, then choose the most accurate response** AP examinations are written to test your precise knowledge of a subject. Some of the responses may be partially correct but there will only be one response that is completely true.
- **Avoid absolute responses** These answers often include the words "always" or "never." For example, the statement "all chlorides are always soluble in water" is incorrect because compounds such as silver chloride and lead (II) chloride are insoluble in water.
- **Mark and skip tough questions** If you are hung up on a question, mark it in the margin of the question book. You can come back to it later if you have time. Make sure you skip that question on your answer sheet too.

TYPES OF MULTIPLE-CHOICE QUESTIONS

There are various kinds of multiple-choice questions. Here are some suggestions for approaching each kind:

CLASSIC/BEST ANSWER QUESTIONS This is the most common type of multiple-choice question. It simply requires you to read the question and select the most correct answer. For example:

1. Which of the following molecules is tetrahedral in shape?
    (A) NH_3
    (B) H_2O
    (C) CH_4
    (D) SF_4
    (E) XeF_4

ANSWER: C. Eliminate (A) and (B) because these do not have 4 atoms attached as in a tetrahedrally shaped molecule. On your booklet or scratch paper, draw the Lewis structures for the remaining molecules. A molecule with tetrahedral shape has 4 atoms attached and no lone pairs. (D) and (E) have lone pairs.

CLASSIFICATION SET In this type of question, there is a list of five lettered headings consisting of possible answers followed by a list of numbered phrases. You must select the heading that is most closely related to each numbered phrase. Headings may be used once, more than once, or not at all.

Questions 1–3 refer to 0.10 M solutions.
    (A) NH_3, ammonia
    (B) NaOH, sodium hydroxide
    (C) CH_3COOH, acetic acid
    (D) $NaNO_3$, sodium nitrate
    (E) CH_3COOK, potassium acetate

1. Has a pH equal to 7.0.

2. Has a pH less than 7.0.

3. Can be mixed with 0.10 M sodium acetate to make a buffer.

For the classification set, answer each question one at a time. Some of the answers can be used more than once or not at all. Begin by knowing what is meant by each phrase in 1 through 3. For example, what is a buffer made of?

1. Look for a neutral solution since the pH equals 7. Salts that have parent acids that are strong will be neutral since their conjugate base will be too weak to undergo hydrolysis. NO_3^- is the conjugate base of the strong acid HNO_3. Answer is (D).

2. Look for an acid or the conjugate acid of a weak base since the pH is less than 7. Answer is (C).

3. A buffer is made of a weak acid and its conjugate base (or a weak base and its conjugate acid). Sodium acetate contains acetate, which is the conjugate base of the weak acid acetic acid. Answer (C).

LIST AND GROUP QUESTIONS In this type of question, there is a list of possible answers, and you must select the answer that contains the correct group of responses.

1. The concentration of a solution can be determined by which of the following methods?
   I.   Titration with a solution of known concentration
   II. Calorimetry
   III. Spectrometric/colorimetric

   (A) I only
   (B) II only
   (C) III only
   (D) I and II only
   (E) I and III only
   (F) I, II, and III

ANSWER E. To approach the question, draw a line through choice II, because calorimetry is used to determine the change in enthalpy of a reaction. Continue to cross out items that are wrong and the responses that contain them. Draw a line through III and answers (D) and (F), which both contain choice II. Now you have narrowed down the possible responses.

NONCALCULATOR COMPUTATIONS These questions require computations without the use of a calculator. Simple mathematics or the choice of the correct algebraic setup will be involved.

> EXAMPLE: A weak acid, HA, has a K_a value of 1.0×10^{-6}. Calculate the pH after the complete reaction of 40.0 mL of 0.0100 M HA with 10.0 mL of 0.0200 M NaOH.

> (A) 2.0
> (B) 6.0
> (C) 7.0
> (D) 8.0
> (E) 14.0

To answer this question, you must consider at what point in the titration the question is referring to. Because the base, NaOH, is twice as concentrated as the acid, HA, it will take half as much NaOH to reach the equivalence point, or 20.0 mL. We are not at the equivalence point. Half the amount of the NaOH, 10.0 mL, needed to reach the equivalence point has been added. Halfway to the equivalence point, the pH equals the pK_a which equals –log (K_a) or 6.0. Answer is (B).

> EXAMPLE: $2C(s) + O_2(g) \rightarrow 2CO(g)$

> Into a 3.0-L container at 25°C are placed 1.2 g of carbon, graphite, and 3.2 g oxygen gas.

If the carbon and the oxygen react completely to form $CO(g)$, what will be the final pressure in the container at $25°C$?

(A) $\dfrac{0.05(0.082)(298)}{3.0}$ atm

(B) $\dfrac{0.10(0.082)(25)}{3.0}$ atm

(C) $\dfrac{0.15(0.082)(298)}{3.0}$ atm

(D) $\dfrac{0.20(0.082)(298)}{3.0}$ atm

(E) $\dfrac{0.10(0.082)(298)}{3.0}$ atm

To answer this question, 1.2 g of carbon is 0.10 mol of carbon (1 mole of carbon equals 12 g). Similarly, 3.2 g of O_2 is 0.10 mol of O_2. The limiting reactant is carbon producing 0.10 mol of CO. O_2 is present in excess. (0.10 mol of CO requires 0.05 mol of O_2 leaving 0.05 mol of O_2.) When the reaction is complete, there is 0.10 mol CO + 0.05 mol O_2 left which equals 0.15 mol total. Answer is (C).

STRATEGIES FOR THE FREE-RESPONSE SECTION

Section II of the AP exam comes with a periodic table, a table of standard reduction potentials, $E°_{red}$, and a table of equations and constants.

- Scan all of the questions in the section you are working in and mark those that you know you can answer correctly. Do these problems first.
- You will be provided with an answer booklet as well as an insert that contains the same questions as the answer booklet, but without the spaces. You can remove the insert for reference. All of your work and answers for each problem must be shown in the answer booklet. No credit will be given for work shown on the insert, but you may write on it.
- Show all of your work. Partial credit will be awarded for problems if the correct work is shown but the answer is not present or is incorrect. In problems involving calculations, circle your final answer.
- Cross out incorrect answers with an "X" rather than spending time erasing.
- Be clear, neat, and organized in your work. If a grader cannot clearly understand your work, you may not receive full credit.
- Some free-response questions have four parts: a, b, c, and d. Attempt to solve each part. Even if your answer to "a" is incorrect, you still may be awarded points for the remaining parts of the question if the work is correct for those parts.
- Units are important in your answer. Keeping track of your units throughout calculations and performing unit cancellation where possible, will help guide you to your answer. Points will be deducted for missing or incorrect units in the answer.
- You do not need to work the questions in order. Make sure you put the number of the question in the corner of each page of your essay

booklet. In addition, questions are sometimes broken into parts, such as (a) and (b). When this is the case, label each part of your response.

STRATEGIES FOR ANSWERING ESSAY QUESTIONS

Essay questions will ask you to explain, compare, and predict. Minor calculations, showing mathematical relationships, or drawing graphs or structures, may also be involved. Usually these are not traditional essay questions. Most free-response questions do not require an introduction or conclusion. Most do not even require a thesis. Many of these questions may be written in a bulleted or short-answer format. Although this may sound easier than writing a traditional essay, it is important that you know the material very well because these are targeted questions. Examination readers want specifics. They are looking for accurate information presented in clear, concise prose. You cannot mask vague information with elegant prose.

To be successful in writing essays for the AP Chemistry exam, be sure to get straight to the answer and use key terms in your explanations. Sometimes, if you ramble on and on, you might accidentally state an incorrect fact. Points will be deducted for incorrect or extraneous information.

Techniques to write essays include chart format, bullet format, and outline format. None of these styles requires that you write complete sentences. In each of these styles, restate the question in simple terms, using your own words. Restate each part (a, b, and so on) separately, not together. In your restatement and response, underline key words or concepts.

- **Fill in a chart to answer the question**. This style is helpful in answering questions about electronic and molecular structure. This method is also used in the example on identification of the set of four quantum numbers for each electron in an element in this chapter.
- **In bullet format, make a list**, using a bullet (■) for each new concept. Leave room between concepts because you may want to come back later to fill them in.
- **Outline format** is more traditional, using Roman numerals, letters, etc. This takes more time to organize ideas, but it does show progression of ideas in a logical sequence. As in the bulleted format, leave room between concepts because you may want to come back later to fill them in.
- Lastly, **the freestyle method** sometimes results in rambling and incomplete answers. Writing in paragraphs does not allow room for additional ideas to be added.

(F) $pH = pK_a + \log ([C_7H_5O_2^-]/[HC_7H_5O_2])$

$5.00 = 4.19 + \log ([\text{mol } C_7H_5O_2^-]/[0.010 \text{ mol}])$

$0.81 = \log([\text{mol } C_7H_5O_2^-]/[0.010 \text{ mol}])$

antilog $0.81 = (\text{mol } C_7H_5O_2^-)/0.010 \text{ mol}$

$0.065 = \text{mol } C_7H_5O_2^-$

Mass of $NaC_7H_5O_2 = 0.065 \times 144 \text{ g/mol} = 9.4 \text{ g}$

1 point for correct values for pK_a and mol of acid
1 point for correct answer

SAMPLE PROBLEM #2 OR 3

You will work two problems here, each worth 20% of Part II.

$$C_2H_4(g) + 3 O_2(g) \rightarrow 2 CO_2(g) + 2H_2O(g)$$

Information about the substances involved in the reaction presented above is tabulated below.

Substance	S^0(J/mol•K)	ΔG^o_f(kJ/mol)
$C_2H_4(g)$	?	68
$O_2(g)$	205	0
$CO_2(g)$	213.6	−394
$H_2O(g)$	189	−229

Bond	Bond Energy (kJ/mol)
O – H	467
O – O	146
O = O	495
C – H	413
C – C	347
C = C	614
C ≡ C	839
C – O	358
C = O	799

(a) Calculate the value for the standard free energy change, ΔG^o, at 25°C for the reaction. What does the sign of ΔG^o indicate about the reaction?

(b) Calculate the value for the standard enthalpy change, ΔH^o, at 25°C for the reaction. What does the sign of ΔH^o indicate about the reaction?

(c) Calculate the value for the standard entropy change, ΔS^o, at 25°C for the reaction. What does the sign of ΔS^o indicate about the reaction?

(d) Calculate the value for the absolute entropy of $C_2H_4(g)$ at 25°C.

SCORING GUIDELINES—SAMPLE PROBLEM #2 OR 3

(A) $\Delta G°$ = 2(–394) + 2(–229) – 68 = –1314 kJ/mol. | 1 point for correct value of $\Delta G°$

(B)

| 1 point for meaning of sign

| 1 point for correct Lewis Structures of Reactants and products

Bonds Broken	Bonds Formed
4 C–H 4(413)	4 C=O 4 (799)
1 C=C 1 (614)	4 O-H 4 (467)
3 O=O 3(495)	–5064 kJ
+3751 kJ	

$\Delta H°$ = 3751 kJ + –5064 kJ = –1313 kJ

The reaction is exothermic since $\Delta H°$ is negative.

| 1 point for correct value of $\Delta H°$

(C) $\Delta G° = \Delta H° - T\Delta S°$; $\Delta S° = (\Delta H° - \Delta G°) / T$

$\Delta S°$ = –1313 – (–1314 kJ) / 298 K = 0.0034 kJ/K

A positive sign of change in entropy indicates that entropy is increasing in the reaction.

Here, it barely changes.

| 1 point for meaning of sign

| 1 point for correct value of $\Delta S°$

| 1 point for meaning of sign

(D) $\Delta S° = 2S°_{CO_2} + 2\,S°_{H_2O} - (S°_{C_2H_4} + 3S°_{O_2})$

$S°_{C_2H_4} = 2\,S°_{CO_2} + 2\,S°_{H_2O} - 3\,S°_{O_2} - \Delta S°$

= 2 (213.6) + 2 (189) – 3(205) – 3.4 = 187 J/K

| 1 point for value of absolute $S°$

SAMPLE PROBLEM #4 REACTIONS—WEIGHTED 10%

You will be given 3 chemical reactions to balance and then you must answer a second question about each reaction. Solutions are aqueous unless otherwise indicated. In all cases, a reaction occurs. If a substance is extensively ionized, such as a strong acid or a completely soluble compound, write it as separate ions. Be sure to cancel out spectator ions or molecules that are unchanged by the reaction.

Each reaction is worth 4 points: 1 point for the correct reactants , 2 points for correct product(s), and 1 point for a correctly balanced equation. For the follow-up question 1 point will be awarded.

1. (i) Equal volumes of equimolar solutions of sodium hydroxide and acetic acid are mixed.
    (ii) Identify a conjugate acid-base pair in the reaction.

(i) $CH_3COOH + OH^- \rightarrow HOH + CH_3COO^-$

1 point for correct reactants
2 points for correct products
Only 1 product point is awarded if water is written as $H^+ + OH^-$ or if CH_3COONa is written as one of the products

(ii) CH_3COOH (acid) and CH_3COO^- (base)   1 point for correct identity of either pair
   OR OH^- (base) and HOH (acid)

SAMPLE PROBLEM #5—WEIGHTED 15%

This mandatory essay could be a lab-based question. There are 22 recommended experiments suggested by the College Board. The question usually asks you to describe a laboratory procedure for one of these experiments or analyze sources of errors. Your response is graded for its accuracy and relevance. You should be clear and well organized in answering.

A student is instructed to identify an unknown solid by conducting four qualitative laboratory tests. The solid substance could either be potassium carbonate, K_2CO_3, or benzoic acid, C_6H_5COOH. Describe each of the four possible lab tests that can be performed. For each test, identify the observations for both potassium carbonate, K_2CO_3, and benzoic acid, C_6H_5COOH.

1) Perform a flame test. The potassium ion in K_2CO_3 will burn a characteristic lavender color. The flame resulting from burning C_6H_5COOH will not be lavender.	1 point for identification of test 1 point for listing observations for both possible substances
2) Dissolve solid in water and perform a conductivity test. Potassium carbonate is a strong electrolyte and the bulb on the tester will be brightly lit. If the unknown is C_6H_5COOH, the bulb will barely be lit since benzoic acid is a weak acid, so it is also a weak conductor of electricity.	1 point for identification of test 1 point for listing observations for both possible substances
3) Dissolve solid in water and measure the pH. A basic pH (>7) reveals that the unknown is K_2CO_3. An acidic pH identifies the unknown as C_6H_5COOH.	1 point for identification of test 1 point for listing observations for both possible substances
4) Add an acid like HCl, which contains a source of hydrogen ion. If the unknown is K_2CO_3, the mixture will bubble (release gas). If the unknown is C_6H_5COOH, there will be no bubbles.	1 point for identification of test 1 point for listing observations for both possible substances

SAMPLE PROBLEM #6—WEIGHTED 15%

This problem usually combines chemical calculations with essay writing. Questions of the type mentioned in Sample Problem #7 have also been found here.

$$A(g) + 2 B(g) \rightarrow C(g) + D(g)$$

Consider the equation above,
   (a) Using the table below, calculate the value of ΔH° for the reaction at 25°C. Show all of your work.

ΔH°_f (kJ/mol)

A	–1275
B	–166

C   –393

D   –286

(b) Sketch the potential energy diagram for the reaction above. Label the activation energy for the forward reaction.

(c) Explain how the magnitude of the standard entropy of the reaction, $\Delta S°$, will change for the reaction at 25°C.

(d) Predict the sign of the standard free energy change, $\Delta G°$, using your answers to parts (a) and (c).

(e) Write the rate law for the reaction using the information in the table below. Explain how you obtained your result.

Trial #	Initial [A] (mol/L)	Initial [B] (mol/L)	Initial Rate of Formation of C (mol L^{-1}s^{-1})
1	0.10	0.10	0.20
2	0.20	0.10	0.40
3	0.40	0.20	6.4

(a) $\Delta H° = [-393 + -286] - [-1275 + 2(-166)] = -679 - (-1607) = 928$ kJ

1 point for correct value of $\Delta H°$

(b)

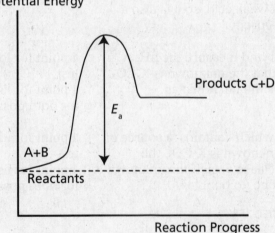

1 point for showing that energy of products > energy of reactants

1 point for labeling the activation energy

(c)   The standard entropy change will decrease because the moles of gas decrease from reactants to products (3 mol → 2 mol).

1 point must say why entropy decreases

(d) $\Delta G° = \Delta H° - T\Delta S°$

= positive – (negative) = positive $\Delta G°$

therefore the reaction is nonspontaneous at all temperatures.

1 point for determination of order in A: must show work

(e) $\dfrac{\text{Rate 2}}{\text{Rate 1}} = \dfrac{k[A]^x[B]^y}{k[A]^x[B]^y}$

$= \dfrac{k[0.20]^x \cancel{[0.10]^y}}{k[0.10]^x \cancel{[0.10]^y}} = \dfrac{0.40}{0.20}$

$= 2^x = 2;\ x = 1$

$\dfrac{\text{Rate 3}}{\text{Rate 2}} = \dfrac{k[A]^x[B]^y}{k[A]^x[B]^y}$

$\dfrac{k[0.40]^x[0.20]^y}{k[0.20]^x[0.10]^y} = \dfrac{6.4}{0.40}$

$= 2^x \times 2^y = 16$

$= 2^1 \times 2^y = 16;\ 2^y = 8;\ y = 3$

1 point for determination of order in B: must show work

ANOTHER POSSIBLE SAMPLE PROBLEM #6

This essay usually asks you to explain chemical phenomena.

Use appropriate chemical principles to account for the following observation. Your answer must include references to both substances.

(a) At 25°C and 1.0 atm, ethane (C_2H_6) is a gas and ethanol (CH_3CH_2OH) is a liquid.

There are usually three more questions, parts (b), (c), and (d), which may or may not be related to (a) or each other, but all are scored in a similar way to (a) below.

(a) Ethane, C_2H_6, has London dispersion forces. Ethanol, CH_3CH_2OH, has hydrogen bonding. Hydrogen bonding is stronger than London dispersion forces.

1 point for indicating the correct intermolecular force for each molecule

1 point for indicating that hydrogen bonding is stronger than dispersion forces

A Diagnostic Test

The purpose of this test is to give you an indication of how well you will perform on the AP Chemistry exam. These questions are representative of the AP Chemistry examination, but bear in mind it is impossible to predict exactly how well you will do on the actual exam. Calculators may not be used for answering questions in the first section of this test. The first section is 50% of your total test grade. Time yourself to finish this part in 90 minutes. There are two types of multiple-choice questions used in this examination. The first type consists of five lettered headings followed by a listing of number phrases. For each phrase you are to select the one heading that is most closely related to it. Headings may be used once, more than once, or not at all. The majority of multiple-choice questions consist of a question or incomplete statements followed by five possible answers. Select the one that is best in each case.

AP CHEMISTRY EXAMINATION
Section I: Multiple-Choice Questions
Time: 90 minutes
Number of Questions: 75

No calculators are to be used in this section; no tables (except Periodic Table) permitted.

Questions 1–3 refer to the following elements:
  (A) Chromium
  (B) Sodium
  (C) Copper
  (D) Phosphorus
  (E) Beryllium

1. Forms colored ions having the formulas $X_2O_7^{2-}$ and XO_4^{2-}

2. Forms an oxide that yields aqueous solutions that are acidic

3. Forms hydrogen gas when it reacts with cold water

Questions 4–7 refer to the following substances:
  (A) Hydrofluoric acid, HF
  (B) Hydrobromic acid, HBr
  (C) Sodium hydroxide, NaOH
  (D) Nitrous acid, HNO_2
  (E) Carbon dioxide, CO_2

4. Will dissolve or etch glass

5. Is a strong acid

6. Will neutralize acids

7. Contributes to the greenhouse effect

20

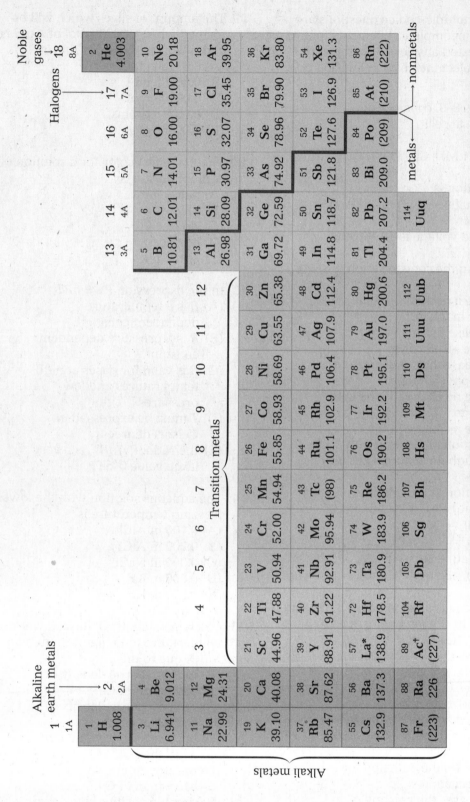

Group numbers 1–18 represent the system recommended by the International Union of Pure and Applied Chemistry.

Most of the multiple-choice questions are questions or incomplete statements followed by five suggested answers or completions. You should select the best one in each case.

For questions 8–9, consider the electrochemical cell:

$$Ag \mid Ag^+(1\ M) \parallel Cu^{2+}(1M) \mid Cu$$

The cell reaction is
$$2Ag^+(aq) + Cu\ (s) \rightarrow Cu^{2+}(aq) + 2Ag(s).$$

The measured voltage is +0.46 volts.

8. Increasing the concentration of silver ions will
    (A) cause a decrease in blue color in the cell.
    (B) increase the cell voltage above +0.46 volts.
    (C) decrease the concentration of copper (II) ions.
    (D) cause no change in the cell voltage.
    (E) cause a change in the direction of electron flow through the external circuit.

9. The reaction at the anode is
    (A) $Ag^+(aq) + e^- \rightarrow Ag(s)$.
    (B) $Ag\ (s) \rightarrow Ag^+(aq)$.
    (C) $Cu(s) \rightarrow Cu^{2+}(aq) + 2e^-$.
    (D) $Cu^{2+}(aq) + 2e^- \rightarrow Cu(s)$.
    (E) $Cu(s) + 2e^- \rightarrow Cu^{2+}(aq)$.

10. A spark coil is used to begin the reaction between hydrogen and oxygen in a balloon. The spark
    (A) allows the gases to obtain the activation energy.
    (B) causes the hydrogen to vaporize, then react.
    (C) causes both the hydrogen and the oxygen to vaporize, then react.
    (D) provides the enthalpy for this reaction.
    (E) increases the entropy of both oxygen and hydrogen.

11. The amount of silver which will be formed when 0.00200 mol of Ag_2S reacts completely with excess zinc is
    (A) 0.00100 mol.
    (B) 0.00200 mol.
    (C) 0.00400 mol.
    (D) 0.00248 grams.
    (E) 0.00456 grams.

12. The flame test color for a solution of sodium nitrate is
    (A) pale yellow.
    (B) blue.
    (C) violet.
    (D) purple.
    (E) crimson.

13. In the expression $PV = nRT$,
    (A) R is a temperature dependent constant.
    (B) R is a pressure dependent constant.
    (C) R is valid for gases at high temperatures and low pressures.
    (D) T must be expressed in Celsius degrees.
    (E) the ratio, PV/nRT, is a very large value @ STP.

14. The aqueous solution with the lowest freezing temperature is
    (A) 0.100 m $AlCl_3$.
    (B) 0.200 m $AlCl_3$.
    (C) 0.300 m NaCl.
    (D) 0.300 m KF.
    (E) 0.350 m KF.

15. Calcium oxide, CaO, has a lower melting temperature than magnesium oxide, MgO, due to the
    (A) higher charge density of Mg^{+2} than of Ca.
    (B) higher charge density of Ca^{+2} than of O^{-2}.
    (C) greater atomic volume of Mg^{+2} than of O^{-2}.
    (D) greater atomic volume of Mg^{+2} than of Ca^{+2}.
    (E) greater positive charge on Ca than on Mg.

16. The general classification for the compound C_3H_7OH is
    (A) an alcohol.
    (B) an aldehyde.
    (C) a carboxylic acid.
    (D) a ketone.
    (E) an ester.

17. A solution of 50.0 mL of 0.0010 M $Ba(NO_3)_2$ is slowly titrated with 50.0 mL of 0.0030 M H_2SO_4. The conductivity of this solution will
    (A) decrease to near zero, then increase.
    (B) decrease to near zero and remain very low.
    (C) increase as the acid is added.
    (D) increase as the acid is added then become constant at a high value.
    (E) increase as the acid is added and then slowly become very low.

18. Which of the following pairs illustrates the Law of Multiple Proportions?
    (A) SO_2, SO_3
    (B) CO_2, CCl_4
    (C) NaCl, NaBr
    (D) NH_4Cl, NH_4Br
    (E) SO_2, CO_2

19. The scientist who is remembered for the alpha particle scattering properties of gold foil, concluding that the nucleus is small, dense, and positively charged, is
    (A) Ernest Rutherford.
    (B) J. J. Thomson.
    (C) John Dalton.
    (D) Robert Boyle.
    (E) Antoine Lavoisier.

20. Isotopic forms of the same element
    (A) differ in the number of neutrons in the nucleus.
    (B) are formed by gaining electrons.
    (C) always have a positive charge.
    (D) are found only in metals.
    (E) have the same number of neutrons.

21. The number of moles of oxygen atoms in one mole of iron (II) phosphate is
    (A) 1.
    (B) 2.
    (C) 3.
    (D) 4.
    (E) 8.

22. Magnesium fluoride, a salt of low solubility in water, has a K_{sp} of 6.4 x 10^{-9}. The concentration of Mg^{2+} ions in this solution would be
    (A) $\sqrt{6.4 \times 10^{-9}}$ M .
    (B) $\sqrt{(6.4 \times 10^{-9}/2)}$ M .
    (C) $\sqrt[3]{(6.4 \times 10^{-9}/3)}$ M .
    (D) $\sqrt[3]{(6.4 \times 10^{-9}/4)}$ M .
    (E) $\sqrt[3]{6.4 \times 10^{-9}}$ M .

Questions 23 and 24 refer to this pH curve for the titration of 50.0 mL of 0.100 M acid with 0.100 M base.

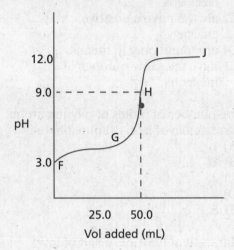

Vol added (mL)

23. This is the curve
    (A) for a strong acid/strong base titration.
    (B) for a strong acid/weak base titration.
    (C) for a weak acid/strong base titration.
    (D) for either a strong acid/strong base or a weak acid/strong base titration.
    (E) for an acid/base titration but it is impossible to describe the strength of the acid and of the base used without more information.

24. Buffering is most effective
    (A) between points F and G.
    (B) between points G and H.
    (C) at point H between points I and J.
    (D) between both F and G and between I and J.
    (E) none of these

25. The systematic (Stock Nomenclature) name for $Ca(OCl)_2$ is
    (A) calcium (II) hypochlorite.
    (B) calcium hypochlorite.
    (C) calcium dioxochloride.
    (D) calcium monoxodichloride.
    (E) calcium dichloride.

26. Bond angle data shows the following angles for three compounds:

    CH_4 109.5°,
    NH_3 107°,
    H_2O 104.5°.

    This trend is basically because
    (A) lone pairs of electrons require more room than bonding pairs.
    (B) hydrogen atoms repel each other more in water than in CH_4 or NH_3.
    (C) oxygen has a higher electronegativity than does N, and C has even less.
    (D) of the attempt of all central atoms to achieve the tetrahedral shape.
    (E) only the positions of nuclei determine molecular shapes and bond angles.

27. The acid $HClO_4$ is given the name
    (A) perchloric acid.
    (B) chloric acid.
    (C) chlorous acid.
    (D) hypochlorous acid.
    (E) hypochlorotetroxide.

28. When 100.0 mL of 2.0 M NH_3 and 100.0 mL of 1.0 M $AgNO_3$ are mixed, but before any reaction occurs, the major species in solution are
    (A) Ag^+, NO_3^-, NH_3, and H_2O.
    (B) Ag^+, NO_3^-, and $Ag(NH_3)^+$.
    (C) $Ag(NH_3)^+$ and $Ag(NH_3)_2^+$.
    (D) NH_3 and $Ag(NH_3)_2^+$.
    (E) NH_3, $Ag(NH_3)_2^+$ and H_2O.

29. H_3PO_4 is a triprotic acid with equilibrium constants of

$K_{a_1} = 7 \times 10^{-3}$,
$K_{a_2} = 6 \times 10^{-8}$,
$K_{a_3} = 5 \times 10^{-13}$.

This information leads to the conclusion that
I H_3PO_4 is a weak acid.
II H_3PO_4 dissociates in three steps to form PO_4^{3-}.
III Only the first dissociation will make an important contribution to the $[H^+]$ when dissolved in water.

Of these three statements
(A) only I is valid.
(B) only II is valid.
(C) only III is valid.
(D) both I and II are valid but not III.
(E) all three statements (I, II, and III) are valid.

30. Burets can usually be read to the nearest ±0.01 mL. If the liquid volume in a buret is recorded as 22.00 mL, the number of significant figures in this value is
(A) 1.
(B) 2.
(C) 3.
(D) 4.
(E) 5.

31. The mass of a sample of $KClO_3$ is determined with five weighings:

1.391 g; 1.392 g; 1.299 g; 1.390 g; 1.388 g.

The number of significant figures in the average of these five readings is
(A) 1.
(B) 2.
(C) 3.
(D) 4.
(E) 5.

32. A weighing tray has a mass of 0.911 g. To this tray is added 3.2 g of NaOH. The total mass of tray plus NaOH is
(A) 4.1 g.
(B) 4.11 g.
(C) 4.111 g.
(D) 4.100 g.
(E) 4.10 g.

33. The density of a new plastic is determined experimentally by first measuring the volume using water displacement; the mass of the dry sample is also found. The Data:

Initial volume of water—13.0 mL

Final volume, water and sample—27.1 mL

Mass of sample —36.123 g

The density of this sample of plastic is best expressed as
(A) 2.5619 g/mL.
(B) 2.562 g/mL.
(C) 2.56 g/mL.
(D) 2.6 g/mL.
(E) 3. g/mL.

34. The density of diethyl ether is 0.714 g/cm^3. The volume occupied by 10.00 g of this liquid is
(A) 1.4 cm^3.
(B) 14. cm^3.
(C) 14.0 cm^3.
(D) 14.01 cm^3.
(E) 71.4 cm^3.

35. When equal volumes of 0.150 M NaOH and 0.150 M $HC_2H_3O_2$ are mixed, the resulting solution has a pH about 9. This is due to
(A) incomplete reaction of acetic acid in water.
(B) an unequal number of moles of OH^- (from the NaOH) and H^+ (from the $HC_2H_3O_2$).
(C) the $C_2H_3O_2^-$ reacting with water to provide more OH^-.
(D) the cation of the acid, which remains in solution at the equivalent point, and is a base.
(E) both the $C_2H_3O_2^-$ reacting with water and the acidic nature of the anion of the acid.

36. An isotope has the atomic number of 8 and a mass number of 17. This element
    (A) is an isotopic form of oxygen.
    (B) has 8 neutrons.
    (C) has 9 protons if it is in ionic form.
    (D) is an isotopic form of fluorine.
    (E) has 17 electrons.

37. The rate of reaction in a collection of gas particles is always much lower than the calculated collision frequency would indicate. This means that only in a small percentage of the collisions does a reaction result. This is because

    I few particles have the required activation energy.
    II few particles have formed the activated complex.
    III few particles have the correct orientation when they collide. Of these three statements, those that are valid are

    (A) only I.
    (B) only II.
    (C) only III.
    (D) both I and II only.
    (E) all three (I, II, and III).

38. The pH of a 0.00100 M HBr solution is
    (A) 0.00100M.
    (B) 3.0.
    (C) 7.0.
    (D) 11.0.
    (E) impossible to determine without more data.

39. Aqueous solutions of lead (II) nitrate and potassium chromate are allowed to react. The results show
    (A) both lead (II) chromate and potassium nitrate precipitate.
    (B) no solid forms from this reaction.
    (C) lead (II) chromate will precipitate from solution.
    (D) potassium nitrate precipitates.
    (E) potassium chromate is insoluble so there is no reaction.

40. A rigid cylinder contains CO_2 gas; some of the carbon dioxide is allowed to escape with temperature adjusted to a constant value. Which of the following applies to the CO_2?
    (A) The pressure of the gas increases.
    (B) The volume of the gas decreases.
    (C) The total number of gas molecules within the cylinder remains unchanged.
    (D) The average molecular speed decreases.
    (E) The distance between CO_2 molecules is increased.

41. Gases are less soluble in water at
    (A) high temperature, low pressure.
    (B) low temperature, high pressure.
    (C) high temperature, high pressure.
    (D) low temperature, low pressure.
    (E) pressure has no effect on the solubility of gases in water.

42. $2NH_3 \rightarrow N_2 + 3H_2$

    The above reaction occurs in a closed system of constant volume and temperature. What is the resultant pressure of the hydrogen if the partial pressure of the ammonia decreases by 0.40 atm.?
    (A) increases by 0.20 atm
    (B) increases by 0.40 atm
    (C) increased by 0.60 atm
    (D) decreased by 0.60 atm
    (E) impossible to calculate without more data

43. Air is pumped into a rigid steel cylinder at constant temperature. The increase in pressure is due to
    (A) increased molecular collisions.
    (B) the greater kinetic energy of the gas particles.
    (C) increase in the size of the individual molecules.
    (D) the greater force of attraction between gas molecules at high pressure.
    (E) the molecular contraction at higher pressure.

44. Real gases depart from ideal at conditions of
    (A) high pressure, low temperature.
    (B) low pressure, low temperature.
    (C) high pressure, high temperature.
    (D) low pressure, low temperature.
    (E) constant pressure and temperature.

45. Gas pressure is due to gas particles
    (A) slowing down at high temperature.
    (B) colliding with container walls.
    (C) neither attracting or repelling each other.
    (D) attracting each other.
    (E) moving in random patterns.

46. $2KClO_3(s) \rightarrow 2KCl(s) + 3O_2(g)$

    According to the above equation, 0.40 mol of solid $KClO_3$ completely decomposes, forming KCl and O_2. The dry gas is collected at STP. The volume of this oxygen gas would be most nearly
    (A) 1.4 L.
    (B) 14 L.
    (C) 140 L.
    (D) 1400 L.
    (E) 14,000 L.

47. A hollow steel cylinder of volume 24 L contains 1.0 mole of Ne and 2.0 mol of Ar. The partial pressure of the Ne is
    (A) 1/2 the total pressure.
    (B) 1/3 the total pressure.
    (C) 3 times that of the Ar.
    (D) 2 times that of the Ar.
    (E) equal to that of the Ar.

48. As the atomic number changes in Period Two of the Periodic Table from alkaline metals to halogens, the atomic radii
    (A) do not change.
    (B) increase.
    (C) decrease.
    (D) decrease, then slightly increase.
    (E) cannot be predicted because there is no trend.

49. Assume that $2A + B \rightarrow C$ is the rate determining step. 3.0 moles of A and 2.0 moles of B are placed in a 1.0-L flask; after five minutes the concentration of C reaches 1.0 M. After five minutes the rate will have
    (A) increased by a factor of 9.
    (B) increased by a factor of 18.
    (C) decreased by a factor of 9.
    (D) decreased by a factor of 10.
    (E) decreased by a factor of 18.

50. For a given reaction at a temperature of 27°C, the rate law is Rate = k [X] [Y]. If the concentration of X and of Y are both 0.40 M, the rate is 4.0×10^{-6} mol/L min. Determine the value of k (the rate constant) at this temperature.
    (A) 2.5×10^{-5} L/mol·min
    (B) 2.5×10^{-3} L/mol·min
    (C) 2.5×10^{-5} mol/L·min
    (D) 1.0×10^{-6} L/mol·min
    (E) 1.0×10^{-5} L/mol·min

51. The major reason an increase in temperature causes an increase in reaction rate is that
    (A) the activation energy changes with temperature.
    (B) the fraction of high energy molecules increases.
    (C) the reactant molecule pressure increases.
    (D) molecules collide with greater frequency.
    (E) catalysts become more effective.

52. Catalysts effectively increase reaction rate by
    (A) increasing the K_{eq}.
    (B) increasing the concentration of the reactant.
    (C) decreasing the concentration of the products.
    (D) lowering the activation energy requirements.
    (E) decreasing the reaction temperature.

53. For all zero-order reactions,
    (A) the reaction rate is independent of time.
    (B) the rate constant equals zero.
    (C) the concentration of reactants does not change over time.
    (D) activation energy is very low.
    (E) the concentration of reactants does not change and the rate is independent of time.

54. It is found that in a certain first order reaction the half-life is 1.4 minutes. The rate constant, k, for that same temperature is about
    (A) 0.35 min^{-1}.
    (B) 0.50 min^{-1}.
    (C) 0.71 min.
    (D) 2.0 min.
    (E) 2.0 min^{-1}.

55. To determine the order with respect to Br$^-$ in the reaction BrO_3^- (aq) + 5Br$^-$(aq) + 6H$^+$(aq)
    → $3H_2O$(l) + $3Br_2$(g) solutions should be prepared which differ in
    (A) [BrO_3^-] and [Br$^-$].
    (B) [Br$^-$].
    (C) [H$^+$].
    (D) [BrO_3^-] and [Br$^-$] and [H$^+$].
    (E) [H_2O] and [Br_2].

56. The first-order rate constant for nuclear unstable ^{60}Co is 0.13 yr^{-1}, and for ^{90}Sr it is 0.24 yr^{-1}.
    (A) The half-life of Sr is longer than that of Co.
    (B) The half-life of Sr is shorter than that of Co.
    (C) The half-lives of Sr and Co are equal.
    (D) The half-lives of Sr and Co cannot be compared from these data.
    (E) There is no simple relationship between half-life and stability.

57. The activation energy for a reaction is +30 kJ/mol; therefore the activation energy for the reverse reaction must be
    (A) –30 kJ / mol.
    (B) greater than 30 kJ / mol.
    (C) less than 10 kJ / mol.
    (D) + 30 kJ / mol.
    (E) a value that cannot be calculated without more data.

58. The rate law expression is rate = $k[X]^2$ [Y] for a certain reaction. Both X and Y are reactant gases. If the volume of the system is reduced to 1/3 of the original volume, the relative new reaction rate will be
    (A) 3 times greater.
    (B) 9 times greater.
    (C) 27 times greater.
    (D) 27 times less.
    (E) 9 times less.

59. For the reaction $X + 2B \rightarrow Z$, the rate law is rate = $k [X] [B]^2$. If the concentration of B increases by a factor of 3 while $[X]$ is held constant, the rate of the reaction will
    (A) increase by 3 times.
    (B) increase by 6 times.
    (C) increase by 9 times.
    (D) increase by 27 times.
    (E) decrease.

60. Spontaneous reactions at all temperatures are favored by values for enthalpy and entropy of

	ΔH	ΔS
(A)	+	+
(B)	–	–
(C)	0	0
(D)	+	–
(E)	–	+

61. A certain liquid boils at $27°C$ at 1.0 atm. pressure, with a heat of vaporization of 60.0 kJ / mol. Determine the entropy change for the boiling of one mole of this liquid.
    (A) +200. J/mol deg.K
    (B) –200. J/mol deg.K
    (C) +2.00 J/mol deg.K
    (D) –2.00 J/mol deg.K
    (E) 0.050 J/mol deg.K

62. The K_a for a weak acid is 5.0×10^{-10} at $25°C$. Determine the value of K_b for the conjugate base of this weak acid.
    (A) 0.50×10^{-5}
    (B) 1.5×10^{-5}
    (C) 2.0×10^{-5}
    (D) 5.0×10^{-5}
    (E) 5.0×10^{-10}

63. Determine the pH of a 0.10 M solution of a base, B, with a K_b of 1.0×10^{-5}.
    (A) 1.0
    (B) 3.0
    (C) 6.0
    (D) 8.0
    (E) 11.0

64. Which of the following is most likely to be a brittle compound with low conductivity as a solid, and have a high melting temperature?
    (A) RbF
    (B) CCl_4
    (C) CS_2
    (D) ICl
    (E) SF_6

65. Of the following, the species that has the correctly predicted largest radius is
    (A) Ar.
    (B) Cl^-.
    (C) Br^-.
    (D) K^+.
    (E) Sr^{2+}.

66. The species with the most polar bond is
    (A) Cl–Cl.
    (B) Si–Si.
    (C) Cr–Br.
    (D) P–Cl.
    (E) Ca–Ca.

67. The molecule whose Lewis structure requires resonance structures to best explain its bonding is
    (A) $BeCl_2$.
    (B) CO_2.
    (C) PCl_5.
    (D) OF_5.
    (E) SO_2.

68. The geometry of a PH_3 molecule is described by the VSEPR model as
    (A) linear.
    (B) trigonal planar.
    (C) tetrahedral.
    (D) bent or angular.
    (E) trigonal pyramidal.

69. The hybridization of an As atom in the AsF_5 molecule is
    (A) sp.
    (B) sp^2.
    (C) sp^3.
    (D) dsp^3.
    (E) d^2sp^3.

70. The following are in order of increasing boiling temperatures:
    (A) $RbCl < CH_3Cl < CH_3OH < CH_4$.
    (B) $CH_4 < CH_3Cl < CH_3OH < RbCl$.
    (C) $CH_3Cl < CH_3OH < RbCl < CH_4$.
    (D) $CH_4 < CH_3OH < CH_3Cl < RbCl$.
    (E) $RbCl < CH_3OH < CH_3Cl < CH_4$.

71. The following liquids are in order of increasing vapor pressure:
    (A) $CH_3CH_2OH < CH_3OCH_3 < CH_3CH_2CH_3$.
    (B) $CH_3CH_2OH < CH_3CH_2CH_3 < CH_3OCH_3$.
    (C) $CH_3CH_2CH_3 < CH_3OCH_3 < CH_3CH_2OH$.
    (D) $CH_3OCH_3 < CH_3CH_2OH < CH_3CH_2CH_3$.
    (E) $CH_3CH_2CH_3 < CH_3CH_2OH < CH_3OCH_3$.

72. The equilibrium system not affected by the pressure change which results from a volume change at constant temperature is
    (A) $2O_3(g) \rightleftharpoons 3O_2(g)$.
    (B) $PCl_5(g) \rightleftharpoons PCl_3(g) + Cl_2(g)$.
    (C) $2NaCl(s) \rightleftharpoons 2Na(s) + Cl_2(g)$.
    (D) $H_2(g) + I_2(g) \rightleftharpoons 2HI(g)$.
    (E) $2NO_2(g) \rightleftharpoons N_2O_4(g)$.

73. Given: $2X + Y \rightleftharpoons 3C + D$ (all are gases)

    Equal numbers of moles of X and Y are added to an empty tank; when equilibrium is achieved,
    (A) $[X] < [Y]$.
    (B) $[D] = [Y]$.
    (C) $[X] = [Y]$.
    (D) $[Y] < [X]$.
    (E) $[X] + [Y] < [C] + [D]$.

74. The K_c for $A + B \rightleftharpoons C$ is 4.0. The K_c for $2C \rightleftharpoons 2A + 2B$ is
    (A) 1 / 16.
    (B) 1 / 4.
    (C) 8.
    (D) 16.
    (E) 32.

75. The solubility of $Zn(OH)_2$ is 2.0×10^{-6} M at a certain temperature. Determine the value of the K_{sp} at this same temperature.
    (A) 2.0×10^{-6}
    (B) 4.0×10^{-6}
    (C) 8.0×10^{-6}
    (D) 1.6×10^{-17}
    (E) 3.2×10^{-17}

A5.5 Standard Reduction Potentials at 25°C (298 K) for Many Common Half-Reactions

Half-Reaction	$e°$ (V)	Half-Reaction	$e°$ (V)
$F_2 + 2e^- \rightarrow 2F^-$	2.87	$Cu^+ + e^- \rightarrow Cu$	0.52
$Ag^{2+} + e^- \rightarrow Ag^+$	1.99	$O_2 + 2H_2O + 4e^- \rightarrow 4OH^-$	0.40
$Co^{3+} + e^- \rightarrow Co^{2+}$	1.82	$Cu^{2+} + 2e^- \rightarrow Cu$	0.34
$H_2O_2 + 2H^+ + 2e^- \rightarrow 2H_2O$	1.78	$Hg_2Cl_2 + 2e^- \rightarrow 2Hg + 2Cl^-$	0.34
$Ce^{4+} + e^- \rightarrow Ce^{3+}$	1.70	$AgCl + e^- \rightarrow Ag + Cl^-$	0.22
$PbO_2 + 4H^+ + SO_4^{2-} + 2e^-$		$SO_4^{2-} + 4H^+ + 2e^- \rightarrow H_2SO_3 + H_2O$	0.20
$\qquad \rightarrow PbSO_4 + 2H_2O$	1.69	$Cu^{2+} + e^- \rightarrow Cu^+$	0.16
$MnO_4^- + 4H^+ + 3e^- \rightarrow MnO_2 + 2H_2O$	1.68	$2H^+ + 2e^- \rightarrow H_2$	0.00
$2e^- + 2H^+ + IO_4^- \rightarrow IO_3^- + H_2O$	1.60	$Fe^{3+} + 3e^- \rightarrow Fe$	−0.036
$MnO_4^- + 8H^+ + 5e^- \rightarrow Mn^{2+} + 4H_2O$	1.51	$Pb^{2+} + 2e^- \rightarrow Pb$	−0.13
$Au^{3+} + 3e^- \rightarrow Au$	1.50	$Sn^{2+} + 2e^- \rightarrow Sn$	−0.14
$PbO_2 + 4H^+ + 2e^- \rightarrow Pb^{2+} + 2H_2O$	1.46	$Ni^{2+} + 2e^- \rightarrow Ni$	−0.23
$Cl_2 + 2e^- \rightarrow 2Cl^-$	1.36	$PbSO_4 + 2e^- \rightarrow Pb + SO_4^{2-}$	−0.35
$Cr_2O_7^{2-} + 14H^+ + 6e^- \rightarrow 2Cr^{3+} + 7H_2O$	1.33	$Cd^{2+} + 2e^- \rightarrow Cd$	−0.40
$O_2 + 4H^+ + 4e^- \rightarrow 2H_2O$	1.23	$Fe^{2+} + 2e^- \rightarrow Fe$	−0.44
$MnO_2 + 4H^+ + 2e^- \rightarrow Mn^{2+} + 2H_2O$	1.21	$Cr^{3+} + e^- \rightarrow Cr^{2+}$	−0.50
$IO_3^- + 6H^+ + 5e^- \rightarrow \frac{1}{2}I_2 + 3H_2O$	1.20	$Cr^{3+} + 3e^- \rightarrow Cr$	−0.73
$Br_2 + 2e^- \rightarrow 2Br^-$	1.09	$Zn^{2+} + 2e^- \rightarrow Zn$	−0.76
$VO_2^+ + 2H^+ + e^- \rightarrow VO^{2+} + H_2O$	1.00	$2H_2O + 2e^- \rightarrow H_2 + 2OH^-$	−0.83
$AuCl_4^- + 3e^- \rightarrow Au + 4Cl^-$	0.99	$Mn^{2+} + 2e^- \rightarrow Mn$	−1.18
$NO_3^- + 4H^+ + 3e^- \rightarrow NO + 2H_2O$	0.96	$Al^{3+} + 3e^- \rightarrow Al$	−1.66
$ClO_2 + e^- \rightarrow ClO_2^-$	0.954	$H_2 + 2e^- \rightarrow 2H^-$	−2.23
$2Hg^{2+} + 2e^- \rightarrow Hg_2^{2+}$	0.91	$Mg^{2+} + 2e^- \rightarrow Mg$	−2.37
$Ag^+ + e^- \rightarrow Ag$	0.80	$La^{3+} + 3e^- \rightarrow La$	−2.37
$Hg_2^{2+} + 2e^- \rightarrow 2Hg$	0.80	$Na^+ + e^- \rightarrow Na$	−2.71
$Fe^{3+} + e^- \rightarrow Fe^{2+}$	0.77	$Ca^{2+} + 2e^- \rightarrow Ca$	−2.76
$O_2 + 2H^+ + 2e^- \rightarrow H_2O_2$	0.68	$Ba^{2+} + 2e^- \rightarrow Ba$	−2.90
$MnO_4^- + e^- \rightarrow MnO_4^{2-}$	0.56	$K^+ + e^- \rightarrow K$	−2.92
$I_2 + 2e^- \rightarrow 2I^-$	0.54	$Li^+ + e^- \rightarrow Li$	−3.05

Advanced Placement Chemistry Equations and Constants

ATOMIC STRUCTURE

$E = hv$

$\lambda = \frac{h}{mv}$

$c = \lambda v$

$p = mv$

$E_n = \frac{-2.178 \times 10^{-18}}{n^2}$ joule

EQUILIBRIUM

$K_a = \frac{[H^+][A^-]}{[HA]}$

$E^\Omega =$

$K_w = [OH^-][H^+] = 1.0 \times 10^{-14}$ at 25°C

$\quad = K_a \times K_b$

$pH = -\log[H^+], \; pOH = -\log[OH^-]$

$14 = pH + pOH$

$pH = pK_a + \log\frac{[A^-]}{[HA]}$

$pOH = pK_b + \log\frac{[HB^+]}{[B]}$

$pK_a = -K_a, \; pK_b = -\log K_b$

$K_p = K_c(RT)^{\Delta n}$

where $\Delta n =$ moles product gas − moles reactant gas

THERMOCHEMISTRY/KINETICS

$\Delta S^\circ = \sum S^\circ$ products $- \sum S^\circ$ reactants

$\Delta H^\circ = \sum \Delta H_f^\circ$ products $- \sum \Delta H_f^\circ$ reactants

$\Delta G^\circ = \sum \Delta G_f^\circ$ products $- \sum \Delta G_f^\circ$ reactants

$\Delta G^\circ = \Delta H^\circ - T\Delta S^\circ$

$\quad = -RT \ln K = -2.303RT \log K$

$\quad = -n\Im E^\circ$

$\Delta G = \Delta G^\circ + RT \ln Q = \Delta G^\circ + 2.303RT \log Q$

$q = mc\Delta T$

$C_p = \frac{\Delta H}{\Delta T}$

$\ln[A]_t - \ln[A]_0 = -kt$

$\frac{1}{[A]_t} - \frac{1}{[A]_0} = kt$

$\ln k = \frac{-E_a}{R}\left(\frac{1}{T}\right) + \ln A$

E = energy

v = frequency

λ = wavelength

p = momentum

υ = velocity

n = principal quantum number

m = mass

Speed of light, $c = 3.0 \times 10^8$ ms^{-1}

Planck's constant, $h = 6.63 \times 10\text{-}34$ Js

Boltzmann's constant, $k = 1.38 \times 10^{-23}$ J K^{-1}

Avogadro's number $= 6.022 \times 10^{23}$ mol^{-1}

Electron charge, $e = -1.602 \times 10^{-19}$ coulomb

1 electron volt per atom $= 96.5$ kJ mol -1

EQUILIBRIUM CONSTANTS

K_a (weak acid)

K_b (weak base)

K_w (water)

K_p (gas pressure)

K_c (molar concentrations)

S° = standard entropy

H° = standard enthalpy

G° = standard free energy

E° = standard reduction potential

T = temperature

n = moles

m = mass

q = heat

c = specific heat capacity

C_p = molar heat capacity at constant pressure

E_a = activation energy

k = rate constant

A = frequency factor

Faraday's constant, $\Im = 96,500$ per mole of electrons

Gas constant, $R \quad = 8.31$ J mol^{-1} K^{-1}

$\quad\quad\quad\quad\quad = 0.0821$ L atm mol^{-1} K^{-1}

$\quad\quad\quad\quad\quad = 8.31$ volt coulomb mol^{-1} K^{-1}

GASES, LIQUIDS, AND SOLUTIONS

$PV = nRT$

$\left(P + \frac{n^2 a}{V^2}\right)(V - nb) = nRT$

$P_A = P_{total} \times X_A$, where $X_A = \frac{\text{moles A}}{\text{total moles}}$

$P_{total} = P_A + P_B + P_C + \ldots$

$n = \frac{m}{M}$

$°K = °C + 273$

$\frac{P_1 V_1}{T_1} = \frac{P_2 V_2}{T_2}$

$D = \frac{m}{V}$

$$u_{rms} = \sqrt{\frac{3kT}{m}} = \sqrt{\frac{3RT}{M}}$$

$$KE \text{ per molecule} = \frac{1}{2}mv^2$$

$$KE \text{ per mole} = \frac{3}{2}RT$$

$$\frac{r_1}{r_2} = \sqrt{\frac{M_2}{M_1}}$$

molarity, M = moles solute per liter solution

molality = moles solute per kilogram solvent

$$\Delta T_f = iK_f \times \text{molality}$$

$$\Delta T_b = iK_b \times \text{molality}$$

$$\pi = MRT$$

$$A = abc$$

OXIDATION-REDUCTION; ELECTROCHEMISTRY

$$Q = \frac{[C]^c[D]^d}{[A]^a[B]^b}, \text{ where } a\,A + b\,B \rightarrow c\,C + d\,D$$

$$I = \frac{q}{t}$$

$$E_{cell} = E^\circ_{cell} - \frac{RT}{n\Im}\ln Q = E^\circ_{cell} - \frac{0.0592}{n}\log Q \,@\, 25°C$$

$$\log K = \frac{nE^\circ}{0.0592}$$

P = pressure
V = volume
T = temperature
n = number of moles
D = density
m = mass
v = velocity

u_{rms} = root-mean-square speed

KE = kinetic energy

r = rate of effusion

M = molar mass

π = osmotic pressure

i = van't Hoff factor

K_f = molal freezing-point depression constant

K_b = molal boiling-point elevation constant

A = absorbance

a = molar absorptivity

b = path length

c = concentration

Q = reaction quotient

I = current (amperes)

q = charge (coulombs)

t = time (seconds)

E° = standard reduction potential

K = equilibrium constant

Gas constant, R = 8.31 J mol^{-1} K^{-1}
  = 0.0821 L atm mol^{-1} K^{-1}
  = 8.31 volt coulomb mol^{-1} K^{-1}

Boltzmann's constant, k = 1.38×10^{-23} J K^{-1}

K_f for H$_2$O = 1.86 K kg mol^{-1}

K_b for H$_2$O = 0.512 K kg mol^{-1}

1 atm = 760 mm Hg
  = 760 torr

STP = 0.000°C and 1.000 atm

Faraday's constant, $\Im$ = 96,500 coulombs per mole of electrons

Section II: Free-Response Questions
Time: 95 minutes
Number of Questions: 6

Section II of the AP Chemistry Examination counts for 50% of the total test grade and involves several parts. You will be required to answer all six questions. Calculators may be used on the first three questions only. A Periodic Table and Equation Tables are provided elsewhere in this study guide and may be used.

Part A: Time—55 minutes

1. Equilibrium (worth 20% of this part of the test)

$$2SO_3(g) \rightleftharpoons O_2(g) + 2SO_2(g)$$

A 3.21 g sample of sulfur trioxide is placed in a 2.25-L cylinder and allowed to reach equilibrium at a constant temperature of 500.K, as shown in the above equation. Analysis shows 1.23×10^{-2} mol of sulfur dioxide at equilibrium.

(a) Write the equilibrium constant expression (K_c) for this system.
(b) Calculate the concentration of all three gases at equilibrium.
(c) Calculate the K_c value for this system.
(d) Calculate the value of K_p at this same temperature.

2. The actual test will have two questions here (each worth 20% of this part of the test).

Perbromic acid, $HBrO_4$, can react with sulfuric acid, H_2SO_4, in the following manner: $HBrO_4 + H_2SO_4 \longrightarrow H_3SO_4^+ + BrO_4^-$.

(a) Indicate the two conjugate acid base pairs in this reaction and label the acids in each pair.
(b) In the SO_4^{2-} ion, sulfur is surrounded by four oxygens in a tetrahedral structure. Indicate a Lewis structure for the $H_3SO_4^+$ ion.
(c) Is a $H_4SO_4^{2+}$ ion stable (can it exist)? Support your answer with a Lewis structure and/or discussion.
(d) Contrast the use of the term 'acid' as used in the Arrhenius model, in the Brønsted-Lowry model, and in the Lewis model of acids.

3. Laboratory Question (worth 20% of this section of the test)

Analysis of an iron (II) solution by titration with a standardized potassium permanganate solution leads to the determination of the percent of iron in the original solution. When performing this laboratory work, the solid of known mass and containing iron (II) is dissolved in 25.0 mL of water.

(a) Explain how each of the following affects the reported percentage of iron in the unknown solid:
   (i)   The student fills the buret with $KMnO_4$ after rinsing with only deionized water.
   (ii) An air bubble appears in the buret tip before titration begins.
   (iii) More than 25.0 mL of water is used to dissolve the solid.
   (iv) Each time the volume of fluid is measured in the buret, measurement is made to the top of the meniscus, rather than the bottom of the meniscus.

(b) Assuming that the accepted value for the percentage of iron in the original sample was 7.77% and that your experiment gave a result of 6.896%, determine the percentage error in your work.

(c) Balance the reaction which occurs between iron (II) ions and the permanganate ion in acidic solution. Identify the reducing agent.

Part B: Time—40 minutes

Calculators may not be used on this part of the test.

4. Reactions (worth 10% of this part of the test)

Write the balanced chemical equation for each of the following three reactions. Then you will also answer a short follow-up question about each reaction. Unless otherwise indicated, all of the solutions are aqueous. Be sure to consider carefully if substances are ions or molecules. Do not show spectator ions. In all cases, a reaction does take place.

a. (i) Ammonia gas is added to boron trichloride gas.
    (ii) Which species acts as the Lewis acid in the reaction? Explain.

b. (i) Iron (III) ions are reduced by iodide ions.
    (ii) What is the oxidation number of the iodine **before** and **after** the reaction?

c. (i) Solid potassium chlorate is heated.
    (ii) In the balanced equation, how many moles of oxygen are produced for every mole of potassium chlorate used?

5. General Discussion (worth 15% of this part of the test)

Explain the following observations:

(a) When a solid is heated at its melting temperature, the temperature does not increase.

(b) The melting temperature of KCl is less than that of NaF.

(c) Liquid bromine and ICl have almost the same molecular mass, but ICl boils at a temperature 40°C higher than Br_2.

(d) The boiling temperature of HCl is lower than that of HF.

(e) When alcohol (ethanol) is poured over your arm, your skin feels cold.

6. General Discussion (worth 15% of this part of the test)

For a given reaction, a proposed rate law is Rate = $k [X]^n [Y]^m$.

(a) Explain how the values for "n" and for "m" can be determined experimentally.

(b) If both "n" and "m" are 1, what is the overall order of this reaction?

(c) How may the value of the rate constant, k, be changed? Support your answer.

(d) Contrast the physical meaning of a first-order reaction with that of a zero-order reaction.

(e) Explain why knowing the rate law expression is essential when proposing a reaction mechanism.

END OF EXAMINATION

ANSWERS TO DIAGNOSTIC TEST

MULTIPLE-CHOICE ANSWERS

Using the table below, score your test.

Determine how many questions you answered correctly and how many you answered incorrectly. You will find explanations of the answers on the following pages.

1. A	2. D	3. B	4. A	5. B
6. C	7. E	8. B	9. C	10. A
11. E	12. A	13. C	14. B	15. A
16. A	17. A	18. A	19. A	20. A
21. E	22. D	23. C	24. A	25. B
26. A	27. A	28. A	29. E	30. D
31. D	32. A	33. C	34. C	35. C
36. A	37. E	38. B	39. C	40. E
41. A	42. C	43. A	44. A	45. B
46. B	47. B	48. C	49. E	50. A
51. B	52. D	53. A	54. B	55. B
56. B	57. E	58. C	59. C	60. E
61. A	62. C	63. E	64. A	65. C
66. C	67. E	68. E	69. D	70. B
71. A	72. D	73. A	74. A	75. E

CALCULATE YOUR SCORE:

Number answered correctly: ‾‾‾‾‾

Adjust ¼ point for guessing penalty:
    Count the number of questions you
    answered incorrectly, multiply by .25, and subtract: – ‾‾‾‾‾
Determine your adjusted score: ‾‾‾‾‾

WHAT YOUR SCORE MEANS:

Each year, since the test is different, the scoring is a little different. But generally, if you scored 25 or more on the multiple-choice questions, you'll most likely get a 3 or better on the test. If you scored 35 or more, you'll probably score a 4 or better. And if you scored a 50 or more, you'll most likely get a 5. Keep in mind that the multiple-choice section is worth 45% of your final grade, and the free-response section is worth 55% of your final grade. To learn more about the scoring for the free-response questions, turn to the last page of this section.

ANSWERS AND EXPLANATIONS

MULTIPLE-CHOICE ANSWERS

1. ANSWER: A

2. ANSWER: D

3.   ANSWER: B   In questions like 1, 2 and 3, it is helpful to consider what you do know about each of the listed responses. In this case you know that chromium is a transition element which suggests that the ion is capable of being multicharged. Transition metals are frequently colored in compounds. You might recall the polyatomic ion list which includes CrO_4^{2-}(chromate ion) and $Cr_2O_7^{2-}$ (dichromate ion). Sodium, like all of the alkali metals, reacts with water to form hydrogen. You have most likely seen this reaction demonstrated in your classroom. Recall also that the oxides of metals form water solutions of hydroxide ions (bases), whereas the oxides of nonmetals form acidic solutions. Note also that an answer can be used more than once in this type of question (*Chemistry* 7th ed. pages 309–318, 588, 943–955 / 8th ed. pages 318–327, 603-604, 954–964).

4   ANSWER: A

5.   ANSWER: B

6.   ANSWER: C

7.   ANSWER: E   The term *descriptive chemistry* hides much information that you "just must know." Of course there is so much of it that it is essential to get some direction on what things are paramount. This includes knowing the six strong acids, the common solubility rules, the colors of several ions (usually encountered in your laboratory work), which ions tend to complex, and a few very specific properties. Knowing that hydrofluoric acid, although a weak acid, will dissolve (etch) glass and that carbon dioxide will not support combustion, are examples of properties that a well-educated chemist—that's you!—is expected to know (*Chemistry* 7th ed. pages 626–631, 140–148, 955–959 / 8th ed. pages 642–647, 144-152, 964–969).

8.   ANSWER: B   Increasing the [Ag^+] will cause a shift toward products, causing an increase in the potential difference of this cell. This would, over time, cause an increase in the concentration of copper (II) ions, making the blue of the copper (II) ions darker. You could also consider this from the Nernst equation. In that case, as [Ag^+] is increased, [Cu^{2+}]/[Ag^+]^2 < 1 such that log Q is a negative number. Subtracting a negative number from the standard potential then gives a larger cell potential (*Chemistry* 7th ed. pages 803–808 / 8th ed. pages 836–842).

9.   ANSWER: C   By definition the anode is the site of oxidation. Oxidation is the loss of electrons. Copper must lose two electrons to become copper (II) ions. Half-reactions involving a decrease in positive charge, like responses (B) and (E), require a gain of electrons (*Chemistry* 7th ed. pages 791–800 / 8th ed. pages 823–833).

10. ANSWER: A   It is necessary for the reactants to gain enough energy to form the activated complex; the difference in the energy of the reactants and the activated complex is called the activation energy

[Section 12.7, especially Figure 12.11). Of course, both hydrogen and oxygen have already vaporized; they exist as gases at room conditions. Enthalpy change represents the difference in energy between the reactants and the products (at constant pressure) [Section 6.2], Entropy change represents the change in the disorder from reactants to products (*Chemistry* 7th ed. pages 552–557 esp. Figure 12.11, 235–242, 755 / 8th ed. pages 563–570 esp. Figure 12.11, 243-251, 779).

11. ANSWER: **E** This is a very straightforward use of the chemical formula and your understanding of moles. Since there are two moles of silver atoms per formula unit of silver sulfide, there are twice as many moles of silver atoms as moles of formula units of the silver sulfide. You can also quickly see that the grams formed would be about 0.4 g if you recall that silver has an atomic mass of about 108 g/mol (*Chemistry* 7th ed. pages 82–85 / 8th ed. pages 81–84).

12. ANSWER: **A** Some flame colors which you should know include lithium: bright red; sodium: pale-yellow; potassium: violet; rubidium: purple; cesium: blue; calcium: orange-red; strontium: crimson-red; and barium: yellowish-green (*Chemistry* 7th ed. pages 284–285 / 8th ed. pages 294–295).

13. ANSWER: **C** "*R*" is the general gas constant and applies as long as the gas behaves ideally. This means that the value of "*R*" and the expression $PV = nRT$ must be adjusted in the nonideal conditions of high temperature and/or low pressure [Section 5.8]. That is what the van der Waals equation is all about. The gas laws are based on Kelvin temperature change (not Celsius), since Kelvin is an index of the motion of the gas particles. The value of PV/nRT for ideal gases is 1 (*Chemistry* 7th ed. pages 208–210 / 8th ed. pages 214–220).

14. ANSWER: **B** Freezing temperature depression is directly related to the number of particles (in this case, ions) in solutions. Since $AlCl_3$ provides four ions per formula unit, the higher concentration of ions for that salt would cause a greater freezing-temperature lowering (*Chemistry* 7th ed. pages 504–507 / 8th ed. pages 516–519).

15. ANSWER: **A** Both CaO and MgO are ionically bonded. Both involve oxygen; therefore the difference in melting temperatures must be due to the difference in attraction developed by Mg compared to that of Ca. Both are 2+ ions, but magnesium ions are smaller, meaning that the charge is more concentrated in the Mg^{2+} ion. The combination of charge and size factors, called the charge density, affects the force of attraction that the ion has; ionic charge / ionic radius = charge density (*Chemistry* 7th ed. pages 309–318 / 8th ed. pages 318-327).

16. ANSWER: **A** The 'general formulas' for these common organic molecules are R–OH—alcohol; R–C=O–H—aldehydes; R–C=O–OH—carboxylic acids; R–C=O–R'—ketones; R–C=O–OH—esters (*Chemistry* 7th ed. pages 1005–1016 esp. Table 22.4 / 8th ed. pages 1014–1025 esp. Table 22.4).

17. **ANSWER: A** The reaction forms $BaSO_4(s)$ which means that ions are removed from solution, thereby making the solution less conductive. However, continued addition of the acid causes the conductivity to increase again with the excess of ions (*Chemistry* 7th ed. pages 140–145 / 8th ed. pages 144–150).

18. **ANSWER: A** The Law of Multiple Proportions is illustrated by two elements which form at least two compounds. It then compares the ratios of the masses of one element with a constant mass of the other element in the two (or more) compounds. These ratios always reduce to simple whole numbers (*Chemistry* 7th ed. pages 41–43 / 8th ed. pages 41–44).

19. **ANSWER: A** In addition to remembering Rutherford for his work with the nucleus, recall Thomson for his work with the electron, Dalton for an early model of the atom, Boyle for the gas law dealing with volume and pressure, and Lavoisier for the conservation of mass in chemical reactions (*Chemistry* 7th ed. pages 39–49 / 8th ed. pages 39–50).

20. **ANSWER: A** Isotopes are found for every element and differ from each other in mass for the same element due only to a different number of neutrons in the nucleus (*Chemistry* 7th ed. pages 49–52, 77–81 / 8th ed. pages 50–52, 77–80).

21. **ANSWER: E** Begin by determining the formula for iron (II) phosphate: $Fe_3(PO_4)_2$. From this you can see that there are two phosphate ions with four oxygens in each, for a total of 8 moles of oxygen for each unit of $Fe_3(PO_4)_2$ (*Chemistry* 7th ed. pages 57–67 / 8th ed. pages 56–67).

22. **ANSWER: D** When solving K_{sp} problems, first write the equation with the solid on the left and the ions on the right: $MgF_2(s) \rightleftharpoons Mg^{2+}$ (aq) + $2F^-$(aq). If s represents the mol/L of the solid that goes into solution, $[Mg^{2+}]$ also equals s, and $2s$ represents the $[F^-]$. Next write the K_{sp} expression in terms of s: $K_{sp} = (s)(2s)^2 = 4s^3 = 6.4 \times 10^{-9}$. Solve for s (*Chemistry* 7th ed. pages 717–719 / 8th ed. pages 743–747).

23. **ANSWER: C** A good clue as to the type of titration curve is the position of the equivalence point, which is the center of the vertical section of the graph showing the very rapid change of pH. In this case, that is in the basic region (above 9). At this point, the amount of added OH^- equals the original amount of acid. The pH exceeds 7 (neutral) due to the hydrolysis of the anion from the acid (*Chemistry* 7th ed. pages 696–697, 700–707 / 8th ed. pages 713–715, 717–725).

24. **ANSWER: A** The leveling-off shown between F and G is caused by buffering. Optimal buffering occurs when $[HA] = [A^-]$, which would be at a volume of 25 mL of base in this case (*Chemistry* 7th ed. pages 693–694, 700–707 / 8th ed. pages 710–712, 717–725).

25. **ANSWER: B** The systematic naming system, the Stock Nomenclature, uses Roman numerals only for transition elements whose

ions display more than one charge (e.g., Cu^{2+} is called copper (II) ion, and Cu^+ is known as copper (I) ion). Calcium is not a transition element; it is found only as a 2+ ion. OCl^- is the hypochlorite ion (*Chemistry* 7th ed. pages 57–67 esp. Table 2.5 / 8th ed. pages 56–67 esp. Table 2.5).

26. ANSWER: **A**   Lone pairs of electrons (unbonded pairs) are less associated with the attractive forces of positive nuclei, therefore will occupy more space than bonded pairs (*Chemistry* 7th ed. pages 369–371 / 8th ed. pages 380–382).

27. ANSWER: **A**   Acids which contain oxygen often give students some difficulty. These names are formed from the root name of the anion with a suffix of -ic or -ous. The -ous form always contains one less oxygen than the -ic form. For example, H_2SO_4 is sulfuric acid and H_2SO_3 is sulfurous acid (*Chemistry* 7th ed. Tables 2.7, 2.8, and Figure 2.25 on page 67 / 8th ed. Tables 2.7, 2.8 on page 66 and Figure 2.24 on page 67).

28. ANSWER: **A**   Before reaction, no combination of NH_3 and Ag^+ has formed; the silver nitrate is found as ions, the ammonia is molecular, and water is also present (*Chemistry* 7th ed. pages 731–734 / 8th ed. pages 759–762).

29. ANSWER: **E**   The low value for K_{a_1} indicates that H_3PO_4 is a weak acid. The three values suggest three steps in dissociation. The large differences in these K_a values indicate that only step 1 will provide a significant $[H^+]$ (*Chemistry* 7th ed. pages 650–655, 682–683 / 8th ed. pages 666–671, 699–700).

30. ANSWER: **D**   Significant figures is a method of indicating to others how accurately you have measured a value; all measured values have some uncertainty in the measurement, (another method is to show the measured value followed by a ± value which indicates how precisely you have measured or how much the measurement is limited by the equipment you have used to obtain it). All of the numerals are significant here. It is the zero which can be difficult. If this value were known only to the nearest 1 mL then it would be correctly written as 22. mL; since 22.00 mL were recorded there must have been a reason to record the values in the tenths and hundredths places. The reason is that the measurement was taken to the hundredths of a mL (*Chemistry* 7th ed. pages 497–504 / 8th ed. pages 509–516).

31. ANSWER: **D**   The average is found by totaling the five readings. When adding (or subtracting), examine the position of the decimal point. In this case, all five values are known to the nearest thousandth, so the total can be known to the nearest thousandths. When this total is divided by five (an exact number in this case), a different rule applies. When multiplying or dividing you must count the number of significant figures; the least precise piece of data used to obtain a result determines the number of significant figures in that result. In this case, that would be four (*Chemistry* 7th ed. pages 10–16 / 8th ed. pages 11–17).

same period (*Chemistry* 7th ed. pages 309–314 esp. Figure 7.34, 875–880 esp. Figure 19.2 / 8th ed. pages 318–323 esp. Figure 7.34, 905–912 esp. Figure 19.2).

49. ANSWER: E   Since this occurs as a one step reaction, we can write that the rate = $k[A]^2[B]$. Initially the rate will be proportional to $(3)^2(2) = 18$. Later, $[A]$ becomes $3 - 2 = 1$, and $[B]$ becomes $2 - 1 = 1$, so the rate is then proportional to $(1)^2(1) = 1$. That means a change in the rate by a factor of 18 to 1, or that the rate is now 1/18 as large as it was originally (*Chemistry* 7th ed. pages 534–552 / 8th ed. pages 547–565).

50. ANSWER: A   This time you need to use the form $k = \text{rate}/[X][Y]$ to determine the value of the rate constant, k.

$k = 4.0 \times 10^{-6}\,\text{mol/L} \cdot \text{min} / (4.0 \times 10^{-1}\,\text{mol/L})(4.0 \times 10^{-1}\,\text{mol/L})$

$= 1/4 \times 10^{-4}\,\text{L} / \text{mol} \cdot \text{min} = .25 \times 10^{-4}\,\text{L} / \text{mol} \cdot \text{min}$

$= 2.5 \times 10^{-5}\,\text{L} \cdot \text{mol}^{-1}\text{min}^{-1}$.

(Note that if set up in this way, the mathematics become rather easy to handle.) Do watch units for rate constants, which, unlike equilibrium constants, traditionally have units assigned (*Chemistry* 7th ed. pages 534–549 / 8th ed. pages 547–565).

51. ANSWER: B   There is almost always a question on kinetics like this on a test. Note that it is not just the greater number of molecular collisions that cause the reaction to occur faster but an increase in the fraction of high energy molecules which can then obtain the activation energy requirement and reaction that cause this rate increase (*Chemistry* 7th ed. pages 552–563 esp. Figure 12.12 / 8th ed. pages 565–575 esp. Figure 12.12).

52. ANSWER: D   Catalysts lower the activation energy barrier by forming a different activated complex (*Chemistry* 7th ed. pages 557–563, esp. Figures 12.15 and 12.16 / 8th ed. pages 570–575 esp. Figures 12.15 and 12.16).

53. ANSWER: A   For a zero-order rate law, rate = $k[A]^0$ (i.e. n = zero). Since any number taken to the zero power is equal to 1, rate = k (the rate constant). This has the physical meaning that the rate does not speed up or slow down over time (like most reactions do); it either takes place at a constant rate or does not take place at all (*Chemistry* 7th ed. pages 538–547, Table 12.6 on page 548, 564–565 / 8th ed. pages 551–560, Table 12.6 on page 561, 576–577).

54. ANSWER: B   Using the general half-life expression, $t_{1/2} = 0.693 / k$, $k = 0.693 / 1.4\,\text{min} = 0.50\,\text{min}^{-1}$. Note that this expression is valid only for first-order reactions, which would include nuclear decay processes. (*Chemistry* 7th ed. pages 538–547 / 8th ed. pages 551–560).

55. ANSWER: B   If you wish to determine how order is affected by a given concentration, then a series of reactions should be run with changes in only the concentration of that substance, with all other concentrations held constant (as well as all other conditions like temperature held constant) (*Chemistry* 7th ed. pages 538–547 esp.

Sample Exercise 12.5 / 8th ed. pages 551–560 esp. Sample Exercise 12.5).

56. ANSWER: **B** This is another application of the half-life expression for first-order rate laws, $t_{1/2} = 0.693/k$. Note that half-life (time) and the rate constant are inversely related; hence the element with the lower rate constant will have the longer half-life (*Chemistry* 7th ed. pages 538–547 / 8th ed. pages 551–560).

57. ANSWER: **E** Since the activation energy for the reverse reaction is the sum of the activation energy for the forward reaction plus the reverse of the enthalpy change for the forward reaction, and we do not know this last value, the value cannot be calculated (*Chemistry* 7th ed. pages 549–552, fig. 12.11 on page 553 / 8th ed. pages 562–565, Figure 12.11 on page 566).

58. ANSWER: **C** If the volume is reduced to 1/3 the original value, then the concentration of these gases will all be three times greater. Using the given rate law and increasing both concentrations by a factor of 3, gives a rate increase of $(3)^2(3) = 27$ times (*Chemistry* 7th ed. pages 534–549 / 8th ed. pages 547–562).

59. ANSWER: **C** Since only the concentration of *B* is altered, you need to see how tripling [*B*] affects the rate: from $r = k[X][B]^2$ you can see that the rate increases by a factor of $(3)^2$, or 9 times (*Chemistry* 7th ed. pages 534–549 / 8th ed. pages 547–562).

60. ANSWER: **E** Use the equation $\Delta G = \Delta H + T\Delta S$, and remember that a negative value for the Gibb's Free Energy is required for a spontaneous reaction. Then determine the signs for ΔH and for ΔS which give a negative ΔG (*T* is always positive, of course) (*Chemistry* 7th ed. pages 749–759 / 8th ed. pages 773–783).

61. ANSWER: **A** For systems at the boiling temperature, the change in state does not offer any free energy. So *G* is zero. This allows you to use $\Delta S = \Delta H / T$ to find the entropy change:

$$\Delta S = +60,000 \, J / mol / \ 300.K = +200. \, J/ mol \cdot deg. \, K.$$

Note the change from kJ to J in this calculation (*Chemistry* 7th ed. pages 759–762 / 8th ed. pages 783–786).

62. ANSWER: **C** From $K_a \times K_b = K_w$; $K_b = 1.0 \times 10^{-14} / 5.0 \times 10^{-10} = 2.0 \times 10^{-5}$ (*Chemistry* 7th ed. pages 655–660 / 8th ed. pages 671–677).

63. ANSWER: **E** Using the general equation $B^- + HOH \rightarrow HB + OH^-$, $[HB][OH^-]/[B^-] = 1.0 \times 10^{-5}$, $[OH^-] = 1.0 \times 10^{-3}$ M, pOH = 3.0. pH = 14.0 − 3.0 = 11.0. OH⁻ pOH = −log 1.0×10^{-3} (*Chemistry* 7th ed. pages 649–650 / 8th ed. pages 665–666).

64. ANSWER: **A** Compounds with these properties are ionically bonded. This suggests elements from the far left side of the Periodic Table (1A or 2A) with elements from the far right side (6A or 7A) (*Chemistry* 7th ed. pages 346–347, 456–459 / 8th ed. pages 357–358, 468–471).

65. **ANSWER: C** Ions with negative charges have gained an electron and are therefore larger than their parent atoms. Between the chloride ions and the bromide ion you have outermost electrons in the third gross energy level versus the fourth gross energy level, hence we would expect Br⁻ to be larger (*Chemistry* 7th ed. pages 338–342 esp. Figure 8.8 / 8th ed. pages 349–353 esp. Figure 8.8).

66. **ANSWER: C** While using actual electronegativities would be helpful for this question, a good generalization is that the further apart the elements are on the periodic table the greater the difference in control they have over the shared electron pair, and the more polar the bond. Of course, if an element from the 1A or 2A Group bonds with something from the 6A or 7A Group, the difference in control over the electron is so great that we call the bond ionic, but that possibility is not one of the answers in this question. Note also that this is a question dealing with the polarity of an individual bond and not with the polarity of an entire molecule (*Chemistry* 7th ed. pages 333–335 / 8th ed. pages 344–346).

67. **ANSWER: E** The need for resonance structures seems to be greatest for molecules in which the same two elements are bonded with different type bonds (e.g. one single, one double bond) in the same molecule. The usual Lewis structure for sulfur dioxide shows that the sulfur and oxygen are bonded with one single bond and, in the other sulfur–oxygen bond, doubly bonded. (Be sure you can draw this Lewis structure.) (*Chemistry* 7th ed. pages 363–367 / 8th ed. pages 374–378)

68. **ANSWER: E** The VSEPR Model is very helpful in describing almost all molecular shapes. Note that in PH_3 the shape may, at first, seem to be tetrahedral, but it does not have a hydrogen in one of the four apex positions, so it is a trigonal pyramid. Molecular shape is determined by the position of the nuclei and not by the position of just electrons, either pairs or single electrons (*Chemistry* 7th ed. pages 367–379 esp. fig. 8.16 / 8th ed. pages 378–390 esp. Figure 8.16).

69. **ANSWER: D** Hybridization is a way of describing the atomic orbitals used to share electrons and thereby form bonds. The careful student will become familiar with the five types of hybrid orbitals indicated by the five responses to this question (*Chemistry* 7th ed. pages 391–403 / 8th ed. pages 404–416).

70. **ANSWER: B** The temperature at which substances boil is a function of the forces between the molecules of that substance (intermolecular forces). Small, nonpolar molecules boil at the lowest temperatures since they have the weakest IMF to overcome. As polarity increases, so does the boiling temperature, from semi-polar to the extremely polar hydrogen 'bond.' Ionic substances have even higher boiling temperatures (*Chemistry* 7th ed. pages 426–429 / 8th ed. pages 440–443).

71. **ANSWER: A** The more easily a substance can obtain the vapor state, the more molecules will be available to exert vapor pressure. The lower the polarity of the molecule, the easier it will vaporize. The CH_3CH_2OH is the most polar of this group, due to the hydrogen 'bonding' at the –OH site; it will have the lowest vapor pressure.

Propane is the most symmetrical of these molecules, the least polar, and will have the highest vapor pressure (*Chemistry* 7th ed. pages 426–429, 459–466 / 8th ed. pages 440–443, 471–479).

72. **ANSWER: D** If you think about equilibrium shifting to attempt to achieve pressure equilibrium, then only when the same number of moles of gas are found on both sides of the reaction will the system not be affected. This is the case in this problem when you have two moles of gas on each side. This effect of a change in pressure on an equilibrium system is described by Le Châtelier's principle (*Chemistry* 7th ed. pages 604–610 esp. Sample Exercise 13.14 / 8th ed. pages 620–626 esp. Sample Exercise 13.14).

73. **ANSWER: A** Since the stoichiometry of this reaction indicates that twice as much X will react as will Y and that you have added equal numbers of moles of these materials initially, then there must be less X remaining after any reaction takes place (*Chemistry* 7th ed. pages 579–582 / 8th ed. pages 594–597).

74. **ANSWER: A** Note that the second equation is the reverse of the first, and has been doubled. Therefore the K_c for the second equation will be the reciprocal of the first squared: $(1 / 4)^2 = 1/16$ (*Chemistry* 7th ed. pages 582–586 esp. Sample Exercise 13.2 and "We Can Summarize These Conclusions About the Equilibrium Expression" / 8th ed. pages 597–601 esp. Sample Exercise 13.2 and "We Can Summarize These Conclusion About the Equilibrium Expression").

75. **ANSWER: E** From the equation $Zn(OH)_2 \rightarrow Zn^{2+}(aq) + 2OH^-(aq)$ you can see that if 2.0×10^{-6} mol/L of the zinc hydroxide dissolves, that will result in the same concentration of zinc ions (2.0×10^{-6} mol/L) and twice that much of hydroxide ions (2.0×10^{-6} mol/L $\times 2 = 4.0 \times 10^{-6}$ mol/L). Then: $K_{sp} = [Zn^{2+}] [OH^-]^2 = (2.0 \times 10^{-6}$ mol/L$)(4.0 \times 10^{-6}$ mol/L$)^2 = 32 \times 10^{-18}$ (usually units are not shown with this kind of value) (*Chemistry* 7th ed. pages 717–724 / 8th ed. pages 744–752).

SECTION II FREE-RESPONSE ANSWERS

Question 1: Answers
(a) $K_C = [O_2] [SO_2]^2 / [SO_3]^2$
(b) $[SO_2] = 1.23 \times 10^{-2}$ mol / 2.25 L = 0.00547 M
   $[O_2]$ = (one half as much as above) = 0.00273 M
   $[SO_3] = 0.0401$ mol – 0.0123 mol / 2.25 L = 0.0124 M
      note: 3.21 g / 80.1 g/mol = 0.401 mol of SO_3 initially
(c) $K_C = (0.00273) (0.00547)^2 / (0.0124)^2 = 5.31 \times 10^{-4}$
(d) $K_p = K_c(RT)^{\Delta n} = 5.31 \times 10^{-4} (0.0821)(500.)^1 = 2.18 \times 10^{-2}$
(*Chemistry* 7th ed. pages 587–588, 593–598 / 8th ed. pages 602–604, 609–614)

Question 2: Answers
(a) The pairs are $HBrO_4$ (acid) / BrO_4^- (conjugate base) and $H_3SO_4^+$ (acid) / H_2SO_4 (base). Note that acid form always has one more H^+ than its conjugate base; this reaction takes place in pure acids; no water is involved.

(b)

$$\left[\begin{array}{c} :\!O\!: \\ \| \\ H\!-\!\ddot{O}\!-\!S\!-\!\ddot{O}\!-\!H \\ | \\ O \\ | \\ H \end{array} \right]^{+}$$

(c)

$$\left[\begin{array}{c} H \\ | \\ :\!O\!: \\ | \\ H\!-\!\ddot{O}\!-\!S\!-\!\ddot{O}\!-\!H \\ | \\ :\!O\!: \\ | \\ H \end{array} \right]^{2+}$$

(d) Arrhenius acids produce H^+ ions in water solutions. Brønsted-Lowry acids donate H^+ (protons) in any solvent; these acids need not be in an aqueous solution. Lewis acids are electron pair acceptors (in any solvent).
(*Chemistry* 7th ed. pages 624–626, 629, 663–665 / 8th ed. pages 640–642, 645, 679–682)

Question 3: Answers

a. (i) The $KMnO_4$ is actually less concentrated than you believe it to be since it has been diluted with the water left in the buret. That means that more of the standardized solution will be needed to react with the iron (II) solution leading you to report that the iron concentration is higher than it actually is.
(ii) If the air bubble stays in the buret during the entire titration, then there is no effect. However, if it is replaced by the $KMnO_4$ solution during the titration, then less solution leaves the buret than you are reporting, leading to a higher than correct percentage of the iron (II).
(iii) If more water is added to the flask that contains the iron salt, it will not affect the number of moles of iron (II) and therefore not affect the reported result.
(iv) If the same point on the meniscus is used as a reference, there will not be any effect on the percentage of iron reported.

b. % Error = |your lab. results – accepted value| / accepted value × 100%
|6.896 – 7.77| / 7.77 × 100% = –11%
Note that the answer is known only to two significant figures.

c. $5Fe^{2+} + MnO_4^- + 8H^{1+} \rightarrow 5Fe^{3+} + Mn^{2+} + 4H_2O$.
Fe^{2+} is undergoing oxidation (the loss of electrons) and so it is the reducing agent.

(*Chemistry* 7th ed. pages 151–152, 158–167 / 8th ed. pages 156–158, 165–168)

Question 4: Answers

Note that physical states need not be included. Do not include spectator ions.

a. (i) $NH_3 + BCl_3 \rightarrow H_3N{:}BCl_3$ (this could be shown as NH_3BCl_3)
(ii) The BCl_3 is the Lewis acid because it accepts the lone pair of electrons from the N in NH_3.

b. (i) $2Fe^{3+} + 2I^- \rightarrow 2Fe^{2+} + I_2$
(ii) The oxidation of iodine before reaction is -1 and after the reaction is zero.

c. (i) $2KClO_3 \rightarrow 2KCl + 3O_2$

(ii) In the balanced equation, 1.5 moles of oxygen are produced for every mole of potassium chlorate used.

(*Chemistry* 7th ed. pages 140–146 / 8th ed. pages 144–150)

Question 5: Answers

(a) The thermal energy added is used to overcome the forces of attraction (IMF) in the solid (not to make the particles move faster, which is an increase in kinetic energy, thereby temperature.)

(b) The distance between K^+ and Cl^- is greater than the distance between Na^+ and F^-, and so the lattice energy is smaller.

(c) Br_2 (i) is nonpolar and has lower IMF than polar ICl.

(d) The bonding electrons are shared much more unevenly in HF, giving rise to a very polar molecule. This leads to stronger dipole–dipole forces in HF (called hydrogen 'bonding,' of course).

(e) An endothermic process causes this. The alcohol evaporates. This requires energy which is lost by your skin, so the temperature of your skin decreases.

(*Chemistry* 7th ed. pages 463–465, 340–345, 435–436, 335–337, 231–236 / 8th ed. pages 475–477, 351–356, 449-459, 346-348, 238-244)

Question 6: Answers

(a) By running a series of reactions while changing the concentration of only substance X, or of only substance Y, and noting how this affects the reaction rate.

(b) It is first order with respect to X, it is first order with respect to Y, and second order overall.

(c) The rate constant may be changed by changing the temperature (but not by changing the concentration of X or of Y). Recall the Arrhenius equation.

(d) For zero-order reactions, the rate is constant {Rate = $k[A]^0 = k(1) = k$}. This means that the rate constant does not change with concentration as it does with a first-order reaction {Rate = $k[A]^1$}.

(e) All reactants in the rate determining step must be part of the rate law expression because each of those reactants controls the reaction rate.

(*Chemistry* 7th ed. pages 534–536, 538–540, 546, 549–552, Table 12.6 on page 548 / 8th ed. pages 547–549, 551-561, 559, 562–565, Table 12.6 on page 561)

SCORING THE DIAGNOSTIC TEST FREE-RESPONSE QUESTIONS

It is difficult to come up with an exact score for this section of test. However, if you compare your answers to the answers in this book, remembering that each part of the test you answer correctly is worth points even if the other parts of the answer are incorrect (see the section titled "Scoring for the Free-Response Questions" on page 13 of this book), you can get a general idea of the percentage of the questions for which you would get credit. If you believe that you got at least one-third of the possible credit, you would probably receive a 3 on this part of the test. If you believe that you would receive close to half or more of the available credit, your score would more likely be a 4 or a 5.

(b)
$$\left[\ \text{H}-\ddot{\text{O}}-\overset{\overset{\displaystyle :\text{O}:}{\|}}{\underset{\underset{\displaystyle\text{H}}{|}}{\underset{\displaystyle\text{O}}{|}}}\text{S}-\ddot{\text{O}}-\text{H}\ \right]^{+}$$

(c)
$$\left[\ \text{H}-\ddot{\text{O}}-\overset{\overset{\displaystyle\text{H}}{|}}{\underset{\underset{\displaystyle\text{H}}{|}}{\underset{\displaystyle :\text{O}:}{|}}}\overset{\displaystyle :\text{O}:}{\text{S}}-\ddot{\text{O}}-\text{H}\ \right]^{2+}$$

(d) Arrhenius acids produce H^+ ions in water solutions. Brønsted-Lowry acids donate H^+ (protons) in any solvent; these acids need not be in an aqueous solution. Lewis acids are electron pair acceptors (in any solvent).
(*Chemistry* 7th ed. pages 624–626, 629, 663–665 / 8th ed. pages 640–642, 645, 679–682)

Question 3: Answers

a. (i) The $KMnO_4$ is actually less concentrated than you believe it to be since it has been diluted with the water left in the buret. That means that more of the standardized solution will be needed to react with the iron (II) solution leading you to report that the iron concentration is higher than it actually is.
 (ii) If the air bubble stays in the buret during the entire titration, then there is no effect. However, if it is replaced by the $KMnO_4$ solution during the titration, then less solution leaves the buret than you are reporting, leading to a higher than correct percentage of the iron (II).
(iii) If more water is added to the flask that contains the iron salt, it will not affect the number of moles of iron (II) and therefore not affect the reported result.
(iv) If the same point on the meniscus is used as a reference, there will not be any effect on the percentage of iron reported.

b. % Error = |your lab. results – accepted value| / accepted value × 100%
|6.896 – 7.77| / 7.77 × 100% = –11%
Note that the answer is known only to two significant figures.

c. $5Fe^{2+} + MnO_4^- + 8H^{1+} \rightarrow 5Fe^{3+} + Mn^{2+} + 4H_2O$.
Fe^{2+} is undergoing oxidation (the loss of electrons) and so it is the reducing agent.
(*Chemistry* 7th ed. pages 151–152, 158–167 / 8th ed. pages 156–158, 165–168)

Question 4: Answers

Note that physical states need not be included. Do not include spectator ions.

a. (i) $NH_3 + BCl_3 \rightarrow H_3N:BCl_3$ (this could be shown as NH_3BCl_3)
(ii) The BCl_3 is the Lewis acid because it accepts the lone pair of electrons from the N in NH_3.

b. (i) $2Fe^{3+} + 2I^- \rightarrow 2Fe^{2+} + I_2$
(ii) The oxidation of iodine before reaction is -1 and after the reaction is zero.

c. (i) $2KClO_3 \rightarrow 2KCl + 3O_2$

(ii) In the balanced equation, 1.5 moles of oxygen are produced for every mole of potassium chlorate used.
(*Chemistry* 7th ed. pages 140–146 / 8th ed. pages 144–150)

Question 5: Answers
(a) The thermal energy added is used to overcome the forces of attraction (IMF) in the solid (not to make the particles move faster, which is an increase in kinetic energy, thereby temperature.)
(b) The distance between K^+ and Cl^- is greater than the distance between Na^+ and F^-, and so the lattice energy is smaller.
(c) Br_2 (i) is nonpolar and has lower IMF than polar ICl.
(d) The bonding electrons are shared much more unevenly in HF, giving rise to a very polar molecule. This leads to stronger dipole–dipole forces in HF (called hydrogen 'bonding,' of course).
(e) An endothermic process causes this. The alcohol evaporates. This requires energy which is lost by your skin, so the temperature of your skin decreases.
(*Chemistry* 7th ed. pages 463–465, 340–345, 435–436, 335–337, 231–236 / 8th ed. pages 475–477, 351–356, 449-459, 346-348, 238-244)

Question 6: Answers
(a) By running a series of reactions while changing the concentration of only substance X, or of only substance Y, and noting how this affects the reaction rate.
(b) It is first order with respect to X, it is first order with respect to Y, and second order overall.
(c) The rate constant may be changed by changing the temperature (but not by changing the concentration of X or of Y). Recall the Arrhenius equation.
(d) For zero-order reactions, the rate is constant {Rate = $k[A]^0 = k(1) = k$}. This means that the rate constant does not change with concentration as it does with a first-order reaction {Rate = $k[A]^1$}.
(e) All reactants in the rate determining step must be part of the rate law expression because each of those reactants controls the reaction rate.
(*Chemistry* 7th ed. pages 534–536, 538–540, 546, 549–552, Table 12.6 on page 548 / 8th ed. pages 547–549, 551-561, 559, 562–565, Table 12.6 on page 561)

SCORING THE DIAGNOSTIC TEST FREE-RESPONSE QUESTIONS

It is difficult to come up with an exact score for this section of test. However, if you compare your answers to the answers in this book, remembering that each part of the test you answer correctly is worth points even if the other parts of the answer are incorrect (see the section titled "Scoring for the Free-Response Questions" on page 13 of this book), you can get a general idea of the percentage of the questions for which you would get credit. If you believe that you got at least one-third of the possible credit, you would probably receive a 3 on this part of the test. If you believe that you would receive close to half or more of the available credit, your score would more likely be a 4 or a 5.

Part II

A Review of AP Chemistry

CHEMICAL FOUNDATIONS

1

In this chapter, you will perform the basic calculations used in chemistry. Chemists use equipment with varying degrees of precision. The measurements recorded in an experiment must reflect the precision of the equipment used. The results of the calculations in an experiment also must reflect the precision of the equipment.

Many of the problems in chemistry use dimensional analysis to convert from one unit to another or to solve stoichiometry problems. Other basic calculations that will be discussed are density and temperature conversions.

You should be able to

- Identify the number of significant figures in a given measurement.
- Perform calculations involving significant figures.
  - ☐ Memorize the rules for counting and performing operations with significant figures.
- Differentiate between accuracy and precision as they apply to measurement.
- Determine the density of solids and liquids and calculate volumes or masses using the given density.
- Convert between units of temperature: degrees Celsius and Fahrenheit, and Kelvin.
- Identify the characteristics of the states of matter: solids, liquids, and gases.
- Identify substances such as elements, compounds, or mixtures.
- Identify methods of separation of mixtures.
- Identify changes as being physical or chemical.

AP tips

Significant figures are important in calculations. One point out of the nine possible in the free-response sections will be deducted for errors in significant figures that are off by ± 1 significant figure.

Units are the key to problem solving. Use conversion factors and cancel out units or problems involving conversions or stoichiometry. Every measured number or number calculated from measurements has a unit except for equilibrium constants, which by convention are reported without units.

UNCERTAINTY IN MEASUREMENT

(Chemistry 7th ed. pages 10–13 / 8th ed. pages 11–14)

SIGNIFICANT FIGURES

The significant figures of a measurement are all of the certain digits in a measurement and the first uncertain digit (estimated number). Students should be able to read measurements to the proper number of significant figures.

EXAMPLE: Figure 1.9 on page 10 in the 7th edition and Figure 1.7 on page 11 in the 8th edition shows the measurement of a volume of liquid using a buret. The *certain digits* in the measurement are the three numbers 20.1. The digit to the right of the one must be estimated by interpolating between the 0.1 mL marks. The measurement with *uncertainty* can be reported as 20.15 mL

PRECISION AND ACCURACY

Accuracy refers to the agreement of a particular value with a true value.

Precision refers to the degree of agreement among several measurements of the same quantity. The degree of precision refers to the number of digits that a measuring device permits one to measure. In a measuring device, all except the last digit, which is estimated, are certain. For example, a balance which measures to the nearest 0.0001 g is more precise than one that measures to the nearest 0.01 g.

$$\text{Percent Error} = \frac{\text{Experimental Value} - \text{Actual Value}}{\text{Actual Value}} \times 100\%$$

EXAMPLE: In an experiment, the density of aluminum is to be determined. Two students perform the experiment three times and obtain the following results.

Trial	Student A	Student B
1	2.45 g/mL	2.69 g/mL
2	2.43 g/mL	2.70 g/mL
3	2.44 g/mL	2.71 g/mL

Describe the accuracy and precision of each student's results. For Student A, calculate the mean and the percent error if the actual value is 2.70 g / mL.

SOLUTION: The values for Student A are precise, but inaccurate. The values for student B are both precise and accurate.

Random error (indeterminate error) means that a measurement has an equal probability of being high or low.

Systematic error (determinate error) occurs in the same direction each time; it is either always high or always low.

EXAMPLE: A balance could have a defect causing it to give a result that is consistently 1.000 g too high.

SIGNIFICANT FIGURES AND CALCULATIONS

It is important to know the uncertainty in the final result in an experiment. The final reported result cannot have more certainty than the least accurate measurement. The number of significant figures in a single value will be determined. Memorize the rules below and use them to answer the examples that follow.

Rules for Counting Significant Figures

1. Nonzero integers always count as significant figures.

2. Zeros: There are three classes of zeros.

    a. Leading zeros precede all the nonzero digits and do not count as significant figures. Example: 0.0025 has 2 significant figures.

    b. Captive zeros are zeros between nonzero numbers. These always count as significant figures. Example: 1.008 has 4 significant figures.

    c. Trailing zeros are zeros at the right end of the number.

        Trailing zeros are only significant if the number contains a decimal point. Example: 1.00×10^2 has three significant figures.

        Trailing zeros are not significant if the number does not contain a decimal point. Example: 100 has one significant figure.

3. Exact numbers, which can arise from counting or definitions such as 1 in = 2.54 cm, never limit the number of significant figures in a calculation.

EXAMPLE: How many significant figures are in each of the following?

100 L

SOLUTION: There is 1 significant figure. Trailing zeros do not count. Only the "1" is significant.

0.001010 L

SOLUTION: There are 4 significant figures. Leading zeros do not count. The numbers "1010" are significant because one zero is captive and the other zero is trailing, but a decimal is present.

Rules for Significant Figures in Calculations

1. For multiplication and division, the number of significant figures in the result is the same as the number with the least number of significant figures in the calculation.

2. For addition and subtraction, the result has the same number of decimal places as the number with the least number of decimal places in the calculation.

3. Rules for rounding:

   In a series of calculations, carry the extra digits to the final result, then round.

   If the digit to be removed

   is less than 5, the preceding digit stays the same. For example, 2.44 rounds to 2.4.

   is greater than or equal to 5, the preceding digit is increased by 1. For example, 2.45 rounds to 2.5.

It is important to calculate the results of mathematical expressions to the proper number of significant figures. Memorize the rules above and apply them to the examples that follow.

EXAMPLE: Perform the following calculations to the correct number of significant figures.

1) 16.8 g + 3.2557 g

SOLUTION: The calculator answer is 20.0557. The correct answer is 20.1g. The answer should have one decimal place.

2) 27 g / 4.148 mL

SOLUTION: The calculator answer is 6.509161041. The correct answer is 6.5 g/ mL. The answer should have two significant figures.

DIMENSIONAL ANALYSIS

(*Chemistry* 7th ed. pages 16–19 / 8th ed. pages 17–20)

Dimensional analysis is used to convert from one unit to another. It is the single most valuable mathematical technique that you will use in general chemistry. The method involves conversion factors to cancel units until you have the proper unit in the proper place. When you are setting up problems using dimensional analysis, you should be more concerned with units than numbers.

EXAMPLE: The density of mercury is 13.6 g/cm³. How many pounds would 1.00 liter of mercury weigh?

SOLUTION:

$$1.00L \times \frac{1000 \text{ cm}^3}{1L} \times \frac{13.6 \text{ g}}{1 \text{ cm}^3} \times \frac{1 \text{ pound}}{453.6 \text{ g}} = 30.0 \text{ pounds}$$

Double check that all of your units cancel properly. If they do, your numerical answer is probably correct. If they don't, your answer is definitely wrong.

DENSITY

(*Chemistry* 7th ed. pages 24–25 / 8th ed. pages 24–26)

Density is the mass of substance per unit volume of substance. Density = mass/volume.

The density of an object can be determined through the *water displacement method*. The object is massed and then submerged in a measured amount of water in a graduated cylinder. The final volume in the graduated cylinder is read. The volume of water displaced by the object is the volume of the object.

EXAMPLE: A sample containing 33.42 g of metal pellets is poured into a graduated cylinder containing 12.7 mL of water, causing the water level in the cylinder to rise to 21.6 mL. Calculate the density of the metal.

SOLUTION:

Volume of metal = (Volume H₂O metal) – Volume H₂O

$$= 21.6 \text{ mL} - 12.7 \text{ mL} = 8.9 \text{ mL}$$

Density of metal = 33.42 g / 8.9 mL = 3.8 g/mL

Note: The answer has 2 significant figures.

TEMPERATURE

(*Chemistry* 7th ed. pages 19–22 / 8th ed. pages 20–23)

You should be able to interconvert among Fahrenheit, Celsius and Kelvin. You should also know the freezing and boiling points of water on each scale.

$$°C \times (9/5) + 32 = °F$$

$$T_K = T_C + 273.15$$

EXAMPLE: The boiling point of water on top of Long's Peak in Colorado (14,255 feet above sea level) is about 86.0⁰C. What is the boiling point in Kelvins and degrees Fahrenheit?

SOLUTION:

$$T_K = T_C + 273.15 = 86.0°C + 273.15 = 359.2 \text{ K}$$

$$T_F = T_C \times (9/5) + 32.0°F = 86.0 \times (9/5) + 32 = 187°F$$

CLASSIFICATION OF MATTER

(*Chemistry* 7th ed. pages 25–28 / 8th ed. pages 26–29)

Matter exists in three states: solid, liquid, and gas. Properties of these three states of matter are listed below.

State of Matter	Shape	Volume
Solid	Fixed	Fixed
Liquid	Not definite	Fixed
Gas	Not fixed; shape of container	Not fixed; volume of container

MIXTURES AND PURE SUBSTANCES

(*Chemistry* 7th ed. page 28 / 8th ed. pages 28–29)

Pure substances, such as elements or compounds, make up mixtures. The following chart summarizes mixtures and the methods for separating them.

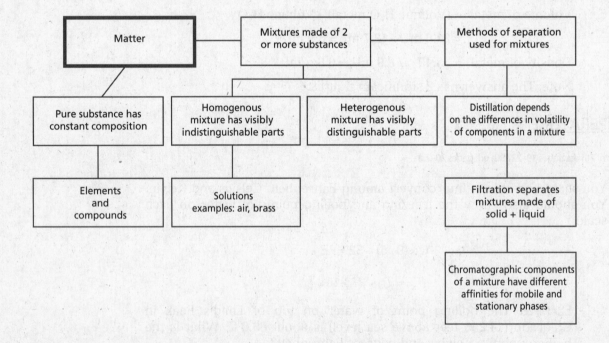

CHANGES IN MATTER

Matter can undergo physical or chemical changes.

Physical changes in matter do not change the original composition of the substance. Changes in state such as boiling or melting are physical changes. Changes involving an alteration in the form of the substance such as grinding or tearing are physical. Physical properties are properties of a substance that can be observed without changing the composition of the substance. During a physical change, bonds are not broken and no reaction between atoms occurs. For example, density, color, and boiling point are physical properties.

Chemical changes in matter change the composition of the original substance by breaking and making bonds between atoms. A new substance is produced when a chemical change occurs. Evidence that a chemical change has occurred includes change in color or odor or the production of a gas or a solid (precipitate). Some examples of chemical properties include flammability and reactivity to air.

MULTIPLE-CHOICE QUESTIONS

No calculators are to be used in this section.

When choosing your answers, it is helpful to make the general observation that numerical answers tend to fall into one of two areas. Quite often the answers will vary greatly from each other. For example, you usually know if an answer of 3.0 grams or 300. grams is more reasonable. Secondly, consider how many significant figures an answer should have when examining all of the answers.

1. Measurements indicate a charge of 0.444 C passes a point in 0.12 seconds. The current (i.e., the rate of charge flow, in C/s) is best expressed as
    (A) 0.27 C/s.
    (B) 0.270 C/s.
    (C) 3.7 C/s.
    (D) 3.70 C/s.
    (E) 3.700 C/s.

2. A given sample contains 2.0 g of hydrogen, 33.1 g of sulfur, and 75.01 g of oxygen. What is the total mass of the sample?
    (A) 110.12 g
    (B) 110.1 g
    (C) 110. g
    (D) 1.1×10^2 g
    (E) 1.1×10^{-2} g

3. What is the percent mass of sulfur in the above mixture (see number 2)?
    (A) 3.01%
    (B) 3.0%
    (C) 3.02%
    (D) 30.1%
    (E) 30%

4. The correct name for Li_3N is
   (A) lithium nitride.
   (B) lithium nitrate.
   (C) trilithium nitride.
   (D) monolithium mononitride.
   (E) lithium nitrogen.

5. The correct name for $Mg(OH)_2$ is
   (A) magnesium hydroxide.
   (B) magnesium(I) hydroxide.
   (C) magnesium(II) hydroxide.
   (D) magnesium(II) hydroxide(I).
   (E) magnesium hydrogen oxide.

6. The correct name for Co_2O_3 is
   (A) cobalt oxide.
   (B) cobalt(II) oxide.
   (C) cobalt oxide(III).
   (D) cobalt(III) oxide.
   (E) dicobalt trioxide.

7. The correct formula for copper(II) phosphate is
   (A) $CuPO_4$.
   (B) CuP.
   (C) $Cu_3(PO_4)_2$.
   (D) Cu_2P.
   (E) Cu_2PO_4.

8. Element 104, rutherfordium, has a half-life of 65 seconds. How much of a 1.00 g sample of element 104 will remain at the end of 2.1667 minutes?
   (A) 0.25 g
   (B) 0.250 g
   (C) 0.50 g
   (D) 0.500 g
   (E) 0.75 g

9. The density of copper is 8.96 g/cm³. What is the mass of 18.88 cm³ of pure copper?
   (A) 1.69 g
   (B) 16.9 g
   (C) 169 g
   (D) 169.0 g
   (E) 1690 g

10. The density of a piece of metal can be determined from mass and water displacement data. A piece of metal with a mass of 15.54 g is placed in a flask with a volume of 50.00 cm³. It is found that 40.54 g of water (d = 0.9971 g/cm³) is needed to fill the flask with the metal in it. The density of the metal is most nearly (all answers in g/cm³):
    (A) 1.66.
    (B) 1.7.
    (C) 9.46.
    (D) 9.5.
    (E) 40.7.

11. Which of the following processes represents a chemical change?
    (A) Water boiling
    (B) Iodine subliming
    (C) Sugar dissolving in water
    (D) Natural gas burning
    (E) Ice melting

12. Which of the following rows has the correct SI unit given for each unit
    of measure (choose the row where all of the SI fundamental units are
    correct).

Mass	Length	Time	Amount
(A) gram	meter	second	mole
(B) milligram	inch	minute	pair
(C) gram	foot	second	dozen
(D) kilogram	meter	hour	mole
(E) kilogram	meter	second	mole

13. A rocket is traveling at 12.0 km/s. What is this speed is miles/hour?
    (A) 4.47×10^2
    (B) 2.68×10^4
    (C) 5.40×10^4
    (D) 6.95×10^4
    (E) 1.20×10^5

14. A pure solid is heated and it decomposes into two substances, one a
    liquid and the other a gas. One can conclude with certainty that:
    (A) The two products are elements.
    (B) One of the products is an element.
    (C) The original solid is not an element.
    (D) The liquid is a compound and the gas is an element.
    (E) Both products are compounds.

15. A student determines the mass of silver in a sample and does four
    determinations. The results are 1.75 g, 1.71 g, 1.85 g and 1.93 g. The
    true value is 1.81 g. Which statement concerning the results is correct?
    (A) High precision and accurate results.
    (B) High precision and poor accuracy.
    (C) Poor precision and poor accuracy.
    (D) Poor precision and accurate results.
    (E) Reasonable precision and poor accuracy.

FREE-RESPONSE QUESTIONS

1(a)   Describe how, in the laboratory, you might determine experimentally
       the density of a solid, such as a sugar cube, which is water soluble.
       Indicate what equipment you might best use in the process.

1(b)   Then describe a second experimental method for determining the
       density of this same object, so that you might verify the results of the
       first method.

2.   Discuss the terms *precision* and *accuracy*. Describe what each term is
     used to indicate.

Answers

MULTIPLE CHOICE

1. **C** Even if you do not understand what is meant by a term (in this case, current) you can often successfully attack a problem by careful attention to units. In the problem, you are seeking an answer in C/s, so divide the 0.444 C by 0.12 s (and keep your answer to two significant figures (*Chemistry* 7th ed. pages 13–14) / 8th ed. pages 14-15).

2. **B** Each of these values is known to one-tenth of a gram (0.1 g) or more, so the total may be known to no more than the nearest 0.1 g. Remember, when adding or subtracting, you do not count the number of significant figures but instead look at the position of the decimal point in each of the data used in the calculation (*Chemistry* 7th ed. pages 13–14 / 8th ed. pages 14-15).

3. **D** Now you are dividing, so the answer is known to the number of significant figures which is the least number of significant figures in data used to obtain the answer (here, three significant figures in the mass of sulfur limits the answer to three significant figures even though you know the total mass to four). Also note that the 100% is known by definition, that is, it is considered an "exact number" (*Chemistry* 7th ed. pages 13–14 / 8th ed. pages 14-15).

4. **A** When naming binary ionic compounds, you are to use the names of the cation followed by the anion. Prefixes are used when considering nonmetal–nonmetal compounds (*Chemistry* 7th ed. pages 57–61 / 8th ed. pages 56-61).

5. **A** Roman numerals are used with transition metal compounds to indicate the charge or oxidation number of the metal ion in this compound. Since transition metal ions can often have more than one charge (oxidation number), this identification is necessary. There is no need for such nomenclature when the magnesium ion used for it can have only a 2+ charge. Note also that even the transition metal ions Ag^+, Zn^{2+}, and Cd^{2+} generally are found with only the oxidation number indicated, and since there is only one form of ion, Roman numerals are not used with these three transition metal ions (*Chemistry* 7th ed. pages 57–61, esp. Table 2.4 / 8th ed. pages 56-61, esp. Table 2.4).

6. **D** As indicated in the discussion for question #5, it is necessary to identify the oxidation number of the cobalt in this oxide since it is not always three (*Chemistry* 7th ed. pages 57–61 / 8th ed. pages 56-61).

7. **C** Both copper with a 2+ and a 1+ charge are common ions. It is necessary to assign copper the 2+ charge in this case and then determine the ratio of Cu^{2+} and PO_4^{3-} to make the unit neutral; hence $Cu_3(PO_4)_2$ (*Chemistry* 7th ed. pages 59-62, esp. Table 2.5 / 8th ed. pages 61-62, esp. Table 2.5).

8. **A** Consider that 2.1667 minutes × 60. s/min = 130. s, and that is equal to two half-lives. So 1.00 g × 1/2 × 1/2 = 0.25 g. Note also that 65s for one half-life, given in the problem, is known only to two significant figures and that is what controls the number of significant figures in this problem (*Chemistry* 7th ed. pages 13–16 / 8th ed. pages 14-17).

9. **C** From density = *m/v*, solving for mass and using mass = (*d*) × (*v*)= 8.96 g/cm^3 × 18.88 cm^3. A 'quick and dirty' estimate using 9 × 20 would given an answer of about 180 g; only three significant figures are allowed by the density value since it is limited to three significant figures, so there is only one possible selection from the possible answers (*Chemistry* 7th ed. pages 24–25 / 8th ed. pages 24-26).

10. **A** The volume of water displaced by the piece of metal is 50.00 cm^3 - 40.66 = 9.34 cm^3. Using mass / volume = 15.54 g / 9.34 cm^3 = 1.66 g / cm^3. Note that 40.54 g of water with *d* = 0.9971 g/cm^3 has a volume of 40.66 cm^3 (*Chemistry* 7th ed. pages 24–25 / 8th ed. pages 24-26).

11. **D** When natural gas burns, at least two new substances are produced, CO_2 and H_2O. Choices (A), (B), and (E) are all examples of phase changes or changes of state, which are physical changes. Choice (C) is an example of a mixture, which can be separated by physical means back into the original substances (*Chemistry* 7th ed. pages 25-28 / 8th ed. pages 25-29).

12. **E** mass   kilogram   kg
    length   meter   m
    time   second   s
    amount   mole   mol
    (*Chemistry* 7th ed. page 9 / 8th ed. page 9)

13. **B** 12.0 km/s ×   (3600s/1 hour) × (1 mile/1.609 km) = 2.68 × 10^4 miles/hour. (*Chemistry,* 7th ed. pages 16-19, and Appendix 6, page A26 / 8th ed. pages 17-20, and Appendix 6)

14. **C** The original solid is pure and therefore it must be a compound because it can be chemically decomposed into simpler substances. You are not given specific information about the composition of the products and thus you can not make any definite conclusions about them (*Chemistry* 7th ed. pages 27-28 / 8th ed. pages 28-29).

15. **D** The average of the measurements is 1.81 g, which is the same as the true value. Therefore, the accuracy is high. The range of measured values is 1.75 g to 1.93 g, which is a difference of 0.18 g, or 10% of the average value, which is poor precision. Precision is a measure of how close the measurements are to each other (*Chemistry* 7th ed. pages 12-13 / 8th ed. pages 13-14).

FREE RESPONSE

1   (a) Since density is a ratio of the mass of the object compared to the volume of the object, you might first determine the mass of the sugar cube using a balance. Greater precision is possible with an analytical balance. Then determine the volume of this regular cube

by measuring its length, width, and depth (they should all be the same, of course!) and calculating the volume $[V = (l) \times (w) \times (d) = (side)^3]$. It is then a simple step to divide the mass by the volume just calculated (*Chemistry* 7th ed. pages 24–25 / 8th ed. pages 24-25).

b) The second method suggests a liquid displacement method for determining volume. However the liquid used may not be water since the object is water soluble. Select a nonpolar liquid to keep this polar solid from dissolving, perhaps, 1,1,1-trichloroethane (*Chemistry* 7th ed. pages 24–25 and 127–129 / 8th ed. pages 24-25 and 130-132).

2. Precision indicates how reproducible a value is (i.e., agreement with other measured data). Accuracy indicates the extent of agreement with an accepted value (i.e., with a standard value) (*Chemistry* 7th ed. pages 10–13 / 8th ed. pages 11-14).

The mass number of an isotope equals the total number of protons and neutrons in an atom which can be obtained by rounding the atomic mass in the periodic table to the nearest whole number. This is not the same as the atomic mass, which is an average of the masses of the isotopes of the particular element.

WRITING SYMBOLS

(*Chemistry* 7th ed. pages 50–51 / 8th ed. pages 51–52)

Mass number → $^{23}_{11}Na$
Atomic number →

> **EXAMPLE:** How many protons, neutrons and electrons are in an atom of sodium – 23?

> **SOLUTION:** There are 11 protons since the atomic number is 11. There are also 11 electrons since atoms are neutral. There are 12 neutrons (mass number – atomic number = 23 – 11).

ISOTOPES

(*Chemistry* 7th ed. pages 78–81 / 8th ed. pages 78–81)

Isotopes are atoms of the same element with different numbers of neutrons and therefore different atomic masses.

CALCULATION OF AVERAGE ATOMIC MASS FROM ISOTOPIC DATA

(*Chemistry* 7th ed. pages 78–81 / 8th ed. pages 78–81)

The average atomic mass of an element can be calculated from the percent abundance and mass of each isotope for that element.

Avg Atomic Mass = Σ(Percent abundance of isotope $\times$ mass value)

> **EXAMPLE:** Element "E" is present with the following mass values and natural abundances.

Isotope	Mass Value (amu)	Percent Abundance
^{10}E	10.01	19.78 %
^{11}E	11.01	80.22 %

What is the average atomic mass of the element, E? What is the element?

(0.1978) 10.01 + (0.8022) 11.01 = 10.812 amu.

The element is boron.

IONS

(*Chemistry* 7th ed. pages 53, 57–58, 62 / 8th ed. pages 52-53, 56–57, 61-62)

An ion is an atom that has lost or gained electrons. A polyatomic ion is a group of atoms bonded together as a single unit that carries a charge.

EXAMPLE: Aluminum forms a cation, a positive ion, by losing three electrons:

$$Al \rightarrow Al^{3+} + 3e^-$$

Oxygen forms an anion, a negative ion, by gaining three electrons:

$$O + 2e^- \rightarrow O^{2-}$$

NOMENCLATURE

(*Chemistry* 7th ed. pages 57–67 / 8th ed. pages 56–67)

When you finish this section you will be able to name, or give formulas for, the following classes of compounds: binary salts, salts with polyatomic ions, binary covalent compounds and acids.

Use the flowcharts that follow to answer the questions below.

Naming Compounds from the Given Formula

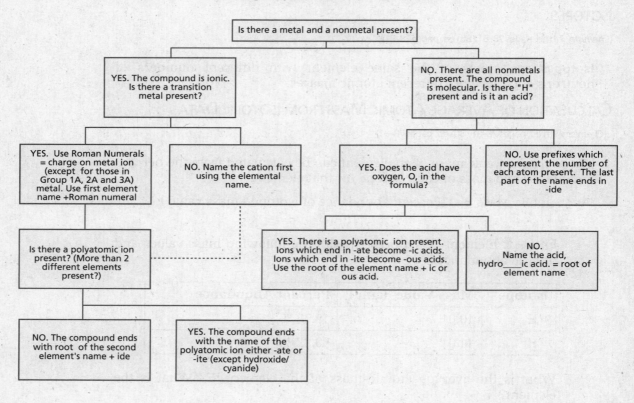

Writing Formulas from the Given Name

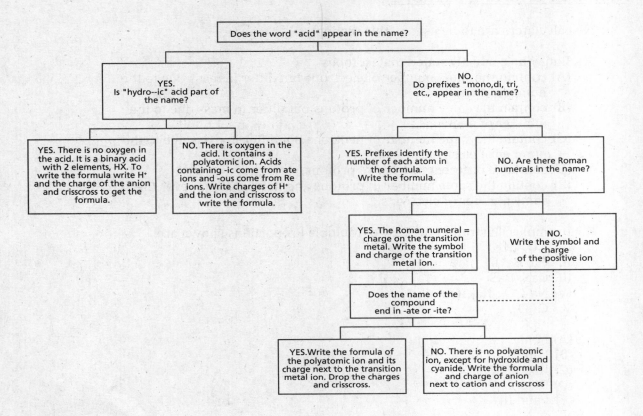

EXAMPLES:

1.   Name each of the following compounds.

      a. CaF_2 d. KIO_3

      b. Cl_2O_7 e. HF (aq)

      c. CuO   f. HNO_2

2. Write the formulas for each of the following compounds.

      a. Tin(IV) oxide   d. Sulfurous acid

      b. Hydrosulfuric acid   e. Potassium chloride

      c. Nickel(II) nitrate   f. Iodine pentafluoride

SOLUTIONS:

1.   a. Calcium fluoride   d. Potassium iodate

      b. Dichlorine heptoxide   e. Hydrofluoric acid

      c. Copper(II) oxide   f. Nitrous acid

2.   a. SnO_2 d. H_2SO_3

      b. H_2S e. KCl

      c. $Ni(NO_3)_2$ f. IF_5

MULTIPLE-CHOICE QUESTIONS

No calculators are to be used in this section.

1. It may be correctly stated that isotopes
    (A) contain the same number of electrons but differ in mass due to the number of neutrons.
    (B) contain the same number of protons but differ in mass due to the number of neutrons.
    (C) contain the same number of neutrons but differ in mass due to the number of electrons.
    (D) contain a different number of protons.
    (E) contain the same number of protons and neutrons but never the same number of electrons.

2. Examples illustrating the Law of Multiple Proportions shown are
    I. CO, CO_2.
    II. Ca, BaO.
    III. CaS, BaS.
    IV. Na_2CO_3, Na_2SO_4.
    V. O_2, O_3.

    (A) I only
    (B) I and V
    (C) III and IV
    (D) III and V
    (E) I and III

3. The chemist who is given credit for being the first to use symbols for elements and who developed a system of formulas for compounds was
    (A) Berzelius.
    (B) Thomson.
    (C) Lavoisier.
    (D) LeChatelier.
    (E) Arrhenius.

4. The chemist given credit for first developing a valid model of the nucleus as a result of experiments with gold foil and alpha particle scattering was
    (A) Boyle.
    (B) Thomson.
    (C) Rutherford.
    (D) Berzelius.
    (E) Arrhenius.

5. The correct name of $Ca_3(PO_4)_2$ is
    (A) tricalcium diphosphate.
    (B) calcium diphosphate.
    (C) calcium (II) phosphate.
    (D) calcium phosphate.
    (E) calcium phosphide.

6. The correct number of atomic particles for $^{51}_{23}V$ is

	electrons	protons	neutrons
(A)	28	28	23
(B)	22	22	29
(C)	23	28	23
(D)	28	23	23
(E)	23	23	28

7. In which of the following is a pair of isotopes represented?

		Atomic Number	Mass Number
(A)	X	24	52
	Y	25	52
(B)	X	26	56
	Y	27	59
(C)	X	30	66
	Y	30	66
(D)	X	28	58
	Y	28	59
(E)	X	16	20
	Y	8	10

8. The correct systematic name for SnI_4 is
   (A) silver iodide.
   (B) tin iodide.
   (C) tin iodate.
   (D) tin (IV) iodide.
   (E) tin iodide(IV).

9. The correct systematic name for N_2O_4 is
   (A) dinitrogen trioxide.
   (B) dinitrogen tetroxide.
   (C) nitrogen(II) oxide.
   (D) nitrogen(IV) oxide.
   (E) nitrogen(II) oxide(IV).

10. The correct formula for silver chlorate is
    (A) $AgCl$.
    (B) Ag_2Cl.
    (C) S_2ClO_3.
    (D) Ag_2ClO_3.
    (E) $AgClO_3$.

11. An ion contains 18 electrons, it has a net charge of 2+, and its nucleus contains 20 neutrons. What is the element?
    (A) argon
    (B) calcium
    (C) chlorine
    (D) potassium
    (E) sulfur

12. A radioactive iodine isotope used in thyroid therapy has a mass
number of 131. The isotope is administered as an iodide salt. How
many protons, neutrons and electrons are present in this isotope?

	protons	neutrons	electrons
(A)	53	131	54
(B)	53	78	77
(C)	53	78	52
(D)	53	78	54
(E)	78	53	79

13. For which of the following pairs are the atoms most likely to form an
ionic bond with each other?
(A) carbon and oxygen
(B) calcium and chlorine
(C) chlorine and oxygen
(D) sodium and magnesium
(E) chlorine and neon

14. The use of mercury salts has been minimized, if not entirely eliminated,
in general chemistry and high school laboratories because of their
potential toxicity and damage to the environment if not disposed of
properly. Which of the following represents the correct chemical
formulas for mercury(I) chloride and mercury(II) sulfide?
(A) $HgCl$ and HgS_2
(B) $HgCl_2$ and HgS_2
(C) Hg_2Cl and HgS_2
(D) Hg_2Cl_2 and HgS
(E) Hg_2Cl_2 and Hg_2S

15. From the list below what do the elements Na, K and Cs have in
common?
(1) metallic
(2) nonmetallic
(3) form cations
(4) form anions
(5) alkali metals
(6) alkaline-earth metals
(7) halogens

(A) 1 and 3
(B) 2 and 4
(C) 1, 4 and 7
(D) 1, 3 and 5
(E) 2, 4 and 6

FREE-RESPONSE QUESTIONS

1. In the late 1880s, J.J. Thomson experimented with cathode rays by
bending them in an applied electromagnetic field. He discovered that
the charge / mass (e/m) ratio was the same no matter what the cathode
was made of.
(a) What did this suggest?

    (b) What were these particles called?
    (c) What model of the atom's structure did he propose?

2. (a) Draw and label the basic parts of Rutherford's famous gold foil experiment.
    (b) Describe his results and the conclusions he reached.
    (c) What would be the results of Rutherford's experiment if J.J. Thomson's model had been correct?

3. Chemist Lawrence Strong once said that the model "must not be confused with the reality that it mirrors." Explain what he meant.

Answers

MULTIPLE CHOICE

1. **B** Isotopes are made of atoms of the same element, differing in mass due only to a different number of neutrons in the nucleus (*Chemistry* 7th ed. pages 78–83 / 8th ed. pages 78–82).

2. **A** The Law of Multiple Proportions deals with the relative masses of two elements that combine in more than one ratio with each other (*Chemistry* 7th ed. pages 43–45 / 8th ed. pages 43–47).

3. **A** (*Chemistry* 7th ed. pages 46–47 / 8th ed. pages 46–48).

4. **C** The summary of the work of many of the early chemists is worth reviewing (*Chemistry* 7th ed. pages 41–48 / 8th ed. pages 41–49).

5. **D** This is a compound made up of a nontransition metal and a polyatomic ion, hence it is named by noting the metal followed by the polyatomic ion (*Chemistry* 7th ed. pages 57–58 / 8th ed. pages 57–58).

6. **E** The atomic number, 23, indicates the number of protons and if this is not an ion, the number of electrons in the atom. The mass number, here 51, gives the total number of protons plus neutrons (51 − 23 = 28) (*Chemistry* 7th ed. pages 49–52 / 8th ed. pages 49–52).

7. **D** Isotopes must have the same atomic number since they have the same number of protons. The mass number differs because of a difference in the number of neutrons (*Chemistry* 7th ed. pages 78–85 / 8th ed. pages 78–84).

8. **D** Tin, symbol Sn, commonly exists in ionic form with either a 2+ or a 4+ charge. It is necessary to indicate which ion is involved in this transition metal–nonmetal compound (*Chemistry* 6th ed. pages 63–65 / 7th ed. pages 58–61).

9. **B** This is a good example of a binary covalent compound of Type III (*Chemistry* 7th ed. pages 57–66 / 8th ed. pages 56–66).

10. **E** Even though it is a transition metal ion, silver ions almost always have a 1+ charge. Therefore there is no need to indicate the charge on this ion with Roman numerals. You do need to know the charge on the silver ion as well as the formula and charge of the chlorate ion (*Chemistry* 7th ed. pages 58–63 / 8th ed. pages 58–63).

11. **B** If the element has a charge of 2+, it must have lost 2 electrons. The number of electrons in the neutral atom = 18+2=20. This number must be equal to the number of protons, which is the atomic number. Calcium has an atomic number of 20 (*Chemistry* 7th ed. pages 50-54 / 8th ed. pages 50-54).

12. **D** Iodine, so the atomic number = number of protons = 53. Since it is administered as the iodide ion, the charge is -1, so the number of electrons is equal to 53 +1 =54. Since the mass number is the number of protons and neutrons, number of neutrons = 131-53 = 78 (*Chemistry* 7th ed. pages 50-54 / 8th ed. pages 50-54).

13. **B** Metals form positive ions and nonmetals form negative ions. This combination of elements is responsible for the makeup of a binary ionic compound. Nonmetals are joined together by covalent bonds. Metal atoms form neither ionic nor covalent bonds with each other (*Chemistry* 7th ed. pages 58-63 / 8th ed. pages 57-63).

14. **D** The mercury(I) ion is unusual in that it occurs in nature as a diatomic chemical species, an ion in this case. The naming of the mercury(II) ion follows the more conventional preferred system of naming transition metal ions that have more than one potential positive charge (*Chemistry* 7th ed. pages 59-61 / 8th ed. pages 58-61).

15. **D** The three elements are members of the alkali metals (Group 1A) and all metals form cations (*Chemistry* 7th ed. pages 55-56 / 8th ed. pages 55-56).

FREE RESPONSE

1. J.J. Thomson was able to determine the ratio of the charge to mass of the electron, showing that cathode rays consist of discrete particles. He also showed that cathode rays always had the same kind of properties. He concluded that atoms of all substances contain the same kind of negative particles, called electrons. From this he developed the plum pudding model (electrons were dispersed in an atom like raisins in a pudding) (*Chemistry* 7th ed. pages 45–49 / 8th ed. pages 46–50).

2. In your diagram, you must show at least (1) a source of alpha particles, (2) a thin sheet of gold foil, and (3) a detecting screen. It is important to indicate that most of the alpha particles were deflected, a few through very large angles, and that this suggested to Rutherford that most of the atom is empty space and that the smaller number of particles experiencing the wide angle deflections suggest a very small and very dense nucleus. If Thomson's model had been correct, none of the alpha particles would have been deflected through wide angles (*Chemistry* 7th ed. pages 48–49 / 8th ed. pages 48–50).

3. A model, as scientists use the term, is neither correct or incorrect, but <u>successful or unsuccessful</u> in matching observations, giving understanding to relationships, and allowing for useful predictions (*Chemistry* 7th ed. pages 348–350 / 8th ed. pages 359–361).

    Note: It is doubtful that a question this philosophical will appear on the AP Chemistry Examination.

3

STOICHIOMETRY

Stoichiometry includes calculations of average atomic mass, the mole, molar mass, percent composition, molarity and empirical formula. The quantities of materials consumed and produced in chemical reactions are also considered. The multiple-choice section of the AP Exam may contain questions that require computations without calculators. The free response portion of the exam may contain one multistep question on stoichiometry.

You should understand

- Moles, mass, representative particles (atoms, molecules, formula units), molar mass, and Avogadro's number.
- Molarity; preparation of solutions.
- The percent composition of an element in a compound.
- Balanced chemical equations; for example, for a given mass of reactant, calculate the amount of product produced.
- Limiting reactants: calculate the amount of product formed when given the amounts of all of the reactants present.
- Reactions in solution: given the molarity and the volume of the reactants, calculate the amount of product produced or the amount of reactant required to react.
- The percent yield of a reaction.

75

THE MOLE

(*Chemistry* 7th ed. pages 82–85 / 8th ed. pages 81–84)

The mole is defined in the table below. Calculations involving moles utilize the conversions in the table below.

1 mol of a monoatomic element = 6.02×10^{23} atoms
1 mol of a molecular compound or diatomic element = 6.02×10^{23} molecules
1 mol of an ionic compound = 6.02×10^{23} formula units

MOLAR MASS

(*Chemistry* 7th ed. pages 86–88 / 8th ed. pages 84–87)

Molar mass is the mass of 1 mole of an element or compound.

> **EXAMPLE:** The principal ore in the production of aluminum cans has a molecular formula of $Al_2O_3 \bullet 2H_2O$. What is the mass in grams of 2.10×10^{24} formula units (f.u.) of $Al_2O_3 \bullet 2H_2O$?

$$2.10 \times 10^{24} \text{ formula units} \times \frac{1 \text{ mol } Al_2O_3 \bullet 2H_2O}{6.02 \times 10^{23} \text{ f.u.}} \times \frac{138 \text{ g}}{1 \text{ mol } Al_2O_3 \bullet 2 H_2O} = 481 \text{ g}$$

MOLARITY

(*Chemistry* 7th ed. pages 133–139 / 8th ed. pages 136–144)

The number of moles of solute in 1 liter of solution is a measure of its molarity or solution concentration.

> **EXAMPLE:** Prepare 2.00 L of 0.250 M NaOH from solid NaOH.

$$2.00 \text{ L} \times \frac{0.250 \text{ mol NaOH}}{\text{L}} \times \frac{40.00 \text{ g NaOH}}{1 \text{ mol}} = \textbf{20.0 g NaOH}$$

Place 20.0 g NaOH in a 2-L volumetric flask; add water to dissolve the NaOH, and fill to the mark with water, mixing several times along the way.

> **EXAMPLE:** Prepare 2.00 L of 0.250 *M* NaOH from 1.00 *M* NaOH.

$$\boxed{M_1 V_1 = M_2 V_2}$$

$$1.00 \text{ } M \text{ } V_1 = 0.250 \text{ } M \times 2.00 \text{ L } \quad V_1 = \textbf{0.500 L}$$

Add 500. mL of 1.00 *M* NaOH stock solution to a 2-L volumetric flask; add deionized water in several increments, mixing until the flask is filled to the mark on the neck of the flask.

PERCENT COMPOSITION

(*Chemistry* 7th ed. pages 89–91 / 8th ed. pages 88–92)

The mass percentages of elements in a compound can be obtained by comparing the mass of each element present in 1 mole of the compound to the total mass of 1 mole of compound.

$$\frac{\text{Mass of element in 1 mol of compound}}{\text{Mass of 1 mol of compound}} \times 100\%$$

EXAMPLE: Calculate the percent of oxygen in $Mg(NO_3)_2$
$$= (16 \times 6/148) \times 100\% = 65\%$$

DETERMINATION OF EMPIRICAL FORMULA BY COMBUSTION ANALYSIS

(*Chemistry* 7th ed. pages 91–93 / 8th ed. pages 90–94)

The empirical formula is the simplest whole number ratio of atoms in a compound.

(*The following problem involves two steps. First, the percent composition of the compound is determined from combustion data. Second, the empirical formula of the compound is determined from the percent composition.*)

EXAMPLE: A compound contains only carbon, hydrogen, and oxygen. Combustion of 10.68 mg of the compound yields 16.01 mg CO_2 and 4.37 mg of H_2O. What is the percent composition of the compound?

SOLUTION: Assume that all the carbon in the compound is converted to CO_2 and determine the mass of carbon present in the 10.68-mg sample.

$$.01601 \text{ g } CO_2 \times \frac{12.01 \text{ g C}}{44.01 \text{ g } CO_2} = 4.369 \text{ mg C}$$

The mass percent of C in this compound is
$$\frac{4.369 \text{ mg C}}{10.68 \text{ mg}} \times 100\% = 40.91\% \text{ C}$$

The same procedure can be used to find the mass percent of hydrogen in the unknown compound. We assume that all the hydrogen present in 10.68 mg of compound was converted to H_2O.

$$0.0437 \text{ g } H_2O \times \frac{2.016 \text{ g H}}{18.02 \text{ g } H_2O} = 0.489 \text{ mg H}$$

The mass percent of H in the compound is

$$\frac{0.489 \text{ mg}}{10.68 \text{ mg}} \times 100\% = 4.58\%\text{H}$$

The unknown compound contains only carbon, hydrogen and oxygen. The remainder must be oxygen.

100.00 % − (40.91% C + 4.58% H) = 54.51% O

EMPIRICAL FORMULA

(*Chemistry* 7th ed. pages 93–95 / 8th ed. pages 93–96)

What is the empirical formula of a compound that contains 40.91% carbon, 4.58% hydrogen and 54.51 % oxygen?

$$40.91 \text{ g C} \times \frac{1 \text{ mol C}}{12.01 \text{ g}} = 3.406 \text{ mol C}$$

$$4.58 \text{ g H} \times \frac{1 \text{ mol H}}{1.008 \text{ g}} = 4.54 \text{ mol H}$$

$$54.51 \text{ g O} \times \frac{1 \text{ mol O}}{16.00 \text{ g}} = 3.407 \text{ mol O}$$

Divide by the smallest number

$$\frac{4.54 \text{ mol H}}{3.406} = 1.33 \text{ mol H} \times 3 = 3.99 \sim 4.00$$

$$C_{\frac{3.406}{3.406}} H_{\frac{4.54}{3.406}} O_{\frac{3.407}{3.406}} = (C_{1.00} H_{1.33} O_{1.00})3 = C_3 H_4 O_3$$

Empirical Formula

Since a whole number ratio is required, we multiply all subscripts by the lowest factor (3) to obtain the whole number. Rounding can only occur when the subscripts are within 0.1 of the nearest whole number.

CONVERSIONS INVOLVED IN STOICHIOMETRIC CALCULATIONS

Type of Conversion	Example	Reference
g → mol (using molar mass)	$\dfrac{18 \text{ g}}{1 \text{ mol } H_2O}$	*Chemistry* 7th ed. page 86/ 8th ed., page 85
mol → mol (using mole ratio – coefficients from balanced chemical equation)	$\dfrac{1 \text{ mol } O_2}{2 \text{ mol } H_2O}$	*Chemistry* 7th ed. pages 102–104/8th ed. pages 102-104
mol → L (using molarity which is a solution's concentration equal to the number of mols of solute in 1 liter of solution)	$\dfrac{6 \text{ mol HCl}}{1L}$	*Chemistry* 7th ed. pages 133–140/ 8th ed. pages 136-144

SOLUTION STOICHIOMETRY

(*Chemistry* 7th ed. pages 151–154 / 8th ed. pages 156–160)

Solution stoichiometry involves calculations for reactions that occur in aqueous solution.

EXAMPLE: The compound with the empirical formula, $C_3H_4O_3$, is a monoprotic acid. A 2.20-g sample of the unknown monoprotic acid is dissolved in 1.0 L of water. A titration required 25.0 mL of 0.500 M NaOH to react completely with all the acid present. What is the molecular formula of the acid? First, find the molar mass of the acid.

SOLUTION: To find the molar mass of the acid, let HA = unknown acid.

Write a balanced molecular equation.

$$HA + NaOH \rightarrow NaA + H_2O$$

Find the moles of HA present from the volume and molarity of NaOH.

$$0.250 \text{ L NaOH} \times \frac{0.500 \text{ mol NaOH}}{1 \text{ L}} \times 1 \text{ mol HA} = 0.0125 \text{ mol HA}$$

To find the molar mass of HA, divide mass (g) of HA which is given by mol of HA.

$$\frac{2.20 \text{ g HA}}{0.0125 \text{ mol HA}} = 176 \text{ g/mol.}$$

To find the molecular formula of the compound if the empirical formula and molar mass are known, use the following equation (*Chemistry* 6th ed. pages 100-102 / 7th ed. pages 95–96):

$\dfrac{\text{Molar mass}}{\text{Empirical formula mass}}$ = Number	*Multiply each subscript in the empirical formula by the number*

Determine the empirical formula mass = 3 (12) + 4(1) + 3(16) = 88 g/mol

Since 176/88 = 2.0, the **molecular formula** is $(C_3H_4O_3)_2$ = $\mathbf{C_6H_8O_6}$

LIMITING REACTANT

(*Chemistry* 7th ed. pages 106–111 / 8th ed. pages 107–113)

In a limiting reactant problem, the amounts of all reactants are given and the amount of product is to be determined. The limiting reactant is completely consumed when the reaction goes to completion. It determines how much product is formed.

EXAMPLE: What mass of precipitate can be produced when 50.0 mL of 0.200 M Al $(NO_3)_3$ is added to 200.0 mL of 0.100 M KOH?

- Always begin with a balanced molecular equation.

$$Al(NO_3)_3 \text{ (aq)} + 3KOH\text{(aq)} \rightarrow Al(OH)_3 \text{ (s)} + 3KNO_3\text{(aq)}$$

- For each reactant, determine the amount of product produced (in mol or g).

$$0.0500 \text{ L} \times \frac{0.200 \text{ mol Al}(NO_3)_3}{L} \times \frac{1 \text{ mol Al}(OH)_3}{1 \text{ mol Al}(NO_3)_3}$$

$$= 0.0100 \text{ mol Al}(OH)_3$$

$$0.200 \text{ L} \times \frac{0.100 \text{ mol KOH}}{L} \times \frac{1 \text{ mol Al}(OH)_3}{3 \text{ mol KOH}}$$

$$= 0.00667 \text{ mol Al}(OH)_3$$

- *The limiting reactant produces the least amount of product.* The limiting reactant in this case is KOH. It produces only 0.00667 mol of product which is less than 0.0100 mol.

The mass of precipitate produced is:

$$0.00667 \text{ mol Al}(OH)_3 \times \frac{78.00 \text{ g Al}(OH)_3}{1 \text{ mol Al}(OH)_3} = \textbf{0.520 g Al(OH)}_3$$

PERCENT YIELD

(*Chemistry* 7th ed. pages 111–113 / 8th ed. pages 113–115)

The *percent yield* of a reaction is the actual yield of a product as a percentage of the theoretical yield. The *actual yield* is the amount of product obtained in an experiment. The *theoretical yield* is the amount of product calculated from the amounts of reactants used.

EXAMPLE: If the reaction above has an 85.3% yield of precipitate, how much aluminum hydroxide is produced?

$$\text{Percent Yield} = \frac{\text{Actual Yield}}{\text{Theoretical Yield}} \times 100\%$$

$$\text{Actual Yield} = \frac{85.3\% \times 0.520 \text{ g}}{100\%} = \textbf{0.444 g Al(OH)}_3$$

MULTIPLE-CHOICE QUESTIONS

No calculators are to be used in this section.

1. There are three naturally occurring isotopes of oxygen: oxygen-16, oxygen-17, oxygen-18. The atomic mass of oxygen is 16.00 amu. From these data it may be concluded that
   (A) most of the oxygen atoms are ^{17}O.
   (B) most of the oxygen atoms are ^{18}O.
   (C) very few of the atoms are ^{17}O and ^{18}O.
   (D) there is equal abundance of all three isotopes in nature.
   (E) more than half of the atoms are ^{18}O.

2. Analysis of a sample of an oxide of chromium is reported as 26 g of chromium and 12 g of oxygen. From these data determine the empirical formula of this compound.
   (A) CrO
   (B) Cr_2O_3
   (C) CrO_3
   (D) CrO_2
   (E) Cr_4O_6

3. Aluminum reacts with sulfuric acid, H_2SO_4, to form aluminum sulfate, $Al_2(SO_4)_3$, and hydrogen gas. Give the sum of all coefficients (all reactants and all products) for this balanced chemical expression.
   (A) 4
   (B) 5
   (C) 6
   (D) 9
   (E) 12

4. Methane reacts with oxygen to form carbon dioxide and water as shown in the following chemical equation: $CH_4 + 2O_2 \rightarrow CO_2 + 2 H_2O$. If 72 g of water form, how much methane must have reacted?
   (A) 16 g
   (B) 32 g
   (C) 36 g
   (D) 84 g
   (E) 72 g

5. Hydrogen reacts with oxygen to form only water. If 16 grams of hydrogen is mixed with 16 grams of oxygen, how much water can form?
   (A) 0.50 grams
   (B) 8.0 grams
   (C) 18 grams
   (D) 72 grams
   (E) 144 grams

6. How many moles of barium sulfide, BaS, will form when 60.0 mL of 1.0 M $Ba(NO_3)_2$ are mixed with 25.0 mL of 0.80 M K_2S solution to form barium sulfide solid?
   (A) 0.020 mol
   (B) 0.040 mol
   (C) 0.060 mol
   (D) 0.080 mol
   (E) 0.10 mol

7. Consider the neutralization reaction $Be(OH)_2 + 2HCl \rightarrow BeCl_2 + 2H_2O$. What volume of 5.00 M HCl is required to react completely with 4.30 g of $Be(OH)_2$? (Molar mass of $Be(OH)_2 = 3.0$ g/mol.)
   (A) 10.0 mL
   (B) 30.0 mL
   (C) 40.0 mL
   (D) 50.0 mL
   (E) 80.0 mL

8. Sucrose, $C_{12}H_{22}O_{11}$, has a molar mass of 342 g/mol. How many moles of molecules of this white crystal are there in 684 g of sucrose?
   (A) 2.00 mole of molecules
   (B) 21.0 mole of molecules
   (C) 42.0 mole of molecules
   (D) 6.02×10^{23} mole of molecules
   (E) 3.01×10^{23} mole of molecules

9. Nitric acid reacts with silver metal: $4HNO_3 + 3Ag \rightarrow NO + 2H_2O + 3AgNO_3$. Calculate the number of grams of NO formed when 10.8 g of Ag reacts with 12.6 g of HNO_3.
   (A) 0.999 g
   (B) 9.0 g
   (C) 12.0g
   (D) 18.0 g
   (E) 30.0 g

10. How many grams of chromium are in 58.5 g of $K_2Cr_2O_7$? (Molar mass $= 294$ g/mol.)
    (A) 200 g of Cr
    (B) 10.4g of Cr
    (C) 15.6 g of Cr
    (D) 20.8 g of Cr
    (E) 208. g of Cr

11. If 7.0 moles of sulfur atoms and 10 moles of oxygen molecules are combined to form the maximum amount of sulfur trioxide, how many moles of which reactant remain unused at the end?
    $$2 S + 3 O_2 \rightarrow 2 SO_3$$
    (A) 0.25 mol O_2
    (B) 0.33 mol O_2
    (C) 0.33 mol S
    (D) 0.67 mol S
    (E) 0.75 mol S

12. What is the percent mass nitrogen in sodium cyanide?
    (A) 14.00%
    (B) 24.41 %
    (C) 28.57 %
    (D) 49.10 %
    (E) 63.12 %

13. A compound of nitrogen and oxygen is 63.64% by mass nitrogen. What is the empirical formula of this compound?
    (A) NO
    (B) NO_2
    (C) NO_3
    (D) N_2O
    (E) N_2O_3

14. When 16 g of methane (CH_4) and 32 g of oxygen (O_2) reacted to produce carbon dioxide and water, 11 g of carbon dioxide was produced. Calculate the percent yield of carbon dioxide in this reaction.
    (A) 5.0%
    (B) 10 %
    (C) 25 %
    (D) 50 %
    (E) 75 %

15. Zinc sulfide reacts with oxygen to yield zinc oxide and sulfur dioxide as follows:

$$2 \, ZnS(s) \, + 3 \, O_2(g) \rightarrow 2 \, ZnO(s) \, + 2 \, SO_2(g)$$

How many moles of ZnO are produced when 32 g of oxygen is allowed to react with an excess of ZnS?
    (A) 0.67
    (B) 1.0
    (C) 1.33
    (D) 2.0
    (E) 2.67

FREE-RESPONSE QUESTIONS

Calculators may be used for this section.

1. A test tube containing $CaCO_3$ is heated until all of the compound decomposes. If the test tube plus calcium carbonate originally weighed 30.08 grams and the loss of mass during the experiment is 4.40 grams, what was the mass of the empty test tube?

2. A 10.0 g sample of an oxide of copper is heated in a stream of pure hydrogen, forming 1.26 g of water.
    (a) Determine the percentage of copper in the compound.
    (b) Determine the empirical formula of the copper oxide. Name it.

ESSAY FOR STOICHIOMETERY

The percentage yield for some reactions is low. Discuss several reasons why this might be so. [Note: While this is not specifically discussed in the chapter, it should direct your attention to practical limitations of chemical reactions often found in your own work in the laboratory.]

Answers

MULTIPLE CHOICE

1. **C** In order for the average mass of naturally occurring oxygen to be 16.00 amu, there cannot be a significant contribution from either oxygen-17 or oxygen-18 (*Chemistry* 7th ed. pages 78–95 / 8th ed. pages 78–97).

2. **B** You have 0.5 mol of Cr (26 g × 1 mol/52 g) and 0.75 mol of O (12 g × 1 mol/16 g). The ratio of mol Cr: mol O is 1/2:3/4 = (1/2 × 4/3) = 2/3 or 2 mol Cr: 3 mol O. While answer E gives the same ratio, it is not the smallest whole number ratio (*Chemistry* 7th ed. pages 93–95 / 8th ed. pages 93–97).

3. **D** The equation is $2Al + 3H_2SO_4 \rightarrow Al_2(SO_4)_3 + 3H_2$. Note that hydrogen is diatomic. Be sure to count the coefficients for the whole formula; it is understood that the coefficient in front of the aluminum sulfate is one (*Chemistry* 7th ed. pages 96–102 / 8th ed. pages 97–103).

4. **B** The equation indicates a mole ratio of water to methane of 2:1. The 72 grams of water is 4.0 moles of water (72 g / 18 g/mol = 4.0 moles), so 2.0 moles of methane are required. Two moles of methane have a mass of 32 grams (16 g/mol × 2) (*Chemistry* 7th ed. pages 102–106 / 8th ed. pages 102–107).

5. **C** Write an equation for the reaction: $2H_2 + O_2 \rightarrow 2H_2O$. The mole ratio can then be seen as 2 mol H_2 for 1 mole O_2. Determine the number of moles of each reactant: 16 g H_2 × 1 mol H_2/2.0 g = 8 mol H_2 and 16 g O_2 × 1 mol O_2/32 g = 0.50 mol O_2.

    Since O_2 is the limiting reagent, it determines the amount of water formed. From the mole ratio in the equation, it follows that 0.50 mol O_2 forms two times that amount of water or 1 mol of water. The molar mass of water is 18 (*Chemistry* 7th ed. pages 106–113 / 8th ed. pages 107–115).

6. **A** The equation, $Ba(NO_3)_2 + K_2S \rightarrow BaS + 2KNO_3$, indicates that with 0.060 mole $Ba(NO_3)_2$ and 0.020 mole K_2S, the potassium sulfide is the limiting reactant (they react in a one-to-one ratio). The equation also shows that for one mole of potassium sulfide, one

mole of barium sulfide forms, hence 0.020 mol of potassium sulfide will allow for the formation of 0.020 mol of the solid BaS (*Chemistry* 7th ed. pages 139–144, 147–148 / 8th ed. pages 144–149, 151-152).

7. **C** The mass of 4.3 g of $Be(OH)_2$ represents 0.10 mol of this base; the equation indicates that twice that amount (0.20 mol) of acid is needed for neutralization to be complete. Since both the concentration and number of moles needed are known for the acid, the volume of the acid can be calculated (0.20 mol / 5.0 mol/L = 0.040 L = 40. mL) (*Chemistry* 7th ed. pages 153–154 / 8th ed. pages 158–160).

8. **A** It takes 342 g to make one mole of the compound and you have twice that amount. Therefore you have 2.00 mole of this substance (684 g / 342 g/mol) (*Chemistry* 7th ed. pages 82–88 / 8th ed. pages 81–87).

9. **A** First determine the number of moles of each reactant to determine which is in excess and which is the limiting reactant: 10.8 g Ag / 108 g/mol = 0.100 mol of Ag; 12.6 g HNO_3 / 63.0 g /mol = 0.40 mol of HNO_3. Since the Ag / HNO_3 ratio (from the equation) is 3 / 4, Ag is the limiting reactant. 0.100 mol of Ag will form one third as much NO (0.100 × 1NO / 3 Ag) or 0.0333 mol of NO. Finally, 0.0333 mol NO × 30.0 g NO / mol = 0.999 g NO (*Chemistry* 7th ed. pages 106–113 / 8th ed. pages 107–115).

10. **D** First determine the number of moles of the salt: 58.5 g / 294 g/mol = 0.200 mol $K_2Cr_2O_7$. Since Cr has an atomic mass of 52.0 and there are two Cr per each unit of $K_2Cr_2O_7$, 0.200 mol × 2 Cr / $K_2Cr_2O_7$ × 52.0 g / mol = 20.8 g of Cr (*Chemistry* 7th ed. pages 83–85 / 8th ed. pages 82–84).

11. **C** According to the balanced equation, for complete reaction the required ratio of moles of oxygen to sulfur is 3/2 or 1.5. In the problem we have a ratio of 10/7 or 1.43. Since this ratio is less than the required ratio, this means the oxygen is limiting; 10 moles of oxygen require 10 × (2/3) mol S = 6.67 mol S used. Therefore, 7.0 – 6.67 = 0.33 mol S remains unused (*Chemistry* 7th ed. pages 106-111 / 8th ed. pages 107-113).

12. **C** Remembering that the formula for sodium cyanide is NaCN, the molar mass is 22.99 + 12.01 + 14.01 = 49.01 g/mol. % mass of N = (14.01/49.01) × 100 = 28.57% (*Chemistry* 7th ed. pages 89-91 / 8th ed. pages 88-90).

13. **D** The % oxygen is 100 − 63.64 = 36.36%

The steps to solving this problem are detailed below:

	N	O
Ratio by mass	63.64	36.36
Divide by molar masses	63.64/14.01	36.36/16.00
Mole ratio	4.54	2.27
Divide by smaller number	2	1

(*Chemistry* 7th ed. pages 91-96 / 8th ed. pages 90-97).

14. **D** First of all, write the balanced equation: $CH_4 + 2O_2 \rightarrow CO_2 + 2H_2O$

16 grams of methane = 1.0 mole

32 grams of oxygen = 1.0 mole. This is the limiting reactant, since by the equation 1 mole of methane requires 2 moles of oxygen, and since there is only 1 mole of oxygen, it will be used up first leaving 0.5 mole of methane unreacted.

Thus, if one mole of oxygen is used up, according to the equation, 0.5 mol CO_2 would be produced. The molar mass of CO_2 is 44.0 g/mol, so the expected yield would be 22 g. The actual yield was 11 g so (11 g/22 g) × 100 = 50% (*Chemistry* 7th ed. pages 111-112 / 8th ed. pages 107-113).

15. **A** According to the equation, 1.0 mol O_2 produces 1.0 × (2/3) mol ZnO. Thus 32 grams of oxygen = 1.0 mol, 2/3 or 0.67 mol ZnO would be produced (*Chemistry* 7th ed. pages 102-106 / 8th ed. pages 102-107).

FREE RESPONSE

1. 20.1 g Since 0.100 mol of CO_2 was lost (4.40g / 44.0 /mol = 0.100 mol), this means that the same number of moles of $CaCO_3$ decomposed. (0.100 mol of $CaCO_3$ is 10.0 g (100 g/mol × 0.100 mol = 10.0 g of $CaCO_3$.) If the test tube and the salt weighed 30.08 g and 10.0 g was the calcium carbonate, then the mass of the test tube must have been 20.1 g (*Chemistry* 7th ed. pages 102–106 / 8th ed. pages 102–107).

2. (a) 88 % copper

$$1.26 \text{ g } H_2O \times \frac{1 \text{ mol}}{18.0 \text{ g } H_2O} = 0.0700 \text{ mol } H_2O \times \frac{1 \text{ mol O}}{1 \text{ mol } H_2O}$$
$$= 0.0700 \text{ mol O atoms}$$

0.0700 mol × 16.0 g/mol = 1.12 g of oxygen in the oxide

10.0 g – 1.12 g = 8.8 g of copper; 8.8 g/10.0 g × 100% = 88% Copper

(b) Cu2O Copper (I) Oxide

Consider the mole ratio of Cu to O:

[8.8 g / 63.55 g/mol = 0.14 mol Cu]

$$\frac{0.14 \text{ mol Cu}}{0.070 \text{ mol O}} = \frac{2 \text{ mol Cu}}{1 \text{ mol O}}$$ Cu_2O is the empirical formula.

This is called copper(I) oxide since the Cu has a charge of +1 in this compound (*Chemistry* 7th ed. pages 102–106, 93–96, 88–91 / 8th ed. pages 102–107, 93-97, 87-91).

ESSAY FOR STOICHIOMETRY

1. The possibility of mechanical loss. Things like spillage, products getting stuck to equipment and not weighed can often occur in the separation and purification processes which are usually necessary to give a pure product.

2. The system may react in such a way that it reaches equilibrium far to the left, i.e., with more reactant than product present at equilibrium. [This is the case with systems that have a low value for the equilibrium constant.]

3. There can also be side reactions that lower the yield by reacting with reactants of the original process thereby preventing them from forming the desired products.

For these reasons the actual yield of products is almost always less than the theoretical yield.

CHEMICAL REACTIONS

Chemical equations represent chemical changes in which atoms in one or more substances are reorganized. Skills involved in this section include balancing chemical reactions, classifying reactions by type, predicting products of chemical reactions, and writing complete and net-ionic equations.

The multiple choice section of the AP Exam will include questions on these topics.

The free-response portion of the exam includes a separate reactions section in which the student is given the names of the reactants combined in three different laboratory situations and asked to provide the balanced net-ionic equation. *The physical states of the reactants and products do not need to be indicated in this section,* although they are extremely helpful in writing the net-ionic equation. If you do not know the products which are worth 2 points, at least write down the correct reactants to obtain 1 of the 5 points for each reaction.

You should be able to
- Classify reactions by type.
- Write a balanced molecular equation, complete ionic equation, and a net ionic equation.
- Balance oxidation-reduction reactions.
- Predict if a precipitate will form using the solubility rules.
- Predict products of reactions given the chemical names of the reactants.
  - Memorize the list of common oxidizers and reducers given in this chapter. Know the chemical formulas of the products they form.
  - This is a separate section in the AP Exam.

89

DESCRIBING REACTIONS IN AQUEOUS SOLUTION

(*Chemistry* 7th ed. pages 145–146 / 8th ed. pages 150–151)

MOLECULAR EQUATION

A molecular equation gives the overall reaction. Although it gives information on stoichiometry, it gives no information on whether or not compounds really exist as ions in solution. For example, $2KCl \text{ (aq)} + Pb(NO_3)_2 \text{ (aq)} \rightarrow PbCl_2 \text{ (s)} + 2KNO_3 \text{ (aq)}$.

COMPLETE IONIC EQUATION

The complete ionic equation gives the equation including all the ions in solution. For example, $2K^+ + 2Cl^- + Pb^{2+} + 2NO_3^- \rightarrow PbCl_2 \text{ (s)} + 2K^+ + 2NO_3^-$ The states are aqueous , unless otherwise indicated.

NET-IONIC EQUATION

The net-ionic equation only gives information on those species that undergo a chemical change. The spectator ions are omitted from the complete ionic equation.

$$Pb^{2+} + 2Cl^- \rightarrow PbCl_2 \text{ (s)}$$

This is the answer for the AP Exam. Remember, it must be balanced on the exam.

TYPES OF REACTIONS

Classification of chemical reactions by type allows the prediction of products if given the reactants.

Main Reaction Types
Combustion, Combination, Complex ion formation
Acid-base
Redox (oxidation–reduction)
Precipitation

COMBUSTION

(*Chemistry* 7th ed. page 1003 / 8th ed. pages 1012)

Combustion is the complete oxidation of any organic compound containing C, H and/or O to yield CO_2 and H_2O.

Ethene is completely burned in oxygen.
$C_2H_4 \text{ (g)} + 3O_2 \text{ (g)} \rightarrow 2CO_2 \text{ (g)} + 2H_2O \text{ (g)}$

Nonmetallic hydrides combine with oxygen to form oxides and water.

Gaseous silane is burned in oxygen.
$SiH_4 \text{ (g)} + 2O_2 \text{ (g)} \rightarrow SiO_2 \text{ (s)} + 2H_2O \text{ (g)}$

Nonmetallic sulfides combine with oxygen to form sulfur dioxide and oxides.

Carbon disulfide vapor is burned in excess oxygen.
$CS_2 \text{ (s)} + 3O_2 \text{ (g)} \rightarrow CO_2 \text{ (g)} + 2SO_2 \text{ (g)}$

COMBINATION

Two or more elements $\rightarrow$ 1 compound

Magnesium ribbon is burned in pure nitrogen gas.
$3Mg \text{ (s)} + N_2 \text{ (g)} \rightarrow Mg_3N_2 \text{ (s)}$

Two or more compounds $\rightarrow$ 1 compound
Metal oxide + water $\rightarrow$ base

Solid lithium oxide is added to water.
$Li_2O \text{ (s)} + H_2O \text{ (l)} \rightarrow 2 Li^+ \text{ (aq)} + 2OH^- \text{ (aq)}$

Nonmetal oxide + water $\rightarrow$ acid (no change in oxidation state)

Sulfur trioxide is bubbled into water.
$SO_3 \text{ (g)} + H_2O \text{ (l)} \rightarrow 2H^+ \text{ (aq)} + SO_4^{2-} \text{ (aq)}$

Metal oxide + nonmetal oxide $\rightarrow$ salt

A mixture of solid calcium oxide and solid tetraphosphorus decaoxide is heated.
$6CaO \text{ (s)} + P_4O_{10} \text{ (s)} \rightarrow 2Ca_3(PO_4)_2 \text{ (s)}$

DECOMPOSITION

Decomposition = 1 compound $\rightarrow$ 2 or more elements or compounds (reverse of combination)

Magnesium carbonate is heated strongly in a crucible.
$MgCO_3 \text{ (s)} \rightarrow MgO \text{ (s)} + CO_2 \text{ (g)}$

COMPLEX-ION FORMATION

(*Chemistry* 7th ed. pages 944, 955, 958–960 / 8th ed. pages 955, 964, 966–969)

Transition metal ions such as Cu^{2+}, Ag^+, and Ni^{2+} react with excess ligands to form coordination complexes.

Examples of Ligands
OH^-, hydroxo
CN^-, cyano
NH_3, ammine
X^-, halo (halides)

Sometimes, # ligands = 2 (charge of transition metal ion)
Charge on complex ion = total charge of all ligands + T.M. ion (T.M. stands for transition metal)

Solid zinc nitrate is treated with excess sodium hydroxide solution.
$$Zn(NO_3)_2 (s) + 4OH^- (aq) \rightarrow Zn(OH)_4{}^{2-} (aq) + 2NO_3{}^- (aq)$$

Complex ions are broken down by the addition of a strong acid which reacts with basic ligands.

An aqueous solution of diamminesilver chloride is treated with dilute nitric acid.
$$Ag(NH_3)_2{}^+ (aq) + Cl^- (aq) + 2H^+ (aq) \rightarrow AgCl (s) + 2NH_4{}^+ (aq)$$

ACID-BASE REACTIONS

(*Chemistry* 7th ed. pages 149–154 / 8th ed. pages 154–161)

Acid + Base $\rightarrow$ Salt + H_2O

Memorize the list of strong acids and strong bases below. All others are weak acids or bases.

Strong Acids	Strong Bases
HCl, HBr, HI	Group I Hydroxides, (LiOH, NaOH . . .)
H_2SO_4 ($HSO_4^- $ = weak)	Group II Hydroxides of Ca, Sr and Ba
HNO_3	
$HClO_4$	

STRONG ACID–STRONG BASE REACTIONS

(*Chemistry* 7th ed. page 149 / 8th ed. pages 154)

Strong acids and bases are assumed to be 100% ionized in aqueous solution; write all reactants in ionic form. Use solubility rules for the salt formed. (See end of this chapter for solubility rules.)

For the examples that follow, all states are aqueous, unless otherwise indicated.

EXAMPLE: Equimolar solutions of hydrochloric acid and sodium hydroxide are mixed.

$$H^+ + Cl^- + Na^+ + OH^- \rightarrow Na^+ + Cl^- + HOH(l)$$

Omit spectator ions to get the net-ionic equation:

$$H^+ + OH^- \rightarrow HOH$$

Note: For most strong acid–strong base reactions, this will be the net-ionic equation, but make sure that the salt formed is soluble.

EXAMPLE: $Ba(OH)_2 + H_2SO_4 \rightarrow BaSO_4 \ (s) + H_2O(l)$

The net-ionic equation for the reaction above is

$$Ba^{2+} + 2OH^- + 2H^+ + SO_4^{2-} \rightarrow BaSO_4(s) + H_2O \ (l)$$

WEAK ACID–STRONG BASE REACTIONS (OR STRONG ACID–WEAK BASE)

(*Chemistry* 7th ed. page 150 / 8th ed. page 155)

Weak acids and bases ionize only slightly. Write the formula for the weak acid or weak base in molecular form (do not separate into ions!).

EXAMPLE: Equimolar solutions of acetic acid and sodium hydroxide are mixed.

$$HC_2H_3O_2(aq) + Na^+ + OH^- \rightarrow Na^+ + C_2H_3O_2^- + HOH(l)$$

Omit spectator ions to get: $HC_2H_3O_2(aq) + OH^- \rightarrow C_2H_3O_2^- + HOH(l)$

All states are aqueous, unless otherwise indicated.

OXIDATION–REDUCTION REACTIONS

(*Chemistry* 7th ed. pages 154–168 / 8th ed. pages 161–168)

Oxidation reactions involve the transfer of electrons. Skills required for this section include the assignment of oxidation numbers and the identification of oxidation–reduction reactions. In an oxidation–reduction reaction, the atoms oxidized will need to be identified as well as the atoms reduced, the oxidizing agent, and the reducing agent.

Rules for Assigning Oxidation States (*Chemistry* 6th ed. pages 167–168 / 7th ed. pages 157–158)

The oxidation state of . . .	Examples
An atom in element is zero.	$Na(s)$, $O_2(g)$
A monatomic ion is the same as its charge.	Na^+
Oxygen is usually –2 in its compounds. Exception: peroxides (containing O_2^{2-}) in which oxygen is 1–.	H_2O, CO_2
Hydrogen is +1 in its covalent compounds. (hydrogen is –1 in binary hydrides, ex. NaH).	H_2O, NH_3
For an electrically neutral compound, the sum of the oxidation states must be zero.	$KMnO_4$ Solution: K = +1, O = -2 $+1 + Mn + 4(–2) = 0$ $–7 + Mn = 0$; Mn = +7
For an ionic species, the sum of the oxidation states must equal the overall charge.	Ex: $Cr_2O_7^{2-}$ Solution: O = –2 , $2 Cr + 7(-2) = –2$; $2Cr = +12$; Cr = +6

EXAMPLE: Is the following reaction an oxidation–reduction reaction? If it is an oxidation–reduction, identify which atoms are oxidized, which atoms are reduced, the oxidizing agent, and the reducing agent (*Chemistry* 7th ed. pages 158–161 / 8th ed. pages 163-165).

$$Zn + Cu(NO_3)_2 \rightarrow Cu + Zn(NO_3)_2$$

SOLUTION: The nitrate ion appears unchanged on both sides of the reaction, so neither element in the ion is oxidized or reduced. *Copper* goes from Cu^{+2} to Cu^0 (It is reduced; its oxidation # goes down. It gains 2 electrons.) *Zinc* goes from Zn^0 to Zn^{2+} (it is oxidized, its oxidation goes up, and it loses 2 electrons.)

Oxidizing agents are reduced and cause something else to be oxidized. Cu^{2+} is the oxidizing agent. *Reducing agents* are oxidized and cause something else to be reduced. Zn is the reducing agent.

NOTE: If there is no change in oxidation numbers in the reaction, then the reaction is not an oxidation–reduction reaction.

STEPS FOR BALANCING OXIDATION–REDUCTION REACTIONS

BALANCE OXIDATION–REDUCTION REACTIONS IN ACIDIC SOLUTION
USING THE HALF-REACTION METHOD
(*Chemistry* 7th ed. pages 162–166 / 8th ed. pages 166–168)

EXAMPLE: $Cr_2O_7^{2-} + Cl^- \rightarrow Cr^{3+} + Cl_2$

STEP 1: *Write separate half-reactions*:

$$Cr_2O_7^{2-} \rightarrow Cr^{3+} \qquad\qquad Cl^- \rightarrow Cl_2$$

STEP 2: *Balance all atoms except H and O:*

$$Cr_2O_7{}^{2-} \rightarrow 2Cr^{3+} \qquad\qquad 2Cl^- \rightarrow Cl_2$$

STEP 3: *Balance oxygen using water:*

$$Cr_2O_7{}^{2-} \rightarrow 2Cr^{3+} + 7H_2O \qquad\qquad 2Cl^- \rightarrow Cl_2$$

STEP 4: *Balance hydrogen with H$^+$:*

$$14H^+ + Cr_2O_7{}^{2-} \rightarrow 2Cr^{3+} + 7H_2O \qquad\qquad 2Cl^- \rightarrow Cl_2$$

STEP 5: *Balance charge using electrons:*

$$6e^- + 14H^+ + Cr_2O_7{}^{2-} \rightarrow 2Cr^{3+} + 7H_2O \qquad 2Cl^- \rightarrow Cl_2 + 2e^-$$

STEP 6: *Equalize electron transfer.* Multiply each reaction by numbers that will allow both reactions to have the same number of electrons exchanged:

$$6e^- + 14H^+ + Cr_2O_7{}^{2-} \rightarrow 2Cr^{3+} + 7H_2O \qquad 6Cl^- \rightarrow 3\,Cl_2 + 6e^-$$

NOTE: 2nd reaction is multiplied by 3.

Add the two half reactions canceling out all the electrons and the formulas which appear on both sides of the equation.

$$\mathbf{14H^+ + Cr_2O_7{}^{2-} + 6Cl^- \rightarrow 2Cr^{3+} + 7H_2O + 3Cl_2}$$

Sum of Charges: $+14 - 2 - 6 = +6 \rightarrow +6 + 0 + 0 = +6$

STEP 7: *Double check that* there is the same number of each kind of atom on both sides and that the sums of all charges are the same on both sides.

BALANCE OXIDATION–REDUCTION REACTIONS IN BASIC SOLUTION USING THE HALF-REACTION METHOD

(Chemistry 7th ed. pages 167–168 / 8th ed. pages 168)

Repeat steps 1–5 above. After Step 5: Add OH$^-$ to both sides of the equation (equal to H$^+$). Form H_2O on the side containing H$^+$ and OH$^-$ ions. Eliminate number of H_2O appearing on both sides. Continue with steps 6 and 7.

EXAMPLE: $NO_2^- + Al \rightarrow NH_3 + AlO_2^-$

STEP 1: Write separate half-reactions.

$$Al \rightarrow AlO_2^- \qquad\qquad NO_2^- \rightarrow NH_3$$

STEP 2: All atoms except H and O are already balanced.

STEPS 3 AND 4: Balance O with H_2O and H with H^+.

$$2H_2O + Al \rightarrow AlO_2^- + 4H^+ \qquad 7H^+ + NO_2^- \rightarrow NH_3 + 2H_2O$$

STEP 5: Balance charge using electrons.

$$6\,e^- + 7H^+ + NO_2^- \rightarrow NH_3 + 2H_2O$$

$$\underline{(2H_2O + Al \rightarrow AlO_2^- + 4H^+ + 3e^-) \times 2}$$

$$7H^+ + NO_2^- + 4H_2O + 2Al \rightarrow 2AlO_2^- + 8H^+ + NH_3 + 2H_2O$$

CANCEL: $NO_2^- + 2H_2O + 2Al \rightarrow 2AlO_2^- + H^+ + NH_3$

New step for basic solution: Add OH^- to both sides equal to the number of H^+ ions:

$$OH^- + NO_2^- + 2H_2O + 2Al \rightarrow 2AlO_2^- + H^+ + NH_3 + OH^-$$

CANCEL: $OH^- + NO_2^- + 2H_2O + 2Al \rightarrow 2AlO_2^- + NH_3 + H_2O$

THE $OH^- + NO_2^- + H_2O + 2Al \rightarrow 2AlO_2^- + NH_3$

REDOX REACTIONS IN THE AP CHEMISTRY FREE-RESPONSE SECTION

Oxidizing Agent		Reducing Agent	
Appears in reactants	Appears in products	Appears in reactants	Appears in products
MnO_4^-, in acid	Mn^{2+}	Halide ions	Halogens, diatomic
MnO_4^- in neutral or basic	MnO_2	Metal element	Metallic ion
MnO_2, in acid	Mn^{2+}	Sulfite	Sulfate
$Cr_2O_7^{2-}$, in acid	Cr^{3+}	Nitrite	Nitrate
HNO_3, concentrated	NO_2	Halogens, dilute base	Hypohalite ion, ClO^-
HNO_3, dilute	NO	Halogens, conc. base	Halite ion, ClO_2^-
H_2SO_4, hot, conc.	SO_2	Metallous ion	Metallic ion
Metallic ions (ex. Pb^{+4})	Metallous ions (Pb^{+2})	H_2O_2	O_2
Halogens, diatomic	Halide ions	$C_2O_4^{2-}$	CO_2
Na_2O_2	NaOH		
$HClO_4$	Cl^-		
H_2O_2	H_2O		

It is important to be very familiar with the list of common oxidizing agents and reducing agents above.

EXAMPLE: A solution of sodium dichromate is added to an acidified solution of sodium iodide.

$$Cr_2O_7{}^{2-} + H^+ + I^- \rightarrow Cr^{3+} + I_2 + H_2O$$

Oxidizing Agent   Reducing Agent

In this reaction, $Cr_2O_7{}^{2-}$ is the oxidizing agent which gets reduced to Cr^{3+}. The reducing agent, I^-, gets oxidized to I_2. H_2O must be added to the products to balance the H^+ in the reactants.

EXAMPLE: Dilute potassium permanganate solution is added to an oxalic acid solution which was acidified with a few drops of sulfuric acid.

$$MnO_4{}^- + C_2O_4{}^{2-} + H^+ \rightarrow Mn^{2+} + CO_2 + H_2O$$

PRECIPITATION REACTIONS

PRECIPITATION REACTIONS INVOLVE THE FORMATION OF A SOLID WHEN TWO SOLUTIONS ARE MIXED

(*Chemistry* 7th ed. pages 149–158 / 8th ed. pages 154–165)

Solubility Rules
1. Most nitrate salts are soluble.
2. Most salts containing Group I ion and ammonium ion, $NH_4{}^+$, are soluble.
3. Most chloride, bromide, and iodide salts are soluble, except Ag^+, Pb^{2+} and $Hg_2{}^{2+}$.
4. Most sulfate salts are soluble, except $BaSO_4$, $PbSO_4$, Hg_2SO_4, and $CaSO_4$.
5. Most hydroxides except Group I and $Ba(OH)_2$, $Sr(OH)_2$, and $Ca(OH)_2$ are only slightly soluble.
6. Most sulfides, carbonates, chromates, and phosphates are only slightly soluble.

PREDICTING PRECIPITATES

When two solutions of ionic compounds are mixed, exchange anions to form products, remembering that each product formed must be neutral. Use the solubility rules above.

EXAMPLE: Solutions of potassium chloride and lead (II) nitrate are mixed:

Molecular Equation:

$$2KCl\ (aq) + Pb(NO_3)_2\ (aq) \rightarrow PbCl_2\ (s)\ +\ 2KNO_3\ (aq)$$

chlorides of Pb insoluble   all nitrates soluble

Complete Ionic Equation:

$$2K^+ + 2Cl^- + Pb^{2+} + 2NO_3^- \rightarrow PbCl_2\ (s) + 2K^+ + 2NO_3^-$$

Net-Ionic Equation: $Pb^{2+} + 2Cl^- \rightarrow PbCl_2\ (s)$

Writing the equation as balanced is the answer expected on the AP Exam.

Multiple-Choice Questions

No calculators are to be used in this section.

1. Consider the equation: $Cl_2(g) + 2KI(aq) \rightarrow I_2(s) + 2KCl(aq)$. Which species is oxidized?
    (A) Cl_2
    (B) K^+
    (C) I^-
    (D) I_2
    (E) Cl^-

2. A solid dissolves easily in benzene (C_6H_6) but it does not dissolve in water. The molecules of the solid can best be described as
    (A) polar molecules.
    (B) nonpolar molecules.
    (C) either polar or nonpolar.
    (D) small with much charge separation.
    (E) large with a high dipole moment.

3. How many grams of NaOH are contained in 200. mL of a 0.400 M solution?
    (A) 3.20 grams of NaOH
    (B) 8.00 grams of NaOH
    (C) 16.0 grams of NaOH
    (D) 40.0 grams of NaOH
    (E) 80.0 grams of NaOH

4. Which of the following solutions contains the largest number of ions?
    (A) 500. mL of 0.10 M $CaCl_2$
    (B) 500. mL of 0.10 M $FeCl_3$
    (C) 700. mL of 0.20 M NaOH
    (D) 400. mL of 0.100 M $Al(NO_3)_3$
    (E) 600. mL of 0.200 M $AlCl_3$

5. There are six strong acids. Which of the following is NOT a strong
   acid?
   (A) HCl
   (B) HF
   (C) HBr
   (D) HI
   (E) HNO_3

6. Spectator ions are those which are in solution but do not react. Identify
   any spectator ions for the reaction of sodium phosphate with calcium
   bromide.
   (A) only $Na^+(aq)$
   (B) only $PO_4^{3-}(aq)$
   (C) $Na^+(aq)$ and $PO_4^{3-}(aq)$
   (D) $Na^+(aq)$ and $Br^-(aq)$
   (E) $Ca^{2+}(aq)$ and $PO_4^{3-}(aq)$

7. A precipitate forms when which of the following solutions are added to
   a solution of NaCl?
   I $AgNO_3$
   II $Pb(NO_3)_2$
   III $Cr(NO_3)_2$

   (A) I only
   (B) I and II
   (C) I and III
   (D) II and III
   (E) I, II, and III

8. Consider the following three equations for chemical reactions:

   $2Na(s) + Cl_2(g) \rightarrow 2NaCl(s)$
   $2NaCl(aq) + Pb(NO_3)_2(aq) \rightarrow PbCl_2(s) + 2NaNO_3(aq)$
   $NaOH(aq) + HBr(aq) \rightarrow H_2O(l) + NaBr(aq)$

   These are examples of:
   (A) three precipitation reactions.
   (B) three acid–base reactions.
   (C) a redox reaction, a precipitation reaction, then an acid–base
       reaction.
   (D) three redox reactions.
   (E) a neutralization reaction, then two precipitation reactions.

9. Which of the following pairs of ions would <u>not</u> form a solid in aqueous
   solution?
   (A) Ba^{2+} and SO_4^{2-}
   (B) Pb^{2+} and Br^-
   (C) Mg^{2+} and SO_4^{2-}
   (D) Pb^{2+} and S^{2-}
   (E) Ca^{2+} and CO_3^{2-}

10. Which of the following ions are likely to form a soluble sulfate in aqueous solution?
    (A) Ba^{2+}
    (B) Pb^{2+}
    (C) Ca^{2+}
    (D) NH_4^{1+}
    (E) Hg_2^{2+}

11. If a solution of sodium bicarbonate is mixed with a solution containing an equal number of moles of nitric acid, then sodium nitrate, water and carbon dioxide are produced. What is the net ionic equation representing the reaction?
    (A) $NaHCO_3(aq) + HNO_3(aq) \rightarrow NaNO_3(aq) + H_2O(l) + CO_2(g)$
    (B) $HCO_3^-(aq) + HNO_3(aq) \rightarrow NO_3^-(aq) + H_2O(l) + CO_2(g)$
    (C) $Na^+(aq) + HCO_3^-(aq) + HNO_3(aq) \rightarrow Na^+(aq) + NO_3^-(aq) + H_2O(l) + CO_2(g)$
    (D) $HCO_3^-(aq) + H^+(aq) \rightarrow H_2O(l) + CO_2(g)$
    (E) $Na^+(aq) + HCO_3^-(aq) + H^+(aq) \rightarrow H_2O(l) + Na^+(aq) + CO_2(g)$

12. Classify the following reactions as one of the following: precipitation, acid-base, or oxidation-reduction (redox).

    Reaction 1 $Ca(OH)_2(aq) + 2\ HNO_3(aq) \rightarrow Ca(NO_3)_2(aq) + 2\ H_2O(l)$
    Reaction 2 $Fe_2O_3(s) + 3\ CO(g) \rightarrow 2\ Fe(s) + 3CO_2(g)$
    Reaction 3 $CuBr_2(aq) + 2\ NaOH(aq) \rightarrow Cu(OH)_2(s) + 2\ NaBr(aq)$
    Reaction 4 $Cl_2(g) + 2NaI(aq) \rightarrow I_2(aq) + 2\ NaCl(aq)$

	Reaction 1	Reaction 2	Reaction 3	Reaction 4
(A)	acid-base	redox	precipitation	redox
(B)	precipitation	redox	acid-base	redox
(C)	redox	precipitation	acid-base	acid-base
(D)	precipitation	precipitation	redox	acid-base
(E)	acid-base	redox	acid-base	redox

13. The oxidation number of N in $Ca(NO_3)_2$ is
    (A) +1
    (B) +2
    (C) +3
    (D) +4
    (E) +5

14. The difference between a strong acid and a weak acid is that
    (A) a strong acid is more concentrated than a weak acid.
    (B) a weak acid is more soluble in water than a strong acid.
    (C) strong acids have more hydrogens per molecule than weak acids.
    (D) weak acids cannot conduct an electric current in solution.
    (E) strong acids are completely dissociated in solution while weak acids are not.

15. How many moles of electrons are transferred between the substance being oxidized and the substance being reduced in the reaction given below?

$$4NH_3(g) + 5 O_2(g) \rightarrow 4 NO(g) + 6 H_2O(l)$$

(A) 5
(B) 10
(C) 16
(D) 20
(E) 25

FREE-RESPONSE QUESTIONS

Every AP Chemistry exam has a question that requires prediction of the product(s). The directions indicate that in all cases a reaction will occur and that you are to write the balanced net-ionic equation for each of the reactions indicated in words. Do not show any spectator ions. Showing physical state is not required. On the actual exam, you are asked to balance three reactions. The scoring is 1 point for the correct formula of the reactants and 2 points for the products, 1 point for properly balancing the equation, and then 1 point for the additional question. Points are subtracted for improper formulas and spectator ions.

1. Write the balanced net-ionic equations for each of the following reactions indicated in words. Remember not to show spectator ions. Showing physical state is not required. Then answer the short question about each reaction.
   a. (i) Hydrogen sulfide gas is bubbled through potassium hydroxide solution.
      (ii) Identify the conjugate acid or base of the hydrogen sulfide in the above reaction. Is this chemical species acting as an acid or base?
   b. (i) Solutions of sodium iodide and lead(II) nitrate are mixed.
      (ii) Identify the spectator ions in this reaction.
   c. (i) Excess ammonia is added to a solution of copper(II) chloride.
      (ii) Identify the Lewis base in the reaction.

2. Consider the reaction $3O_2(g) + 2PbS(s) \rightarrow 2PbO(s) + 2SO_2(g)$. Show the two half-reactions and identify the reducing agent for this reaction.

Answers

MULTIPLE CHOICE

1. **C** We need only look at reactants when asked to identify the species oxidized or reduced. KI is an ionic compound and we can observe that the oxidation number of K^+ is the same in both KI and KCl. Whenever an element is found in a compound on one side of the equation and as a free element on the other side of the equation, there has to be a change in oxidation number for that species.

By definition, oxidation is the loss of electrons. Writing the two half-reactions [$Cl_2 + 2e^- \rightarrow 2Cl^-$ and $2I^- -2e^- \rightarrow I_2$] shows that it is the two iodide ions that lose two electrons to become diatomic iodine (*Chemistry* 7th ed. pages 154–162 / 8th ed. pages 161–166).

2. **B** The phrase "like dissolves like" refers to the polarity of the solute and of the solvent. Benzene is a nonpolar solvent (from the formula it can be observed that only nonpolar C-C and C–H bonds are involved), so the solid is most likely nonpolar. Water is a polar solvent (*Chemistry* 7th ed. pages 127–129 / 8th ed. pages 130–132).

3. **A** There are 0.400 moles in each liter of this solution, and you are using 0.200 liter of the solution; there are 40.0 grams in each mol of NaOH, hence 0.400 mol/L NaOH × 0.200 L × 40.0 g NaOH / mol NaOH = 3.20 grams of NaOH (*Chemistry* 7th ed. pages 133–134 / 8th ed. pages 136–138).

4. **E** There are three factors which must be considered in this problem, the number of ions per formula unit, the concentration of the solution, and the volume of the solution used. For example, in the first solution there are 0.15 moles of ions present: 0.500L × 0.100 mol of formula units of $CaCl_2$ / L × 3 ions/ $CaCl_2$ = 0.15 moles of ions. In response E: 0.600L × 0.200 mol of formula units of $AlCl_3$ × 4 ions/ $AlCl_3$ = 0.48 mol. of ions (*Chemistry* 7th ed. pages 133–135 / 7th ed. pages 136–140).

5. **B** Even though hydrofluoric acid will dissolve glass (!), it does not ionize significantly. It is important for you to know the six strong acids (then you know that all others are weak!). The six strong acids are $HClO_4$, HCl, HBr, HI, H_2SO_4, and HNO_3. These are the acids that are 100% ionized in water. You can assume that all other acids are weak (*Chemistry* 7th ed. pages 129–131, 626–627, Appendix 5.1 / 8th ed. pages 132–134, 642-644, Appendix 5.1).

6. **D** The total unbalanced expression for this reaction is: $Na^+ + PO_4^{3-} + Ca^{2+} + Br^- \rightarrow Ca_3(PO_4)_2 + Na^+ + Br^-$. In this reaction, the only insoluble compound is calcium phosphate, leaving the other two ions in solution unreacted (i.e., NaBr is soluble in water). You must know the solubility rules to write equations successfully for reactions (*Chemistry* 7th ed. pages 145–146, 149 / 8th ed. pages 149–151, 154).

7. **B** The solubility rules indicate that all chlorides are soluble except for AgCl, $PbCl_2$ and Hg_2Cl_2. Again, you must know the solubility rules (*Chemistry* 7th ed. pages 142–143 / 8th ed. pages 147–148).

8. **C** Being able to classify reactions will help in predicting products. In the first reaction, Na is oxidized and Cl_2 is reduced (a metal–nonmetal reaction can always be assumed to be a redox reaction); in the second reaction, insoluble lead(II) chloride forms; in the third reaction, a base and an acid form water and a salt (*Chemistry* 7th ed. pages 147–151, 158–162 / 8th ed. pages 152–156, 164-166).

9. **C** Most sulfates are soluble. Exceptions are barium, lead(II), mercury(I), and calcium sulfates (*Chemistry* 7th ed. pages 142–143 / 8th ed. pages 147–148).

10. **D** Salts containing the ammonium ion are soluble (*Chemistry* 7th ed. pages 142–143 / 8th ed. pages 147–148).

11. **D** Sodium bicarbonate and sodium nitrate are both strong electrolytes (salts) and therefore are written as ions in aqueous solution. Nitric acid is a strong acid and is completely ionized in aqueous solution. Water and carbon dioxide are molecular species and have very little ionization occurring, so they stay written in the molecular form. The two spectator ions, Na^+ and NO_3^- are canceled from both sides of the equation (*Chemistry* 7th ed. pages 144-146 / 8th ed. pages 148-151).

12. **A** (*Chemistry* 7th ed. pages 149, 154–162 / 8th ed. pages 144–150, 154–155, 161–166).

13. **E** Ca ion is always +2 and oxygen is -2. If you remember that the nitrate ion is -1, then $3(-2) + n = -1$, and when solving for n, n = +5 (*Chemistry* 7th ed. pages 155–158 / 8th ed. pages 162–164).

14. **E** This is the definition of a strong electrolyte, which is why strong acids are labeled as "strong" (*Chemistry* 7th ed. page 129–132 / 8th ed. pages 132–136).

15. **D** The change in oxidation number of the nitrogen is from –3 in the NH_3 to +2 in the nitric oxide molecule. This requires 5 electrons/nitrogen atom, or 5 moles electrons/mol of nitrogen. Since there are 4 moles of nitrogen atoms, the total moles of electrons needed is 4 × 5 or 20 mol electrons (*Chemistry* 7th ed. pages 162–168 / 8th ed. pages 166–168).

FREE RESPONSE

1. a.   (i) $H_2S + 2OH^- \rightarrow S^{2-} + 2H_2O$

        (ii) The conjugate base of H_2S is the sulfide ion, S^{2-}.

    b.   (i) $Pb^{2+} + 2I^- \rightarrow PbI_2$

        (ii) The spectator ions are the Na^{1+} and the nitrate ion.

    c.   (i) $4NH_3 + Cu^{2+} \rightarrow Cu(NH_3)_4^{2+}$

        (ii) The ammonia is acting as a Lewis base, since it is donating its pair of electrons to form the metal-ligand bond.

This is a complex ion formation reaction. Know that ammonia complexes with Cd, Cu, and Zn ions in a 4 to 1 ratio, and with Ag^+ in a 2 to 1 ratio. Also watch for the formation of $Al(OH)^{4-}$ (*Chemistry* 7th ed. pages 140–146, 955–956 / 8th ed. pages 144–151, 964–965).

2. $O_2 \rightarrow 2O^{2-}$

Since the oxidation state of each oxygen decreases from zero to 2-, the oxygen is reduced.

$S^{2-} \rightarrow S^{4+}$

Sulfur is oxidized (electrons are lost as the substance becomes more positive). The reducing agent for this reaction is therefore sulfur in lead(II) sulfide (*Chemistry* 7th ed. pages 154–162 / 8th ed. pages 161-166).

5

GASES

In this section, the laws describing gas behavior are explored. Mathematical laws relating the properties of pressure, volume, temperature, and moles of gas are considered. Densities and molar masses of ideal gases will be determined from the ideal gas law. Stoichiometric calculations will be performed for reactions involving gases. Pressures of gases in a mixture will be examined. The behavior of ideal gases will be explained by the kinetic molecular theory. Real gases will be compared to ideal gases.

Questions about gas behavior and gas laws can be either qualitative or quantitative on the multiple choice section.

The free-response portion of the exam can include questions about an experiment involving gases such as the determination of the molar mass or molar volume of a gas. Essay questions about ideal or real gas behavior or the kinetic molecular theory may also appear.

You should be able to
- Perform calculations with gas laws: Boyle's, Charles', Avogadro's, Combined, and Ideal.
- Perform calculations with the ideal gas law to find the density or molar mass of the gas.
- Interpret or draw graphical relationships between gas variables.
- Perform stoichiometric calculations for reactions which produce gases.
- Perform calculations with molar volume.
- Perform calculations with Dalton's Law for a mixture of gases.
- Perform calculations for gases collected over water.
- Perform calculations with rates of effusion to find the molar mass.
- Perform calculations with root mean square velocity.
- Compare real gases to ideal gases.

105

GAS LAWS

(*Chemistry* 7th ed. pages 181–190 / 8th ed. pages 183–194)

The table below summarizes the gas laws that can be derived from the ideal gas law. A typical problem involves solving for the missing variable when given a set of initial and final conditions for a gas sample. Perform all gas calculations involving temperature using the Kelvin scale, not degrees Celsius. Temperature (K) = °C + 273.15.

Gas Law	Equations	Definition
Boyle's law	$P = (nRT)\,1/V$ $P = (\text{constant})\,1/V$ $\boxed{P_1V_1 = P_2V_2}$	The product of the pressure and the volume is a constant, k, for a trapped sample of gas at constant temperature
Charles' law	$V = (nR/P)\,T$ $V = (\text{constant})T$ $\dfrac{V_1}{T_1} = \dfrac{V_2}{T_2}$	The volume of a gas at constant pressure increases linearly with the temperature of the gas.
Avogadro's law	$V = (RT/P)\,n$ $V = (\text{constant})\,n$ $\dfrac{V_1}{n_1} = \dfrac{V_2}{n_2}$	Equal volumes of gases at the same temperature and pressure contain the same number of moles of gas.
Ideal Gas law	$\boxed{PV = nRT}$ $R = $ universal gas constant $= 0.08206\ \text{L} \times \text{atm}\,/(\text{K} \times \text{mol})$	An equation of state for a gas, where the state of a gas is its condition at a given time.
Dalton's law	$\boxed{P_{total} = P_1 + P_2 + P_3 + \dots}$ $\boxed{P_{total} = n_{tot}RT/V}$	For a mixture of gases, the sum of the pressures of the individual gases is equal to the total pressure. The total pressure depends on the total moles of gas present.

GRAPHICAL RELATIONSHIPS FOR GAS LAWS

(*Chemistry* 7th ed. pages 181–184 / 8th ed. pages 183–187)

BOYLE'S LAW Pressure is inversely related to volume. A graph of P vs. $1/V$ is a straight line.

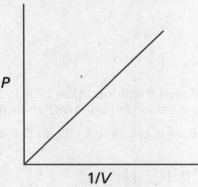

CHARLES' LAW Volume is directly proportional to temperature. A graph of V vs. T is linear. Extrapolation of this graph for all gases to zero volume results in the same temperature, $-273.2°C$ (0 K), which is absolute zero.

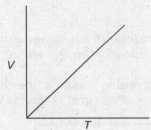

EXAMPLE: A sample tube containing 103.6 mL of CO gas at 20.6 torr is connected to an evacuated 1.13 liter flask. What will the pressure be when the CO is allowed into the flask?

$$P_1V_1 = P_2V_2; \; P_2 = \frac{P_1V_1}{V_2} = \frac{(20.6 \text{ torr})(0.1036 \text{ L})}{1.47 \text{ L}} = 1.45 \text{ torr}$$

Note: Milliliters need to be converted to liters so that the units of volume cancel out.

EXAMPLE: A quantity of gas at 27.0°C is heated in a closed vessel until the pressure is doubled. To what temperature is the gas heated?

Pressure is directly proportional to temperature. If the pressure is doubled, the temperature will also double. 27.0°C + 273.2 = 300.2 K. The answer is 600.4 K (or 327.2°C).

GAS STOICHIOMETRY

(*Chemistry* 7th ed. pages 190–194 / 8th ed. pages 194–199)

If the pressure, volume, and temperature of ideal gases are known, stoichiometric calculations can be performed.

MOLAR VOLUME

(*Chemistry* 7th ed. pages 190–191 / 8th ed. pages 194–195)

One mole of an ideal gas at 0°C and 1 atm occupies 22.4 L. Standard temperature and pressure (STP) are the conditions 0°C and 1 atm.

> EXAMPLE: What mass of helium is required to fill a 1.5-L balloon at STP?

$$1.5 \text{ L} \times \frac{1 \text{ mol He}}{22.4 \text{ L}} \times \frac{4.0 \text{ g He}}{1 \text{ mol}} = 0.27 \text{ g He}$$

MOLAR MASS OF A GAS

(*Chemistry* 7th ed. pages 193–194 / 8th ed. pages 198–199)

The molar mass of a gas can be calculated from its measured density. The density of a gas can also be calculated from the molar mass of the gas.

The equation relating density and molar mass can be derived from $PV = nRT$. Since n = grams/molar mass, PV = (grams /molar mass) RT.

$$\boxed{\text{Molar Mass} = \frac{dRT}{P} = \frac{(\text{mass})RT}{V \times P}}$$

> EXAMPLE: A sample of gas weighing 0.800 g occupies a 256 mL flask at 100°C and 750.0 torr. Determine the molar mass of the gas.

$$\text{Molar mass} = \frac{0.800 \text{ g} \times \dfrac{0.08206 \text{ L atm}}{\text{mol K}} \times 373 \text{ K}}{0.256 \text{ L} \times 0.986 \text{ atm}} = 97.0 \text{ g/mol}$$

DALTON'S LAW

(*Chemistry* 7th ed. pages 194–199 / 7th ed. pages 199–205)

For a mixture of gases in a container, the total pressure is the sum of the pressures that each gas would exert if it were alone.

$$\boxed{P_{\text{total}} = P_1 + P_2 + P_3 + ...}$$

The partial pressure of a gas, P_{gas}, is the pressure that a particular gas would exert if it were alone in the container. The subscripts refer to the individual gases (gas$_1$, gas$_2$, and so on). The partial pressure of each gas can be calculated from the ideal gas law:

$$P_1 = n_1RT/V$$

The total pressure of the mixture, P, can be represented as

$$P_{total} = n_{tot}RT/V$$

The mole fraction is the ratio of the number of moles of a given component in a mixture to the total number of moles in the mixture.

$$x_1 = n_1/n_{tot}$$

EXAMPLE: A mixture of 1.00 g of H_2 and 1.00 g of He is placed in a 1.00 L container at 27°C. Calculate the mole fraction and partial pressure of each gas. Calculate the total pressure in the container.

$$n_{H_2} = 1.00 \text{ g } H_2 \times \frac{1 \text{ mol } H_2}{2.02 \text{ g } H_2} = 0.496 \text{ mol } H_2$$

$$n_{He} = 1.00 \text{ g He} \times \frac{1 \text{ mol He}}{4.00 \text{ g}} = 0.250 \text{ mol He}$$

$$X_{H2} = \frac{0.496 \text{ mol}}{0.496 \text{ mol} + 0.250 \text{ mol}} = 0.665; \quad X_{He} = \frac{0.250 \text{ mol}}{0.496 \text{ mol} + 0.250 \text{ mol}} = 0.335$$

$$P_{H_2} = \frac{n_{H_2}RT}{V} = \frac{0.496 \text{ mol} \times \frac{0.08206 \text{ L} \times \text{atm}}{\text{mol} \times \text{K}} \times 300 \text{ K}}{1.00 \text{ l}} = 12.2 \text{ atm}$$

$$P_{He} = \frac{n_{He}RT}{V} = \frac{0.250 \text{ mol} \times \frac{0.08206 \text{ L} \times \text{atm}}{\text{mol} \times \text{K}} \times 300 \text{ K}}{1.0 \text{ L}} = 6.15 \text{ atm}$$

$$P_{total} = P_{He} + P_{H_2} = 12.2 \text{ atm} + 6.15 \text{ atm} = 18.4 \text{ atm}$$

GAS COLLECTION OVER WATER

(*Chemistry* 7th ed. pages 197–199 / 8th ed. pages 202–205)

A gas can be collected by displacement of water. A mixture of gases results due to a mixture of water vapor and the gas being collected.

EXAMPLE: A sample weighing 0.986 g contains zinc and some impurities. Excess hydrochloric acid is added and reacts with the zinc but not the impurities. Determine the percentage of zinc in the sample if 240.0 mL hydrogen gas are collected over water at 30.0°C and 1.032 atm.

Begin with the balanced equation: $Zn + 2\ HCl \rightarrow ZnCl_2 + H_2$.

Using Dalton's law and the vapor pressure of water at the specified temperature:

$$P_{total} = P_{atm} = P_{H_2} + P_{H_2O}; \quad P_{H_2} = P_{atm} - P_{H_2O}$$

$$P_{H_2} = 1.032\ atm - 0.042\ atm = 0.990\ atm$$

$$n_{H_2} = \frac{P_{H_2} \times V}{RT} = \frac{0.990\ atm \times 0.2400\ L}{\dfrac{0.08206\ L\ atm \times 303.2\ K}{mol\ K}} = 0.00955\ mol\ H_2$$

$$0.00955\ mol\ H_2 \times \frac{1\ mol\ Zn}{1\ mol\ H_2} \times \frac{65.4\ g\ Zn}{1\ mol\ Zn} = 0.625\ g\ Zn$$

$$\%Zn = \frac{g\ Zn \times 100\%}{g\ sample} = \frac{0.625\ g}{0.986\ g} \times 100\% = 63.3\%\ Zn$$

KINETIC MOLECULAR THEORY

(*Chemistry* 7th ed. pages 199–206 / 8th ed. pages 205–213)

The kinetic molecular theory is a model that attempts to explain the properties of an ideal gas. The postulates of the kinetic molecular theory are

1. The volume of the individual particles of a gas can be assumed to be negligible.
2. The gas particles are in constant motion. The pressure exerted by the gas is due to the collisions of the gases with the walls of the container.
3. The gases are not attracted to one another.
4. The average kinetic energy of a gas is directly proportional to the Kelvin temperature.

$$(KE)_{average} = 3/2\ RT = \frac{1}{2}\ mv^2_{avg}$$

EXAMPLE: Three identical flasks are filled with three different gases.

Flask A: CO at 760 torr and 0°C

Flask B: N_2 at 250 torr and 0°C

Flask C: H_2 at 100 torr and 0°C

In which flask will the molecules have the greatest average kinetic energy? In which flask will the molecules have the greatest average velocity?

All molecules will have the same average kinetic energy since they are all at the same temperature. Flask C will have the greatest average velocity since hydrogen has the lowest molar mass. At constant T, the lightest molecules are fastest, on average.

ROOT-MEAN-SQUARE VELOCITY

(*Chemistry* 7th ed. pages 204–205 / 8th ed. pages 211–212)

The root-mean-square velocity is the square root of the squares of the individual velocities of the gas particles.

$$u_{rms} = (3RT/M)^{1/2}$$

8.3145 J/K × mol

R = Ideal gas constant

I = Kelvin temperature

M = Molar mass of the gas

EXAMPLE: Which one of the following molecules has the largest root-mean-square velocity? H_2, N_2, CH_4, C_2H_6

Hydrogen, the substance with the lowest molar mass, has the largest root-mean-square velocity.

EFFUSION AND DIFFUSION

(*Chemistry* 7th ed. pages 206–208 / 8th ed. pages 213–214)

Diffusion is the mixing of gases. Effusion is the passage of a gas through a small opening into an evacuated chamber.

$$\text{For gases at the same temperature,}$$
$$\frac{\text{Rate of effusion for gas}_1}{\text{Rate of effusion for gas}_2} = \frac{(M_2)^{1/2}}{(M_1)^{1/2}}$$

EXAMPLE: The rate of effusion of a particular gas was measured and found to be 24.0 mL/min. Under the same conditions, the rate of pure methane (CH_4) gas is 47.8 mL/min. What is the molar mass of the unknown gas?

$$\frac{24.0}{47.8} = \frac{(16.04)^{1/2}}{(M1)^{1/2}} = 0.502$$

$16.04 = (0.502)^2 \times M_1$; $M_1 = 16.04 / 0.252 = 63.7$ g/mol

REAL GASES

(*Chemistry* 7th ed. pages 208–210 / 8th ed. pages 214–217)

The ideal gas model fails at high pressure and low temperature. Real gas molecules take up space and experience attractive forces between molecules. At high pressure there is less empty space between molecules, and the volume of molecules becomes more significant. An ideal gas could be compressed to zero volume, but for a real gas, as the pressure doubles, the volume of empty space cannot continue to be halved. As the temperature decreases, the molecules have less kinetic energy to overcome the attractive forces between gas molecules; these attractive forces may cause the gas to condense.

The van der Waals equation modifies the assumptions of the kinetic molecular theory to fit the behavior of real gases.

$$[P_{obs} + a(n/V)^2] \times (V - nb) = nRT$$

EXAMPLE: Which of the following gases would you expect to have the largest value of the van der Waals constant, b: H_2, N_2, CH_4, C_2H_6, or C_3H_8?

The van der Waals constant, b, is a measure of the size of the molecule. C_3H_8 has the largest molar volume and should have the largest value of b.

EXAMPLE: Which of the following gases would you expect to have the largest value of the van der Waals constant, a: H_2, CO_2, CH_4, or N_2?

CO_2 has the largest value for a, which measures intermolecular attractions. All the molecules are nonpolar so the only force present is an induced dipole or London force which increases as the number of electrons and protons in the molecule increases. (This will be discussed in Chapter 8.)

MULTIPLE-CHOICE QUESTIONS

No calculators are to be used in this section.

1. Ideal gases vary from real gases at conditions of
   (A) high temperature and low pressure.
   (B) high temperature and high pressure.
   (C) low temperature and low pressure.
   (D) low temperature and high pressure.
   (E) both high density and low pressure.

2. What is the volume of 3.00 mol of gas @ STP?
    (A) 22.4 L
    (B) 3 × 22.4 L
    (C) 3 × 22.4 L × 760
    (D) 3 × 22.4 LL × 273 / 760
    (E) It cannot be determined without knowing which gas is involved.

3. An ideal gas of volume 189. mL is collected over water at 30°C and 777 torr. The vapor pressure of water is 32 torr @ 30°C. What pressure is exerted by the dry gas under these conditions?
    (A) 320 torr
    (B) 745 torr
    (C) 777 torr
    (D) 32 / 77 torr
    (E) 32 x 777 torr

4. A 14.0-L cylinder contains 5.60 g N_2, 79.9 g Ar and 6.40 g O_2. What is the total pressure in atm at 27°C? (R = the ideal gas constant.)
    (A) 20 R
    (B) 26 R
    (C) 30 R
    (D) 60 R
    (E) 120 R

5. In a closed inflexible system, 7.0 mol CO_2, 7.0 mol Ar, 7.0 mol N_2, and 4.0 mol Ne are trapped, with a total pressure of 10.0 atm. What is the partial pressure exerted by the neon gas?
    (A) 1.6 atm
    (B) 4.0 atm
    (C) 10.0 atm
    (D) 21.0 atm
    (E) 29.0 atm

6. Consider the reaction: $C_2H_6(g) + 7/2 \, O_2(g) \rightarrow 2CO_2(g) + 3H_2O(g)$. If 6.0 g ethane, $C_2H_6(g)$ burn (as shown above), what volume of $CO_2(g)$ will be formed at STP?
    (A) 0.20 L
    (B) 0.40 L
    (C) 2.2 L
    (D) 9.0 L
    (E) 22.4 L

7. Cl_2 and F_2 combine to form a gaseous product; one volume of Cl_2 reacts with three volumes of F_2 yielding two volumes of product. Assuming constant conditions of temperature and pressure, what is the formula of the product?
    (A) ClF
    (B) Cl_2F_2
    (C) ClF_2
    (D) Cl_2F
    (E) ClF_3

8. Decreasing the temperature of an ideal gas from 80°C to 40°C causes the average kinetic energy to
   (A) decrease by a factor of two.
   (B) decrease by a factor of four.
   (C) increase by a factor of two.
   (D) increase by a factor of four.
   (E) decrease by less than a factor of two.

9. The effusion rate of helium gas, He, compared with that of methane gas, CH_4, is
   (A) four times greater for helium.
   (B) four times less for helium.
   (C) twice as great for helium.
   (D) twice as great for methane.
   (E) sixteen times as great for helium.

10. The average speed of the molecules of a gas is proportional to the
   (A) volume of the container.
   (B) reciprocal of absolute temperature, $(1 / T)$.
   (C) absolute temperature.
   (D) square root of the absolute temperature.
   (E) square of the absolute temperature.

11. Nitrogen gas diffuses through a porous barrier at the rate of 5.80 liters per minute. An unknown gas under the same conditions diffuses through the same barrier at a rate of only 1.80 liters per minute. What is the molar mass of the unknown gas?
   (A) 45.1
   (B) 50.3
   (C) 90.3
   (D) 145
   (E) 291

12. A 5.0 L vessel contains 2.0 moles of helium and 3.0 moles of hydrogen at a pressure of 10 atm. Maintaining a constant temperature, an additional 3.0 moles of hydrogen are added. What is the partial pressure of hydrogen gas in the vessel at the end? (Assume that the gases behave ideally.)
   (A) 6.0 atm
   (B) 8.0 atm
   (C) 10.0 atm
   (D) 12.0 atm
   (E) 20.0 atm

Questions 13-15 refer to Figure 1 and Figure 2 below. On the right is a key to identify the gases in the problems.

Figure 1   **Figure 2**   **Key to Gas Symbols**

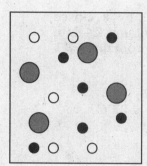   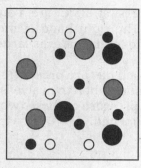   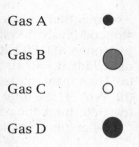

13. In Figure 1 there are three gases present, gas A, B and C. The number of spheres represents the number of moles of each gas present. If the total pressure within the vessel is 5.00 atm, what is the partial pressure of gas B?
    (A) 0.500 atm
    (B) 1.00 atm
    (C) 1.33 atm
    (D) 1.50 atm
    (E) 1.67 atm

14. Suppose that another gas D is added to the vessel in Figure 1, keeping the temperature and the volume constant. This is represented by Figure 2 above. What happens to the partial pressure of gas B?
    (A) increases
    (B) decreases
    (C) stays the same

15. What happens to the total pressure for the system described in Question 14? (In other words, compare the total pressure in Figure 1 versus Figure 2).
    (A) increases
    (B) decreases
    (C) remains the same

FREE-RESPONSE QUESTIONS

Calculators may be used for this section.

1. (a) Compare the temperature of freshly made coffee made at lower altitudes to coffee made at higher altitudes. Support your ANSWER with discussion.
   (b) The van der Waals equation, $(P + a/V_2)(V - b) = nrt$, is a necessary form of $PV = nRT$ under certain conditions.
      i.   What are these conditions? Explain your ANSWER.

      ii. What are the physical properties which the constants "*a*" and "*b*" are designed to describe? Explain your ANSWER.

  (c) Explain on a molecular level why gases exert pressure.

  (d) How does gaseous pressure relate to changes in volume? Explain.

2. Assume that two cylinders at 27°C are connected by a closed stopcock (valve) system. The right-hand cylinder contains 2.4 L of hydrogen at 0.600 atm; the left cylinder is larger and contains 6.8 L of helium at 1.40 atm.

  (a) How many moles of each gas are present?

  (b) What is the total pressure when the valve is open?

  (c) Determine the partial pressure of these two gases at 27°C when the stopcock is opened.

Answers

MULTIPLE CHOICE

1. **D** At low temperature and high pressure, the molecules are closer together and therefore the forces between molecules become more important as they are stronger. The actual (finite) volume of individual molecules also becomes more important as more of the total space is actually occupied by finite molecular volume (*Chemistry* 7th ed. pages 208–210 / 8th ed. pages 214–217).

2. **B** The molar volume of all gases at STP is about 22.4 L, so three moles would occupy 22.4 L /mol × 3 moles (*Chemistry* 7th ed. pages 208–210 / 8th ed. pages 214–217).

3. **B** The total pressure = 777 torr. Of this, 32 torr is due to the water vapor, hence 777–32 = 745 torr of pressure are allocated to the dry gas (*Chemistry* 7th ed. pages 194–199, 459–466 / 8th ed. pages 199–205, 471-478).

4. **C** Use $P_{Total} = N_{Total}RT / V$ to determine the pressure. From the mass of each of the gases you can find 0.200 mol of N_2, 1.00 mol of Ar, and 0.200 mol of O_2 to give a total number of moles (n) of 1.4 mol. Therefore $P_{Total} = R$ × 1.40 mol × 300 K / 14.0 L = 30 R. Note: You do not need a calculator to divide 1.4 by 14 and then multiply by 300! (*Chemistry* 7th ed. pages 186–190 / 8th ed. pages 188–194).

5. **A** The total number of moles is 25 (7.0 + 7.0 + 7.0 + 4.0 = 25), hence the Ne is 4 /25 of the total amount of gas and exerts 4 /25 of the total pressure, or 4/25 x 10 atm = 1.6 atm (*Chemistry* 7th ed. pages 194–199 / 8th ed. pages 199–205).

6. **D** 6.0 g of ethane / 30. g/mol yields 0.20 mol of ethane, which forms twice that number of moles of carbon dioxide. Since one mol of gas occupies 22.4 L at STP, 22.4 L/mol × 0.20 mol × 2 CO_2/C_2H_6 = 8.96 L of carbon dioxide (rounded to two significant

figures = 9.0 L)   (*Chemistry* 7th ed. pages 190–194 / 8th ed. pages 194–199).

7. **E** This problem assumes you understand Avogadro's Law, "At constant temperature and pressure, the volume of a gas is directly proportional to the number of moles of the gas". To apply that to this problem, 1 volume Cl_2 + 3 volumes F_2 → 2 volumes Cl_xF_{3x}. The simplest formula for the product becomes ClF_3 (*Chemistry* 7th ed. pages 181–186 / 8th ed. pages 183–188).

8. **E** Be careful here to note the difference between the kinds of temperature scales and what they mean. Even though the Celsius temperature is half as much, the average kinetic energy is proportional to the Kelvin temperature. In this case that ratio is only 353 / 313 = 1.13 so the average kinetic energy decreases only by a factor of 1.13 (*Chemistry* 7th ed. pages 198–206 / 8th ed. pages 203–212).

9. **C** Effusion rates are inversely proportional to the square root of the molecular masses of the gases (*Chemistry* 7th ed. pages 206–208 / 8th ed. pages 212–214).

10. **D** In this case the question is about the average speed (not the energy) of the molecules. Review root-mean-square velocity (*Chemistry* 7th ed. pages 204–206 / 8th ed. pages 211–212).

11. **E** The unknown gas diffuses more slowly; therefore, its molar mass must be greater than that of nitrogen, N_2, 28.0 g mol^{-1}.

$$(5.80/1.80) = \sqrt{M/28.0}$$

$$M = 291 \text{ g mol}^{-1}$$

(*Chemistry* 7th ed. pages 206-208 / 8th ed. pages 212–214).

12. **D** The initial partial pressure of the hydrogen = 10 atm x (3.0/5.0) = 6.0 atm. The amount of hydrogen present is doubled from 3.0 moles to 6.0 moles, therefore the partial pressure will increase proportionally to 12.0 atm. It does not matter what other gases are added or what other gases are present. (*Chemistry* 7th ed. pages 194–197 / 8th ed. pages 199–203).

13. **C** The partial pressure = mole fraction × total pressure. In this case, partial pressure of gas B = (4/15)(5.00 atm) = 1.33 atm

14. **C** Since the amount of gas B remains the same, there would be no change in the partial pressure of this gas.

15. **A** Since the total number of moles of gas in the vessel has now increased, the total pressure will also increase. (Reference for questions 13-15: *Chemistry* 7th ed. pages 194-198 / 8th ed. pages 199-203).

FREE RESPONSE

1. (a) At low altitudes, the amount of air above the surface of the earth, and therefore the total atmospheric pressure, would be greater. Water boils at a higher temperature under such conditions since a higher vapor pressure is required for the liquid to become a gas (*Chemistry* 7th ed. pages 469–470 / 8th ed. pages 481–482).

   (b) In this question, be careful not to confuse "conditions" with "physical properties" or "characteristics of the gaseous molecules." The conditions under which real gases do not act like ideal gases and therefore do not obey the simpler $PV = nRT$ are those of high pressure and low temperature. This is because when the gas molecules are closer together, the forces of attraction between the molecules (accounted for by the constant "*a*") become more important. Also, the finite volume of individual molecules (accounted for by constant "*b*") becomes a larger part of the total volume occupied by the gas (*Chemistry* 7th ed. pages 186–190 / 8th ed. pages 188–194).

   (c) Gases exert pressure because the molecules are moving and strike the sides of the container exerting force on it; this force per unit of area is what we call pressure (*Chemistry* 7th ed. pages 179–180 / 8th ed. pages 181–182).

   (d) Gaseous pressure is inversely proportional to the volume of the container. If only the volume of the container is less, for example, the molecules have less room to move around before striking the sides and therefore strike the sides of the container more often. This assumes that the temperature is held constant (*Chemistry* 7th ed. pages 181–184 / 8th ed. pages 183–186).

   Note: What assumption is made in your ANSWER to D?

2. Solving this problem has three steps. (1) To determine the pressure exerted by each gas, first calculate the number of moles of each ($n = PV / RT$), (2) and then use the total number of moles of gas and $p = nRT/V$ to calculate the total pressure when the total volume is 9.20 L. (3) Finally, the partial pressure of each gas is the total pressure × the mol fraction of that gas (*Chemistry* 7th ed. pages 186–190 and 194–195 / 8th ed. pages 188–194 and 199-200).

   For hydrogen: (0.600 atm × 2.40 L) / (R × 300 K) = 0.0585 mol of H_2

   For helium: (1.40 atm × 6.80 L) / (R × 300 K) = 0.397 mol of He

   Total number of moles = 0.0585 + 0.397 mol = 0.456 mol

   Total pressure is then (0.456 mol × R × 300 K) / (9.20 L) = 1.22 atm

   Use the mol fractions: (0.0585 / 0.456) × 1.22 atm = 0.157 atm H_2

   (0.397 / 0.456) × 1.22 atm = 1.06 atm He

6

THERMOCHEMISTRY AND THERMODYNAMICS

Thermochemistry describes the heat flow of a chemical reaction or physical change. Changes in enthalpy are calculated through calorimetry, Hess's Law, and standard heats of formation. The study of thermodynamics allows one to predict whether a process will occur or not. The dependence of Gibbs free energy on temperature, entropy, and enthalpy will be identified. The laws of thermodynamics will be explored.

In the free-response portion of the exam, one of the questions in Part B usually involves entropy, enthalpy, and free energy. A lab question regarding the use of a calorimeter to determine the heat of a reaction may be asked in the essay portion of the test.

You should be able to
- Perform stoichiometric calculations with the enthalpy of the reaction.
- Perform calculations with specific heat.
- Discuss how a calorimeter is used and perform related calculations.
- Draw, label, and perform associated calculations for heating curves involving specific heat and changes in enthalpy for phase changes.
- Perform calculations with Hess' Law.
- Perform calculations with standard heats of formation, ΔH°_f.
- Write reactions representing standard heats of formation, ΔH°_f.
- Compare the absolute entropies, S°, of elements and compounds.
- Perform calculations with entropy S° and ΔS°.
- Perform calculations with free energy, ΔG°_f.

119

AP tips

Mathematical equations will be provided in the free-response portion of the test, but not in the multiple choice. It is best to memorize the equations.

Questions on enthalpy will be revisited in the chapter on bonding. Bond energies can be used to estimate the change in enthalpy of a reaction.

FIRST LAW OF THERMODYNAMICS

(*Chemistry* 7th ed. pages 231–232 / 8th ed. pages 238–239)

The first law of thermodynamics states that energy can be converted from one form to another, but can neither be created nor destroyed: The energy of the universe is constant.

ENTHALPY

(*Chemistry* 7th ed. pages 235–236 / 8th ed. pages 243–244)

At constant pressure, the change in enthalpy (ΔH°) of a system is equal to the energy flow as heat. Enthalpy is a state function, so the change in H is independent of the pathway.

EXOTHERMIC

A reaction that is exothermic gives off heat and the change in enthalpy is negative ($\Delta H^\circ < 0$). The temperature of the surroundings increases during an exothermic process.

ENDOTHERMIC

A reaction that is endothermic absorbs heat and the change in enthalpy is positive ($\Delta H^\circ > 0$). The temperature of the surroundings decreases in an endothermic process.

EXAMPLE: How much heat is released when 4.03 g of hydrogen is reacted with excess oxygen?

$$2H_2(g) + O_2(g) \rightarrow 2H_2O(l) \qquad \Delta H^\circ = -572 \text{ kJ}$$

$$4.03 \text{ g H}_2 \times \frac{1 \text{ mol H}_2}{2.02 \text{ g}} \times \frac{-572 \text{ kJ}}{2 \text{ mol H}_2} = -572 \text{ kJ}$$

Note that conversion factors can be written using the relationship between the enthalpy, ΔH°, and the moles of reactants and products. 572 kJ of heat is produced when 1 mole of O_2 reacts with 2 moles of H_2 to produce 2 mol of H_2O.

CALORIMETER

(*Chemistry* 7th ed. pages 237–242 / 8th ed. pages 244–249)

A calorimeter is a device used to experimentally determine the heat energy change of a chemical reaction. Calorimetry can be carried out under constant pressure (measures ΔH) or constant volume (measures ΔE).

SPECIFIC HEAT

(*Chemistry* 7th ed. pages 237–238 / 8th ed. pages 245–246)

The amount of heat lost or gained in a reaction depends on
1. The change in temperature during the reaction.
2. The amount of substance present.
3. The heat capacity (*C*) of a substance which is the amount of energy required to change the temperature of an object by 1°C.
    C = heat absorbed / increase in temperature = J / °C
    a. Specific heat capacity, *s*, is the amount of energy required to change the temperature of 1.0 g of a substance by 1°C (J/ g °C).
    b. Molar heat capacity is the amount of energy required to change the temperature of 1 mole of a substance by 1°C.

EXAMPLE: The same amount of heat energy is applied to 1.0 g of each of solid aluminum, solid iron, mercury liquid, and carbon graphite. Which one of these substances will reach the highest temperature?

The answer is Hg(l). Mercury has the lowest specific heat of the four mentioned substances (see table in the textbook), which means it takes a lesser amount of heat to raise 1 gram of mercury 1°C.

The heat released by a reaction can be determined by

$$q \;=\; m \times s \times \Delta T$$

EXAMPLE: A 110.g sample of copper (specific heat capacity 0.20 J / (g° C) is heated to 82.4°C and then placed in a container of water at 22.3° C. The final temperature of the water and the copper is 24.9° C. What was the mass of the water in the original container, assuming that all the heat lost by the copper is gained by the water?

The heat lost by copper is equal to:

$$110. \text{ g Cu} \times \frac{0.20 \text{ J}}{\text{g°C}} \times (82.4°C - 24.9°C) = 1265 \text{ J} = 1.3\text{kJ}$$

The heat lost by the copper is equal to the heat gained by the water.

−(heat lost by copper) = (heat gained by water)

$$\text{Mass of water} = \frac{q}{s \times \Delta T} = \frac{1265 \text{ J}}{\dfrac{4.184 \text{ J} \times 2.6°C}{g°C}} = 120 \text{ g water}$$

AP tips

Students sometimes have trouble with the sign of ΔT. Change in temperature, ΔT, is usually calculated $T_{final} - T_{initial}$. When the heat lost by one substance equals the heat gained by the other substance, ΔT for the substance losing heat needs to be calculated $T_{initial} - T_{final}$ in order for the heat gained to equal the heat lost. (Otherwise you will end up with $-q = q$.)

HEATING CURVES

(*Chemistry* 7th ed. pages 463–464 / 8th ed. pages 475–476)

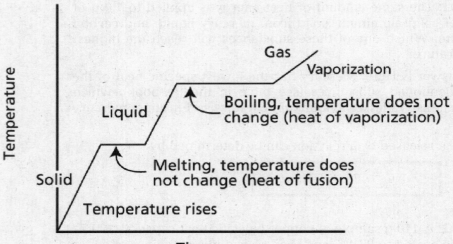

When a system is heated, energy is transferred into it. In response to the energy it receives, the system changes, for example, by increasing its temperature. A plot of the temperature versus time is called **the heating curve**. One such heating curve is shown above. The physical states of the substance and the phase transitions are identified along the curve.

The temperature of the system usually increases when energy is applied. However, when the energy supplied is used for phase transition, a change in the physical state, the temperature (average kinetic energy) remains constant because the potential energy of the system is being increased as the molecules are rearranged in the phase change.

HEAT OF FUSION

(*Chemistry* 7th ed. page 464 / 8th ed. page 476)

The heat of fusion, ΔH_{fus} is the enthalpy change that occurs in melting a solid at its melting point.

EXAMPLE: What quantity of heat is required to melt 1.00 kg of ice at its melting point? For ice, $\Delta H_{fus} = 6.0$ kJ/mol

$$1.00 \text{ kg} \times \frac{1000 \text{ g}}{1 \text{ kg}} \times \frac{1 \text{ mol}}{18.0 \text{ g}} \times \frac{6.0 \text{ kJ}}{1 \text{ mol}} = 333 \text{ kJ}$$

HEAT OF VAPORIZATION

(*Chemistry* 7th ed. page 458 / 8th ed. pages 470–471)

The heat of vaporization, ΔH_{vap} , is the energy required to vaporize one mole of a liquid at a pressure of one atmosphere.

EXAMPLE: What quantity of heat is required to vaporize 130 g of water? For liquid, water, $\Delta H_{vap} = 43.9$ kJ/mol.

$$130 \text{ g} \times \frac{1 \text{ mol}}{18.0 \text{ g}} \times \frac{43.9 \text{ kJ}}{1 \text{ mol}} = 317 \text{ kJ}$$

EXAMPLE: The substance, X, has the following properties:

	Specific Heat
ΔH_{vap} =20. kJ/mol	Solid = 3.0 J/ (g °C)
ΔH_{fus} = 5.0 kJ/mol	Liquid 2.5 J/ (g °C)
Boiling point 75°C	Gas 1.0 J/ (g °C)
Melting point –15°C	

Calculate the energy required to convert 250. g of substance X from a solid at –50.0°C to a gas at 100°C. Assume that X has a molar mass of 75.00 g/mol.

There are 5 steps involved.

1. Heating the solid from –50.0°C to –15°C.

2. Melting the solid at –15°C.

3. Heating the liquid to its boiling point from –15°C to 75°C.

4. Boiling the liquid at 75°C.

5. Heating the gas to 100°C

The energy used = sum of energies from individual steps.

There are 3.33 mol of X in 250. g of X.

Step 1: $q = m \times s \times \Delta T$; 250. g × [3.0 J/(g °C)] × 35°C =   26 kJ

Step 2: mol × ΔH_{fus} = 3.33 mol × 5.0 kJ/mol =   17 kJ

Step 3: $q = m \times s \times \Delta T$; 250.0g × [2.5 J/(g °C)] × 90°C =   56 kJ

Step 4: mol × ΔH_{vap} = 3.33 mol × 20. kJ/mol =   67 kJ

Step 5: $q = m \times s \times \Delta T$; 250.0g × [1.0 J/(g °C)] × 25 °C = __6.2 kJ__

                                                                                                      172 kJ

The total energy required is the sum of the energies of the individual steps.

HESS'S LAW

(*Chemistry* 7th ed. pages 242–246 / 8th ed. pages 249–255)

Hess's Law states that enthalpy change, in going from a particular set of reactants to a particular set of products, is the same whether the reaction takes place in one step or in a series of steps. Enthalpy is a state function.

EXAMPLE: Given the following reactions and $\Delta H°$ values,

$$\Delta H°$$

Si(s) + 2H$_2$(g) → SiH$_4$ (g)   34 kJ/mol

Si(s) + O$_2$(g) → SiO$_2$(s)   –911 kJ/mol

H$_2$(g) + ½ O$_2$(g) → H$_2$O(g)   –242 kJ/mol

Calculate ΔH for

SiH$_4$ (g) + 2O$_2$(g) -→ SiO$_2$(s) + 2 H$_2$O (g) $\Delta H° = ?$

The idea is to manipulate the three equations so that they add up to the overall reaction. There are two ways that the equations can be manipulated.

1. Reverse the reaction. The sign of ΔH is also reversed.

2. Multiply the coefficients in a balanced reaction by an integer. The value of ΔH is multiplied by the same integer.

SOLUTION: Start by finding a substance that appears only once in all of the reactions, SiH$_4$ or H$_2$O. Look to see where that substance is in the final reaction and in what amount. Modify the reaction so that the substance appears where it should be and in the correct amount. SiH_4 needs to be in the reactants, so the first reaction is reversed. H_2O stays in the products, but there are two of them in the final reaction, so, the third equation needs to be multiplied by two.

$SiH_4(g) \rightarrow \cancel{Si}(s) + 2H_2(g)$ – (34 kJ/mol) (Note: This was reversed.)

$\cancel{Si}(s) + O_2(g) \rightarrow SiO_2(s)$ – 911 kJ/mol

(Note: The 2nd reaction stays the same so that Si can cancel out.)

$\underline{2H_2(g) + O_2(g) \rightarrow 2H_2O(g)}$ $\underline{2(-242kJ/mol)}$

$SiH_4(g) + 2O_2(g) \rightarrow SiO_2(s) + 2 H_2O(g)$

$$\Delta H^\circ = -1429 \text{ kJ}$$

STANDARD ENTHALPIES OF FORMATION, ΔH°_F

(*Chemistry* 7th ed. pages 246–252 / 8th ed. pages 255–261)

The standard enthalpy of formation of a compound is the change in enthalpy that accompanies the formation of one mole of a compound from its elements with all substances in their standard states.

The degree symbol on ΔH° indicates that the process occurred under standard conditions of 25°C, 1 atm and 1M solutions.

By definition, the standard heat of formation for elements in their standard states equals zero.

EXAMPLE: Which of the following will have standard heats of formation equal to zero?

$H_2(g)$, $Hg(s)$, $CO_2(g)$, $H_2O(l)$, $Br_2(l)$

The only elements present in their standard states are $H_2(g)$ and $Br_2(l)$. Mercury is a liquid under standard conditions. Carbon dioxide and water are compounds.

EXAMPLE: Write the balanced molecular equation representing the ΔH_f^0, standard heat of formation reaction of ethanol, $C_2H_5OH(l)$.

$2C(s) + 3H_2(g) + 1/2 O_2(g) \rightarrow C_2H_5OH(l)$

Note, that 1 mole of product is produced according to the definition of the standard heat of formation. All of the reactants are elements in their standard states.

To calculate the enthalpy change for a given reaction, the enthalpies of formation of the reactants are subtracted from the enthalpies of formation of the products.

$$\Delta H^\circ_{reaction} = \Sigma \Delta H^\circ_{f(products)} - \Sigma \Delta H^\circ_{f(reactants)}$$

EXAMPLE: Using the standard heats of formation, $\Delta H°_f$,calculate the change in enthalpy, ΔH^0, for the following reaction. The standard heats of formation for hydrogen peroxide and water are –187 kJ/mol and –285 kJ/mol, respectively.

$$2H_2O_2(l) \rightarrow 2\ H_2O(l) + O_2(g)$$

$$\Delta H°_{reaction} = [2\ (–285\ kJ) + 0] – 2\ (–187\ kJ) = –196\ kJ$$

EXAMPLE: Calculate the heat of formation of gaseous carbon monoxide based on the reaction below.

$$2CO(g) + C(s) \rightarrow C_3O_2(g) \qquad \Delta H° = 127.3\ kJ$$

$\Delta H°_f$ for $C_3O_2(g)$ is –93.7 kJ/mol .

$$\Delta H°_{reaction} = \Sigma\ \Delta H°_{f(products)} – \Sigma\ \Delta H°_{f(reactants)}$$

$$\Delta H°_{reaction} = \Delta H°_f\ \text{for}\ C_3O_2(g) – [2\ \Delta H°_f\ \text{for}\ CO(g) – \Delta H°_f\ \text{for}\ C(s)]$$

Rearrange the equation and solve for missing variable; note $\Delta H°_f$ for C(s) is zero.

$$\Delta H°_f\ \text{for}\ CO(g) = [\Delta H°_f\ \text{for}\ C_3O_2(g) - \Delta H°_{reaction}] / 2$$

$$= (–93.7\ kJ – 127.3\ kJ) / 2 = –110.5\ kJ\ /mol$$

ENTROPY

(*Chemistry* 7th ed. pages 752–755 / 8th ed. pages 776–779)

Entropy, *S*, is the driving force for a spontaneous process which readily occurs without intervention. Entropy measures molecular randomness or disorder.

Positional entropy describes the number of arrangements or positions available to a system in a given state. Positional entropy increases in going from a solid to liquid to gas. In the solid state, molecules are much closer together than in the gaseous state, with very few positions available for them.

Entropy during a process is said to be increasing when the value of $\Delta S > 0$; The value of ΔS is positive.

Entropy during a process is said to be decreasing when the value of $\Delta S < 0$; The value of ΔS is negative.

EXAMPLE: Which of the following pairs is likely to have the higher positional entropy per mole at a given temperature?

1) Solid CO_2 or gaseous CO_2 ?

2) N_2 gas at 1.0 atm and N_2 gas at 0.001 atm?

1) Gaseous CO_2 has more positional entropy than solid CO_2 since there are more positions for the molecules the gaseous state to move to than in the solid state.

2) N_2 gas at 0.001 atm has more positional entropy than N_2 gas at 1.0 atm, because at a lower pressure, there is more volume for the molecules to move than at a higher pressure.

EXAMPLE: Describe the change in entropy when solid salt is added to water.

The entropy increases when salt in its solid state is dissolved in water. The ions fixed in a crystal lattice in the solid are free to move about in the water due to solute–solvent interactions.

SECOND LAW OF THERMODYNAMICS

(*Chemistry* 7th ed. page 755 / 8th ed. page 779)

The second law of thermodynamics states that in any spontaneous process there is always an increase in the entropy of the universe. The change in entropy of the universe is equal to the change in the entropy of the system and the change in entropy of the surroundings.

$$\Delta S_{univ} = \Delta S_{sys} + \Delta S_{surr}$$

$$\Delta S_{surr} = -\Delta H/T$$

The sign of ΔS_{surr} depends on the direction of the heat flow. ΔS_{surr} is positive at constant temperature when the reaction is exothermic, since heat flows to the surroundings increasing the random motions and the entropy of the surroundings. The opposite is true for an endothermic reaction at constant pressure.

The magnitude of ΔS_{surr} depends on temperature. At a low temperature, the production of heat effects a much greater percent change in the randomness of the surroundings than it does at high temperature.

ENTROPY CHANGES IN A CHEMICAL REACTION

(*Chemistry* 7th ed. pages 762–766 / 8th ed. pages 786–790)

The change in entropy, ΔS, for a chemical reaction can be predicted without calculation.

For a chemical reaction involving only gaseous reactants and products, entropy is related to the total number of moles of gas on either side of the equation. If the moles of gas increase from reactants to products in a chemical reaction, the entropy is increasing. If the moles of gas decrease in a chemical reaction, the entropy is decreasing.

For a chemical reaction involving solid, liquids and gases, the production of a gas will, in general, increase the entropy of the reaction much more than an increase in the number of moles of liquids or solids.

EXAMPLE: Predict the sign of ΔS° for the following reaction:

$N_2(g) + 3H_2(g) \rightarrow 2NH_3(g)$

The entropy decreases as the reaction proceeds from reactants to products because the number of moles of gas decreases from four total moles to two moles. The sign of $\Delta S°$ is negative.

The change in entropy for a reaction, $\Delta S°$, can also be calculated using tabulated thermodynamic values in the appendix of the textbook.

$$\Delta S°_{reaction} = \Sigma\, S°_{(products)} - \Sigma\, S°_{(reactants)}$$

EXAMPLE: Calculate $\Delta S°$ for the following reaction:

$N_2(g) + 3H_2(g) \rightarrow 2NH_3(g)$

$\Delta S° = 2(193) - [1(192) + 3(131)] = -199$ J/K.

The sign of $\Delta S°$ is negative, confirming the prediction in the previous example.

FREE ENERGY

(*Chemistry* 7th ed. pages 766–773 / 8th ed. pages 790–797)

Free energy, G, is a thermodynamic function whose value describes whether or not a process is spontaneous. Gibbs free energy is dependent on the change in enthalpy, change in entropy, and temperature of the system.

$$\Delta G = \Delta H - T\,\Delta S;\ T \text{ is the Kelvin temperature}$$

Using the chart below, you can predict if a reaction will occur without the exact value for ΔH and ΔS.

The Dependence of Spontaneity on Temperature		
ΔS	ΔH	ΔG
+	–	Spontaneous at all temperatures
+	+	Spontaneous at high temperatures
–	–	Spontaneous at low temperatures
–	+	Process not spontaneous at any temperature

FREE ENERGY AND CHEMICAL REACTIONS

(*Chemistry* 7th ed. pages 766–770 / 8th ed. pages 790–794)

The standard free energy change, $\Delta G°$, is the change in free energy that will occur if the reactants in their standard states are converted to the products in their standard states. This value cannot be measured directly, but it can be calculated from other measured quantities such as the equilibrium constant and the standard cell potential.

The standard free energy change, ΔG, can be calculated from the changes in enthalpy and entropy.

$$\Delta G° = \Delta H° - T\,\Delta S°$$

EXAMPLE: Consider the reaction $2\ POCl_3(g) \rightarrow 2PCl_3(g) + O_2(g)$.

The value of $\Delta S°$ is 179 J/K. The value of $\Delta H°$ is 542 kJ. At what temperature is this reaction spontaneous? Assume that $\Delta H°$ and $\Delta S°$ do not depend on temperature.

When the process is spontaneous, solve for the temperature when $\Delta G° = 0$.

$\Delta G° = \Delta H° - T\,\Delta S°$; $T = \Delta H° / \Delta S° = 542$ kJ/ (0.179 kJ/K) = 3030 K.

The standard free energy change, ΔG, is a state function and can be calculated by

$$\Delta G°_{reaction} = \Sigma\ \Delta G°_{f(products)} - \Sigma\ \Delta G°_{f(reactants)}$$

THE DEPENDENCE OF FREE ENERGY ON PRESSURE

(*Chemistry* 7th ed. pages 770–772 / 8th ed. pages 794–796)

The free energy of a reaction is dependent on the pressure of a gas or the concentration of species in solution.

Consider the reaction: $N_2(g) + 3H_2(g) \rightarrow 2NH_3(g)$

$$\Delta G° = \Delta G°_{reaction} + RT \ln Q$$

$$\Delta G = \Delta G°_{reaction} + RT \ln \left[\frac{(P_{NH_3})^2}{(P_{N_2})(P_{H_2})^3} \right]$$

R = Gas law constant = 8.3145 J/K

$\Delta G°$ = the free energy change with all reactants and products at a pressure of 1 atm and 25°C.

ΔG = the free energy change for the reaction for the specified pressures of reactants and products.

Q = the reaction quotient and will be discussed in the equilibrium chapter in this book.

FREE ENERGY AND EQUILIBRIUM

(*Chemistry* 7th ed. pages 774–778 / 8th ed. pages 798–802)

$$\Delta G° = -RT \ln K;\ K = \text{equilibrium constant}$$

This will be discussed in the chapter on equilibrium in this book.

FREE ENERGY AND CELL POTENTIAL

(*Chemistry* 7th ed. pages 800–803 / 8th ed. pages 800–803)

$$\Delta G^{\circ} = -nFE^{\circ}; \ E^{\circ} \text{ is the standard cell potential}$$

This will be discussed in the chapter on electrochemistry in this book.

MULTIPLE-CHOICE QUESTIONS

No calculators are to be used in this section.

1. The standard enthalpy of formation for nitrogen dioxide is the enthalpy change of the reaction
   (A) $1/2 \ N_2O_4(g) \rightarrow NO_2(g)$.
   (B) $1/2 \ N_2(g) + O_2(g) \rightarrow NO_2(g)$.
   (C) $N_2(g) + 2O_2(g) \rightarrow N_2O_4(g)$.
   (D) $N_2(g) + 2O_2(g) \rightarrow 2NO_2(g)$.
   (E) $NO(g) + 1/2 \ O_2(g) \rightarrow NO_2(g)$.

2. Which of the following has a standard enthalpy of formation which is not zero?
   (A) $Na(s)$
   (B) $Hg(l)$
   (C) $H_2O(l)$
   (D) $N_2(g)$
   (E) $C(s)$

3. For endothermic reactions at constant pressure
   (A) $\Delta H < 0$.
   (B) $\Delta H > 0$.
   (C) $\Delta G < 0$.
   (D) $\Delta S > 0$.
   (E) $\Delta S < 0$.

4. At a certain temperature $C_{(s)} + O_{2(g)} \rightarrow CO_{2(g)}$ has a ΔG of -339.4 kJ/mol. This means that at this temperature
   (A) the system is at equilibrium.
   (B) gaseous carbon dioxide is unstable.
   (C) gaseous carbon dioxide spontaneously forms.
   (D) this system has a high reaction rate.
   (E) the system will not react.

For questions 5, 6, and 7, consider four separate situations:

I with $\Delta H = +$ and $\Delta S = +$

II with $\Delta H = +$ and $\Delta S = -$

III with $\Delta H = -$ and $\Delta S = -$

IV with $\Delta H = -$ and $\Delta S = +$

Assume that both ΔH and ΔS are temperature independent.

5. Which processes are spontaneous at all temperatures?
   (A) I only
   (B) II only
   (C) III only
   (D) IV only
   (E) I and IV

6. The process which must be nonspontaneous at all values of temperature is
   (A) I.
   (B) II.
   (C) III.
   (D) IV.
   (E) Impossible to identify without more data.

7. Which of these four processes is improbable at a low temperature but becomes more probable as the temperature rises?
   (A) I
   (B) II
   (C) III
   (D) IV
   (E) Impossible to identify without more data

8. In which of the following four processes is there an increase in entropy?
   I $2SO_2(g) + O_2(g) \rightarrow SO_3(g)$
   II $H_2O(g) \rightarrow H_2O(s)$
   III $Hg(l) \rightarrow Hg(g)$
   IV $H_2O_2(l) \rightarrow H_2O(l) + 1/2\ O_2(g)$

   (A) all of the above
   (B) I, II
   (C) I, IV
   (D) III, IV
   (E) II, III, IV

For questions 9 and 10 refer to the following exothermic reactions involving gases.

I $CH_4 + 2O_2 \rightarrow CO_2 + 2H_2O$

II $1/2\ C + 1/2\ O_2 \rightarrow 1/2\ CO_2$

III $CH_4 + O_2 \rightarrow C + 2H_2O$

IV $C + O_2 \rightarrow CO_2$

9. According to the data given, which reaction liberates the most energy?
(A) I
(B) II
(C) III
(D) IV
(E) This cannot be determined without more data.

10. According to the data given, which reaction liberates the least energy?
(A) I
(B) II
(C) III
(D) IV
(E) This cannot be determined without more data.

11. What is the enthalpy change for the following reaction under standard conditions?

$CS_2(l) + 3\ O_2(g) \rightarrow CO_2(g) + 2\ SO_2(g)$

$H_f^\circ\ CS_2(l) = +\ 88$ kJ/mol

$H_f^\circ\ CO_2(g) = -394$ kJ/mol

$H_f^\circ\ SO_2(g) = -297$ kJ/mol

(A) −900 kJ
(B) −779 kJ
(C) −603 kJ
(D) −1076 kJ
(E) +1173 kJ

12. Which statement is **false**?
(A) All standard heats of formation cancel in a balanced chemical equation.
(B) The standard heat of formation of oxygen, O_2 gas is zero.
(C) The heat of reaction can be calculated from the standard heats of formation.
(D) The heat of reaction at constant external pressure is called the enthalpy change.
(E) The standard heat of formation of liquid mercury is zero.

13. A 57 gram block of metal at 92°C is dropped into an insulated flask containing approximately 45.0 grams of ice and 30.0 grams of water at 0°C. After the system had reached equilibrium it was determined that 9.5 grams of the ice had melted. What is the specific heat of the metal? (Heat of fusion of water = 333 J/g)
    (A) 0.22 J/°C/g
    (B) 0.32 J/°C/g
    (C) 0.60 J/°C/g
    (D) 0.79 J/°C/g
    (E) 0.92 J/°C/g

14. A reaction takes place within a system. As a result, the entropy of the system decreases – the system becomes more ordered. Which of the following statements *must* be true?
    I.   The reaction is exothermic.
    II. The entropy S of the universe increases.
    III. The entropy of the surroundings increases.
    IV. The Gibbs free energy G of the system decreases.

    (A) I only
    (B) II only
    (C) I and II
    (D) I and III
    (E) I, II, III and IV

15. When propane burns in air, heat is released:

    $$C_3H_8(g) + 5\,O_2(g) \rightarrow 3\,CO_2(g) + 4\,H_2O(g)$$

    What are the signs of H, S, and G for this process as illustrated by the above equation?

	ΔH	ΔS	ΔG
(A)	–	+	+
(B)	–	+	–
(C)	–	–	–
(D)	–	–	+
(E)	+	+	–

FREE-RESPONSE QUESTIONS

1. Ozone can be prepared from molecular oxygen by subjecting oxygen to an electrical discharge. The standard free energy of formation of ozone is +163.4 kJ/mol.
    (a) Write the chemical equation for the formation of ozone.
    (b) What is the sign of the entropy change for this reaction? Justify your answer.
    (c) Comment on the stability of ozone based on these data.
    (d) As temperature increases, would you expect ozone to become more or less stable? Justify your answer.

2. The molar heats of fusion and vaporization of benzene are 10.9 kJ/mol and 31.0 kJ/mol, respectively. The melting temperature of benzene is 5.5° C and it boils at 80.1°C.
   (a) Calculate the entropy changes for solid → liquid, and for liquid → vapor for benzene.
   (b) Would you expect the ΔS for these two changes to be about the same?
   (c) Comment on the physical significance of the difference in these two values.
   (d) Why are the values for heat of vaporization usually so much greater than the heats of fusion?

Answers

MULTIPLE CHOICE

1. **B** This question should determine if you understand how to apply the definition of the term "standard enthalpy of formation." Like may terms in chemistry, this has a very specific meaning: It is the energy involved in forming one mole of a compound from its elements in their standard or free state at 25°C and 1 atm. pressure. As applied here, you must write the equation showing the formation of $NO_2(g)$ as the product from the elements nitrogen and oxygen (both are diatomic elements under these conditions), and balance it so that only one mole of the product compound is formed (*Chemistry* 7th ed. pages 246–251 / 8th ed. pages 255–256).

2. **C** By definition, standard enthalpy of formation is the energy involved when one mole of a compound is formed from its elements in their standard states at 25°C and 1 atm. In this question, only water is a compound (*Chemistry* 7th ed. pages 246–247 / 8th ed. pages 255–256).

3. **B** In an endothermic reaction, heat is gained by the system. The convention to show this is to indicate that the enthalpy change is positive (*Chemistry* 7th ed. pages 242–244 / 8th ed. pages 249–253).

4. **C** A negative Gibbs free energy value indicates a spontaneous reaction; CO_2 forms. Note that it says nothing about the rate of reaction. Kinetics will be the topic of a later chapter (*Chemistry* 7th ed. pages 759–762, 773–776 / 8th ed. pages 783-786, 797-800).

5. **D** From $\Delta G = \Delta H - T\Delta S$. For example, if ΔH is negative and ΔS is positive the reaction must be spontaneous because ΔG is negative in all such cases (*Chemistry* 7th ed. pages 759–768 / 8th ed. pages 783–792).

6. **B** From $\Delta H - T\Delta S = \Delta G$. If ΔH is positive and $T\Delta S$ is negative, then ΔG is positive in all cases, so the reaction is always nonspontaneous (*Chemistry* 7th ed. pages 759–768 / 8th ed. pages 783–792).

7. **A** From $\Delta H - T\Delta S = \Delta G$, if the temperature is high, the $T\Delta S$ factor is large and "overcomes" the influence of ΔH (+) to make ΔG negative; hence it becomes spontaneous at the higher temperatures (*Chemistry* 7th ed. pages 759–768 / 8th ed. pages 783–792).

8. **D** If there is an increase in entropy, then the products must be more disordered than the reactants. When a liquid becomes a gas (process III) the disorder increases; when one mole of a liquid becomes one mole of a different liquid and half a mole of gas (process IV), the entropy increases (*Chemistry* 7th ed. pages 762–766 / 8th ed. pages 786–790).

9. **A** Since II is exactly half of IV, it must be less exothermic. Note also that if you add III to IV you get I, hence I must be the most exothermic of these four processes (each of the parts must be less than the whole (*Chemistry* 7th ed. pages 242–246 / 8th ed. pages 249–255).

10. **E** From these data there is no certain way to determine if II or III is less energetic (*Chemistry* 7th ed. pages 242–246 / 8th ed. pages 249–255).

11. **D** Enthalpy change = [(-394 + (2 × -297)) – (+88)] = –1076 kJ This is a basic Hess's law problem. Remind yourself that the heat of formation of any element in its standard state is defined as zero, which is why there is no value listed for oxygen gas (*Chemistry* 7th ed. pages 242-252 / 8th ed. pages 251-261).

12. **A** It would be exceedingly unusual to have all of the standard heats of formation cancel in a chemical reaction when applying Hess's Law. Choices B, C, D and E are all standard definitions or procedures used when applying Hess's Law (*Chemistry* 7th ed. pages 246-249 / 8th ed. pages 255-258).

13. **C** Heat required to melt the ice = 9.5 g × 333J/g = 3164 J

Heat lost by the metal block = mass of block x (specific heat) × ΔT

= 57g × (specific heat) × 92°C

= specific heat × 5244 °C-g

Since heat list by metal block – heat gained by ice,

3164J = specific heat × 5244 °C-g; and solving for specific heat,

specific heat = 0.603 J/°C-g

(*Chemistry* 7th ed. pages 237-242 / 8th ed. pages 244-251).

14. **E** The entropy of the universe must increase and the free energy of the system must decrease, as these are expressions of the second law. If the entropy of the system decreases, the entropy of the surroundings must increase. If the entropy of the surrounds is to increase, the reaction must be exothermic. Thus all four

statements are true (*Chemistry* 7th ed. pages 749-756, 759-766 / 8th ed. pages 773-776, 783-790).

15. **B** Since the reaction occurs we know it is spontaneous, which means ΔG must be negative. The reaction is exothermic, meaning that ΔH is negative and since there is an increase in the number of molecules of gas, there is an increase in the randomness, meaning at ΔS is positive (*Chemistry* 7th ed. pages 766-773 / 8th ed. pages 783-787).

Free Response

1. (a) $3O_2(g) \rightarrow 2O_3(g)$

   (b) ΔS is (–) since order is increasing (this is to say, entropy is decreasing) as 3 moles of O_2 form 2 mol of O_3.

   (c) Note that the reaction in (A) is for the formation of ozone and that ΔG is a large positive value. This means that the reverse reaction must have a large negative ΔG and that ozone will spontaneously decompose, hence ozone is less stable than oxygen (*Chemistry* 7th ed. pages 752–755, 762–766 / 8th ed. pages 776–779, 786–790).

   (d) As temperature increases, it is expected that O_3 is less stable. From $\Delta G = \Delta H - T\Delta S$. As T increases, the value of ΔG becomes more positive and therefore the reaction is less likely to occur.

2. When a liquid boils at its boiling temperature or freezes at its freezing temperature, no useful work can be done by the process, i.e., ΔG is zero. Therefore, under these conditions,

   $\Delta S = \Delta H/T$.

   In melting, $\Delta S = 10.9 \times 10^3$ J/mol $/(5.5 + 273$ K$) = +39.1$ J/K•mol.

   In boiling, $\Delta S = +87.8$ J/K•mol.

   Vaporization involves a much greater change in disorder than melting (gases are very disordered compared to liquids, whereas liquids and solids vary less in disorder, distance between molecules, and number of possible positions for molecules); hence $\Delta S_{vap} > \Delta S_{fus}$ (*Chemistry* 7th ed. pages 762–766 / 8th ed. pages 762–790).

7

ATOMIC STRUCTURE AND PERIODICITY

In this chapter, coverage of the electromagnetic spectrum will lead to the electronic structure of atoms. Properties such as the sizes of atoms and ionization energy can be predicted from electronic structures. Bond formation between atoms can also be rationalized from the electronic structure of atoms in the next chapter.

Questions on electronic structure, periodicity, and bonding are often found in the free-response essay section as well as in the multiple-choice section of the AP Chemistry exam.

You should be able to:

- Identify characteristics of and perform calculations with frequency and wavelength.
- Know the relationship between types of electromagnetic radiation and energy; for example, gamma rays are the most damaging.
- Know what exhibits continuous and line spectra.
- Know what each of the four quantum numbers n, ℓ, m_ℓ and m_s represents.
- Identify the four quantum numbers for an electron in an atom.
- Write complete and abbreviated electron configurations as well as orbital diagrams for an atom or ion of an element.
- Identify the number and location of the valence electrons in an atom.
- Apply the trends in atomic properties such as atomic radii, ionization energy, electronegativity, electron affinity, and ionic size.

ELECTROMAGNETIC RADIATION

(*Chemistry* 7th ed. pages 275–276 / 8th ed. pages 285–286)

Electromagnetic radiation provides an important means of energy transfer. Light from the sun reaches the earth in the form of radiation that includes ultraviolet and visible radiation. Electromagnetic radiation, from the shortest to the longest wavelength, includes gamma rays, x-rays, ultraviolet, visible, infrared, microwaves, and radio waves. All types of electromagnetic radiation exhibit both wavelike and particle-like behavior and travel at the speed of light in a vacuum.

Chemists have used the interaction of light with matter to study the structure of the atom.

CHARACTERISTICS OF WAVES

(*Chemistry* 7th ed. pages 275–276 / 7th ed. pages 285–286)

Wavelength, λ, is the distance between two consecutive peaks or troughs in a wave.

Frequency, υ, is the number of waves (cycles) per second that passes a given point in space.

Wavelength and frequency are related by the equation:

$$\upsilon = c / \lambda$$

in which c is the speed of light in a vacuum, 2.9979×10^8 m/s. Wavelength is measured in meters, and frequency is measured in 1/s, or hertz (Hz).

ENERGY OF ELECTROMAGNETIC RADIATION

(*Chemistry* 7th ed. pages 275–277 / 8th ed. pages 285–287)

Energy is quantized and can occur only in whole number multiples of $h\upsilon$. Each small packet of energy is called a quantum. A system can transfer energy only in discrete quanta. The change in energy for a system can be represented by

$$\Delta E = nh\upsilon$$

where n is an integer (1,2,3,...), h is Planck's constant which equals 6.626×10^{-34} Js, and υ is the frequency.

Einstein viewed electromagnetic energy as a stream of particles called photons. And the energy of each photon can be given by

$$E_{photon} = hc/\lambda$$

EXAMPLE: The laser in an audio compact disc player uses light with a wavelength of 7.80×10^2 nm. Calculate the frequency of this radiation. What is the energy of this radiation per photon?

$$\upsilon = \frac{c}{\lambda} = \frac{2.998 \times 10^8 \text{ m/s}}{7.80 \times 10^{-7} \text{ m}} = 3.84 \times 10^{14} \text{ s}^{-1}$$

$E_{\text{photon}} = hc/\lambda$

$= (6.626 \times 10^{-34} \text{ Js})(2.998 \times 10^8 \text{ m/s}) / 7.80 \times 10^{-7} \text{ m}$

$= 25.5 \text{ J} \times 10^{-19}$

ATOMIC SPECTRUM

(*Chemistry* 7th ed. pages 284–285 / 8th ed. pages 294–295)

A continuous spectrum contains all wavelengths of visible light. A line spectrum contains only a few lines, each corresponding to a discrete wavelength. When atoms are excited by adding energy, the excess energy is given off by emitting light of various wavelengths to produce an emission spectrum or line spectrum.

BOHR MODEL

(*Chemistry* 7th ed. page 285 / 8th ed. page 295)

Bohr proposed that the electron in the hydrogen atom moves around the nucleus only in certain allowed circular orbits and the line spectra of elements are the result of electrons moving between these allowed orbits. It has often been described as a solar system model. Since a moving electron radiates energy, this model was unstable using classical mechanics and the model was shown to be incorrect.

QUANTUM MECHANICAL MODEL

(*Chemistry* 7th ed. pages 290–293 / 8th ed. pages 300–303)

Our current model of the atom is the quantum mechanical model. In this model, we treat the electrons as a wave and do not believe it is possible to describe the path of an electron. An orbital is a three-dimensional electron density map in which there is a 90% probability of finding the electron. The Heisenberg Uncertainty Principle is an important part of the model. It states that we cannot know both the momentum and location of an electron.

QUANTUM NUMBERS

(*Chemistry* 7th ed. pages 293–294 / 8th ed. pages 303–304)

Quantum numbers characterize various properties of the orbitals. There are four quantum numbers. The Pauli Exclusion Principle states that no two electrons can have the same set of four quantum numbers.

The principal quantum number, n, is related to the size and energy of the orbital. As n increases, the orbital becomes larger, which also means it is higher in energy.

The angular momentum quantum number, ℓ, is related to the shape of atomic orbitals.

The magnetic quantum number, m_ℓ is related to the orientation of the orbital in space relative to the other orbitals in the atom.

The electron spin quantum number, ms, has only two values: +1/2 and −1/2. This quantum number is necessary to describe the fact that the electron has two possible orientations, or spins, when placed in an external magnetic field.

The four quantum numbers are summarized in the following table.

Quantum Numbers			
Name	Designation	Possible range of values	Property of the orbital
Principal quantum number	n	Integers > 0 (1,2,3,…)	Related to size and energy of the orbital
Angular momentum quantum number	ℓ	Integers from 0 to $n-1$	Related to the shape of the orbital
Magnetic quantum number	m_ℓ	Integers from $-\ell$ to $+\ell$	Related to the position of the orbital in space in relation to other orbitals
Electron spin quantum number	m_s	+½ or −½	Related to the spin of the electron, which can be only one of two values

For each value of the angular quantum number, ℓ, the designated atomic orbitals are as follows:

The Angular Momentum Quantum Number, ℓ, and the Letters Used to Designate Atomic Orbitals					
Values of ℓ	0	1	2	3	4
Letter used	s	p	d	f	g
Number of orbitals	1	3	5	7	9
Total number of electrons in sublevel	2	6	10	14	18

The principal quantum number, n, is the number which appears in front of the types of orbital s, p, d or f. In general, for periods (rows) 1, 2 and 3, the principal quantum number, n, is the same as the row number. For periods 3, 4, 5, 6, and 7, the principal quantum number written before s and p orbitals is the same as the row number for the main group of elements, those in columns 1A through 8A. For the transition elements, the value of n which is written in front of the d orbital is 1 number less than the row number. For example, scandium, in period 4, has a value of n equal to 3 for its d electron. In the sixth row of the periodic table, the f-transition elements occur after lanthanum. For these elements, the value of n which is placed before the f is 2 less than the row in which they occur.

ORBITAL SHAPES AND ENERGIES

(*Chemistry* 7th ed. pages 295–298 / 8th ed. pages 305–308)

s ORBITALS

The s orbitals have a spherical distribution of electron density. The s orbitals become larger as the value of n increases.

Nodes are areas in which there is zero probability of finding an electron. The 2s has one node which separates areas of high probability, the 3s has two nodes, and so forth.

p ORBITALS

There are three p orbitals for each value of $n > 1$. The p orbitals have two lobes which are separated by a node at the nucleus. The p orbitals are labeled n_{px}, n_{py}, and n_{pz}, according to the axis along which the lobe lies.

Degenerate orbitals are orbitals with the same value of n which have the same energy. The three p orbitals in the same energy level are degenerate or equal in energy.

d AND f ORBITALS

The d orbitals first occur when $n = 3$ ($\ell = 2$), but in the fourth period of the periodic table. The f orbitals first occur when $n = 4$ and in the sixth period.

> **EXAMPLE:** Which of the following orbital designations is incorrect? 1s, 1p, 2d, 4f?
>
> 1p is incorrect: $n = 1$, for a p orbital, ℓ must equal 1. If $n = 1$, $\ell = n - 1$ or 0 is the only possible value.
>
> 2d is incorrect: $n = 2$, for a d orbital, ℓ must equal 2. If $n = 2$, the largest value possible of ℓ is $n - 1$ or 1.

> **EXAMPLE:** Give the maximum number of electrons in an atom that can have the quantum number $n = 3$.
>
> ℓ can be 0, 1, or 2. Thus we have s ($2e^-$) p ($6e^-$), and d ($10e^-$). The total number of electrons to fill these orbitals is 18.
>
> For any energy level, $2n^2$ = max number electrons for any energy level. For the above example, $2(3)^2 = 18$.

> **EXAMPLE:** Give the maximum number of electrons in an atom that can have the quantum numbers $n = 2$ and $\ell = 1$.

If $\ell = 1$, then it is a p-type sublevel, so there are three orbitals in that sublevel, each of which can hold two electrons or a total of six electrons is possible in the sublevel.

AUFBAU PRINCIPLE AND ELECTRON CONFIGURATIONS

(*Chemistry* 7th ed. pages 302–309 / 8th ed. pages 312–318)

The Aufbau principle is a scheme used to reproduce the electron configuration of the ground states of atoms by successively filling sublevels with electrons in a specific order.

The electron configuration for an atom of an element represents all of the electrons in the atom and it shows in which energy levels and orbitals the electrons reside.

EXAMPLE: Write an electron configuration for oxygen.

The electron configuration for oxygen is $1s^2 2s^2 2p^4$.

To write an electron configuration for an atom of an element, follow the Aufbau principle and add electrons to the lowest energy orbital, the 1s. Using the Pauli Exclusion Principle, each orbital holds two electrons, we fill up the 1s with two electrons of opposite spin, then the 2s, to get $1s^2 2s^2$. We have used four electrons so far and oxygen has eight total electrons. The remaining four electrons go into the three 2p orbitals. Hund's Rule requires that we place one electron in each separate p orbital with parallel spin before pairing electrons. Write the electron configuration as $1s^2 2s^2 2p^4$.

EXAMPLE: Write the abbreviated electron configuration for oxygen.

The abbreviated electron configuration for oxygen is [He] $2s^2 2p^4$.

To write the abbreviated electron configuration, write the noble gas symbol which comes before the element in brackets followed by the remaining electrons.

VALENCE ELECTRONS

(*Chemistry* 7th ed. page 304 / 8th ed. page 313)

Valence electrons are the electrons in the outermost principal quantum level of an atom. In the example with oxygen, the valence electrons are in the second quantum (energy) level. In the second energy level, there are four valence electrons, two in the 2s and two in the 2p. Coincidentally, the valence electrons of a main group element are the same as the group number, except for helium which has two valence electrons. The two electrons in the 1s orbital are referred to as the core (inner) electrons. Pages 306 and 307 of the 7th edition of *Chemistry* and pages 315 and 316 of the 8th edition display helpful diagrams which show how the orbitals are filled in various parts of the periodic table.

AP tip

Questions about trends in the periodic table often appear in the essay portion of the free-response section. It is important to be clear and concise in your writing. Different essay writing styles are discussed in this book in the test preparation chapter.

There are exceptions to electron configurations. One exception is chromium. Instead of having the expected electron configuration $[Ar]4s^2 3d^4$, the configuration is $[Ar]4s^1 3d^5$. This configuration is energetically more favorable (lower in energy) than the configuration with only four d electrons. This also occurs for molybdenum. A similar situation occurs for Cu and all of the transition elements beneath it. Instead of having the electron configuration $[Ar]4s^2 3d^9$, the configuration for copper is $[Ar]4s^1 3d^{10}$. The filled d sublevel is lower in energy than the partially filled sublevel.

ORBITAL DIAGRAMS

(*Chemistry* 7th ed. pages 302–304 / 8th ed. pages 312–313)

The orbital diagram displays the same information as the electron configuration. In addition, the spin of the electron is also represented.

EXAMPLE: Give the orbital diagram for oxygen.

$$\underset{1s}{\uparrow\downarrow} \quad \underset{2s}{\uparrow\downarrow} \quad \underset{2p}{\underline{\uparrow\downarrow \; \uparrow \; \uparrow}}$$

The first arrow in the 1s orbital represents an electron spinning in a particular direction. The second electron occupying the same orbital must spin in the opposite direction since, according to the Pauli exclusion principle, only two electrons with opposite spin can occupy the same orbital. When the 1s orbital is filled, the 2s orbital fills in the same manner as the 1s. There are three degenerate 2p orbitals, so one electron must go into each of the 2p orbitals. The fourth 2p electron becomes paired with the first 2p electron.

EXAMPLE: Give an acceptable set of four quantum numbers for each electron in oxygen:

	$\uparrow\downarrow$	$\uparrow\downarrow$	$\uparrow\downarrow$	$\uparrow$	$\uparrow$
	1s	2s		2p	
n	1 1	2 2	2 2	2	2
ℓ	0 0	0 0	1 1	1	1
m_ℓ	0 0	0 0	−1 −1	0	+1
m_s	+½ −½	+½ −½	+½ −½	+½ −½	+½ −½

Begin with the orbital diagram for oxygen. Write the set of four quantum numbers under each electron (arrow). The value of n is obtained from the integers in front of the type of orbital, s or p. For the s orbitals, $\ell = 0$ and for p, $\ell = 1$.

The values for m_ℓ can be from $+\ell$ to $-\ell$. For $\ell = 0$, $m_l = 0$. For $\ell = 1$, m_l can be -1, 0, $+1$. Each p orbital is assigned one of the following values -1, 0, $+1$.

The value for m_s is either $+ 1/2$ or $- 1/2$. If the arrows are pointing in the same direction, then the value of m_s must be the same.

For example, an acceptable set of four quantum numbers for the first 1s electron is $n = 1$, $\ell = 0$, $m_l = 0$, $m_s = +1/2$. Another acceptable set would be $n = 1$, $\ell = 0$, $m_\ell = 0$, $m_s = -1/2$.

PERIODIC TRENDS IN ATOMIC PROPERTIES

(*Chemistry* 7th ed. pages 309–314 / 8th ed. pages 318–323)

ATOMIC RADII

(*Chemistry* 7th ed. pages 313–314 / 8th ed. pages 322–323)

Atomic radii decrease in going from left to right across a period from the alkali metals to the halogens. This can be explained by an increasing effective nuclear charge, which causes the electrons to be drawn closer to the nucleus.

Atomic radius increases down a group because of the increases in the orbital sizes in successive principal quantum levels. Also, there is an increase in the shielding of the nucleus by the core electrons, which decreases the effective nuclear charge felt by the valence electrons.

> **EXAMPLE:** Arrange the following atoms in order of increasing atomic radius.
>
> C, Al, F, Si
>
> F < C < Si < Al ; F and C are in the second period, which means that they have two shells or energy levels. Within a period, the radius decreases as the atomic number increases. This means F will be smaller than C. Al and Si are in the third period and have one more energy level, so they will have a larger radius than atoms in period two. Since Si has more protons than Al, it has a smaller radius.

IONIZATION ENERGY

(*Chemistry* 7th ed. pages 309–312 / 8th ed. pages 318–321)

Ionization energy is the energy required to remove an electron from a gaseous atom or ion.

The first ionization energy is the energy required to remove the highest energy electron (the one bound least tightly) which is one of the valence electrons.

$$X(g) + energy \rightarrow X^{1+}(g) + e^-$$

For lithium, the first ionization energy is the energy required to remove an electron from the 2s orbital. The configuration for lithium is $1s^22s^1$ and the valence electron is the 2s electron.

In general, the first ionization energy increases across a period from left to right. This trend can be related to the trend in atomic radii. The smaller the atom, the greater is its first ionization energy. In going across a period from left to right, atomic radii decrease as the charge on the nucleus increases. There are irregularities in this trend in ionization energy. The first ionization energy decreases from beryllium to boron because the electrons in the 2s orbital are much more effective at shielding the electrons in the 2p orbital than they are at shielding each other. In all atoms except for hydrogen, the 2p orbital is slightly higher in energy than the 2s orbital. The higher energy 2p electron of boron is easier to remove than one of the 2s electrons of beryllium.

The first ionization energy decreases in going from nitrogen to oxygen due to repulsion of paired electrons in the p^4 configuration of oxygen. (Nitrogen has a valence electron configuration $[He]2s^22p^3$ and oxygen has a valence electron configuration of $[He]2s^22p^4$) The first ionization energy decreases down a group with increasing atomic number because the valence electrons become easier to remove as the atoms become larger going down a group.

The second ionization energy is the energy required to remove the second outermost electron. This value is larger than the first ionization energy. The first ionization energy removes an electron from a neutral atom. The second ionization energy removes an electron from a positive ion.

$$X^{1+}(g) + energy \rightarrow X^{2+}(g) + e^-$$

The increase in positive charge binds the electrons more tightly, and the ionization energy increases as each successive electron is removed. Also, a slightly larger increase in ionization energy occurs when removing an electron from a higher orbital, such as a 3p orbital vs. a 3s orbital. Note a large increase occurs in ionization energy when a core electron from a full energy level is removed.

> **EXAMPLE:** Explain why the difference between the third ionization energy and fourth ionization energy for aluminum is so large in comparison with the first and second and second and third ionization energies.
>
> Aluminum has the valence electron configuration $[Ne]3s^23p^1$. The first electron removed is the 3p; the second electron removed is a 3s, the third electron removed is the remaining 3s electron. The fourth electron, a core electron (a 2p), requires much more energy for removal than the valence electrons, the 3s and the 3p electrons, because it is more tightly bound.
>
> **EXAMPLE:** Arrange the following atoms in order of increasing first ionization energy. I, Rb, Na
>
> Na will have the smaller radius. Rb < I < Na. Na has the largest IE_1 because the electron is removed from the 2s sublevel. Both Rb and I have valence electrons in the 5th energy level, but I has

53 protons and a smaller radius than Rb with only 37 protons which makes I smaller than Rb. The smaller the atom, the more energy required to remove the valence electron.

ELECTRON AFFINITY

(Chemistry 7th ed. pages 312–313 / 8th ed. pages 321–322)

Electron affinity is the energy change associated with the addition of an electron to a gaseous atom.

$$X(g) + e^- \rightarrow X^-(g)$$

In general, electron affinity becomes increasingly negative or exothermic as we proceed across a period with increasing atomic number. The greater the charge on the nucleus of the atom, the greater is the affinity for electrons. The addition to a noble gas, however, would require the added electron to reside in a new, higher energy level which is energetically unfavorable, so that stable, isolated X- ions do not exist. Electron affinities are not measured for noble gases.

Electron affinities become slightly more positive, less exothermic, going down a group. The electron–nucleus attraction decreases down a group due to increasing atomic size and shielding. But the added electron experiences less electron–electron repulsion since the orbital to which the electron is being added is increasingly spread out. One exception to this trend is fluorine. The value of electron affinity for F is more positive than that of Cl due to the small size of the F atom and the greater electron–electron repulsion in the smaller 2p sublevel.

> **EXAMPLE:** Arrange the following elements from the least to the most exothermic electron affinity. Cl, Al, P.

> From the least to the most exothermic is Al, P, Cl. In general, the greater the nuclear charge on the atom, the more exothermic the electron affinity. Atoms become smaller across the period as the atomic number increases.

MULTIPLE-CHOICE QUESTIONS

No calculators are to be used in this section.

1. Which of the following sets of quantum numbers is unacceptable?

	n	ℓ	m_l	m_s
(A)	4	3	-2	+1/2
(B)	3	0	+1	-1/2
(C)	3	0	0	-1/2
(D)	3	1	1	+1/2
(E)	2	0	0	-1/2

2. The differentiating electron in gallium, Ga, is the single $4p^1$ electron.
    A valid set of quantum numbers for this electron is

	n	ℓ	m_l	m_s
(A)	3	2	1	+1/2
(B)	3	1	−1	+1/2
(C)	4	0	0	+1/2
(D)	4	−1	+1	+1/2
(E)	4	1	−1	+1/2

3. In a given atom, no two electrons can have the same set of four
    quantum numbers. This statement is knows as the
    (A) Pauli exclusion principle.
    (B) Hund's rule.
    (C) Einstein principle.
    (D) Heisenberg uncertainty principle.
    (E) Bohr law.

4. If the angular momentum (second) quantum number ℓ equals 3, the
    total number of allowed orbitals in just that subenergy level is
    (A) 1.
    (B) 3.
    (C) 5.
    (D) 7.
    (E) 9.

5. The element with the ground state electron configuration
    (spectroscopic notation) of $[Ar]\, 3d^7\, 4s^2$ is
    (A) Mg.
    (B) K.
    (C) Ar.
    (D) Co.
    (E) Ni.

6. The electron configuration for the element antimony, Sb, #51, is
    (A) $[Na]\, 3s^2\, 2d^{10}\, 3p^3$.
    (B) $[Ar]\, 4s^2\, 3d^{10}\, 4p^5$.
    (C) $[Ar]\, 4s^2\, 3d^{10}\, 4p^3$.
    (D) $[Kr]\, 5s^2\, 4d^{10}\, 5p^3$.
    (E) $[Kr]\, 5s^2\, 4d^{10}\, 5p^5$.

7. Atomic radii decrease from left to right across a period because of
    (A) an increase in effective nuclear charge.
    (B) an increase in gross energy level (n).
    (C) an increase in subenergy level (ℓ).
    (D) an increase in shielding.
    (E) a decrease in effective nuclear charge.

8. The number of unpaired electrons (no counterspinning electron) in the chromium atom is
   (A) 1.
   (B) 2.
   (C) 4.
   (D) 5.
   (E) 6.

9. The correct ordering of atoms in progressively decreasing ionization energy is
   (A) F > O > C > Li > Na.
   (B) Na > Li > C > O > F.
   (C) F > O > C > Na > Li.
   (D) C > O > F > Li > Na.
   (E) O > F > C > Na > Li.

10. Electron affinity is the
    (A) energy required to remove an electron from an atom in the gaseous state.
    (B) energy released when an electron is gained by an atom in its standard state.
    (C) maximum energy required to remove an electron from an atom in its standard state.
    (D) energy involved when an electron is gained by an atom in the gaseous state.
    (E) energy gained by an electron when it is absorbed by another electron.

11. In which sequence are the regions of the electromagnetic spectrum arranged in increasing energy per photon?
    (A) radio waves, infra red, violet, microwaves, x-rays
    (B) infrared, orange, indigo, green, ultraviolet
    (C) γ-rays, x-rays, ultraviolet, red, microwaves
    (D) microwaves, infrared, orange, violet, γ-rays
    (E) radio waves, red, infrared, ultraviolet, x-rays

12. The following diagrams represent electron orbitals of the hydrogen atom. Which labels correctly characterize each orbital?

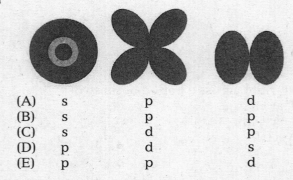

    (A)   s   p   d
    (B)   s   p   p
    (C)   s   d   p
    (D)   p   d   s
    (E)   p   p   d

13. According to the aufbau principle, which set of orbitals is filled after the 4d set?
    (A) 5s
    (B) 6s
    (C) 6p
    (D) 5p
    (E) 5d

14. The longest wavelength of electromagnetic radiation that will cause the emission of a photoelectron from the surface of a metal is 300 nm. What is the energy per photon of this radiation?
    (A) 5.52×10^{19}J
    (B) 1.81×10^{-20}J
    (C) 6.63×10^{-19}J
    (D) 1.81×10^{48}J
    (E) 8.33×10^{-16}J

15. What statement about the principal quantum number, n, is *false*?
    (A) In part, it determines the energy of the electron.
    (B) In part, it determines the most probable distance of the electron from the nucleus.
    (C) It can never be 0.
    (D) It corresponds to rows in the periodic table.
    (E) It can never be larger than the quantum number ℓ.

FREE-RESPONSE QUESTIONS

1. One of the emission lines that provides the green color in fireworks, which comes from barium salts, has a wavelength of 505 nm.
    (a) Determine the frequency of this green line.
    (b) Determine the energy associated with this green line.
    (c) Arrange the visible spectrum (from red to violet) in order of increasing:
        i. wavelength
        ii. frequency
        iii. energy

2. You have studied the Bohr model of the atom and the quantum mechanical model.
    (a) What phenomena was Bohr able to explain using this model?
    (b) What was one of major shortcomings of Bohr's model?
    (c) Give two differences in the quantum mechanical model from the Bohr model.

3. The emission spectrum of hydrogen consists of several series of sharp emission lines. Why is the emission spectrum of hydrogen made of discrete wavelengths rather than a continuum of wavelengths? Discuss this in terms of features of the electronic energies of the hydrogen atom.

Answers

MULTIPLE CHOICE

1. **B** Quantum number assignments follow set patterns or rules. The angular momentum quantum number ℓ (also called the second quantum number by some chemists) is limited to a maximum number of whole number values, from 0 ... to ... (n − 1). The magnetic quantum number m_ℓ is limited to integers from −ℓ to zero to +ℓ. Finally, the spin quantum number can have only one of two values, +1/2 or −1/2 (*Chemistry* 7th ed. pages 293–294 / 8th ed. pages 303–304).

2. **E** Since the last part of the electron configuration for Ga is $4p^1$, you can assign $n = 4$ from the 4p orbital; ℓ follows the pattern: $\ell = 0$ for all s orbitals, $\ell = 1$ for all p orbitals, $\ell = 2$ for all d orbitals, etc. If $\ell = 1$, as it does in this case, then m can be only -1, 0, or $+1$ (representing the three orientations of a "p" orbital in space) (*Chemistry* 7th ed. pages 293–294 / 8th ed. pages 303–304).

3. **A** This is also another way of saying that no more than two electrons are allowed in the same orbital in the same atom (*Chemistry* 7th ed. pages 296–298 / 8th ed. pages 306–308).

4. **D** If $\ell = 3$, then m_ℓ can have values of -3, -2, -1, 0, $+1$, $+2$, $+3$. That is seven possible values, one for each of the possible orbitals allowed (*Chemistry* 7th ed. pages 293–296 / 8th ed. pages 303–306).

5. **D** From the 4 in the $4s^2$ notation, you know that you are operating in the fourth period of the Periodic Table. Since the 3d sub-energy level fills next and has seven electrons in it, you must be dealing with the seventh element in the transition metals level, hence cobalt, Co (*Chemistry* 7th ed. pages 302–306 / 8th ed. pages 312–315).

6. **D** First find Sb on the Periodic Table, and note that it is in the 5A family (also called the 15 family). This grouping from 3A to 8A is filling the "p" orbital subenergy level; hence Sb must end in p^3. But which "p" subenergy level? Now note that this element is in the fifth period; hence it ends in $5p^3$ (*Chemistry* 7th ed. pages 309–313, esp. Exercise 7.10 / 8th ed. pages 318–322, esp. Exercise 7.10).

7. **A** While adding electrons as you proceed from left to right might seem to add to the atomic radii, remember that you are also adding protons to the nucleus while adding electrons to the same gross energy level; so the increase in nuclear charge is the determining factor in this trend (*Chemistry* 7th ed. pages 295–298 / 8th ed. pages 305–306).

8. **E** Remember that chromium is an exception to the general rules of electron configuration and that it ends in $4s^1 3d^5$, so it has six unpaired electrons (*Chemistry* 7th ed. pages 302–309 / 8th ed. pages 312–318).

9. **A** The trends in decreasing ionization are similar to those in increasing in atomic radii. From left to right in one period, the atomic radius decreases and the ionization energy increases from top to bottom in a family; the ionization energy decreases as you move to greater energy levels (*Chemistry* 7th ed. pages 309–313, esp. Exercises 7.8 and 7.9 / 8th ed. pages 318–322, esp. Exercises 7.8 and 7.9).

10. **D** This is a basic definition you need to know (*Chemistry* 7th ed. pages 312–313 / 8th ed. pages 321–322).

11. **D** Review a diagram of the electromagnetic spectrum. Energy is directly related to the frequency of radiation and inversely

proportional to the wavelength (*Chemistry* 7th ed. page 276 / 8th ed. page 286).

12. **C** (*Chemistry* 7th ed. pages 295-297 / 8th ed. pages 305 -307)

13. **D** (*Chemistry* 7th ed. pages 302–309 / 8th ed. pages 312–318)

14. **C** Frequency × wavelength = speed of light = 3.0×10^8 m/s

frequency = $(3.0 \times 10^8 \text{m/s})/(300 \times 10^{-9}\text{m}) = 1.00 \times 10^{15}/\text{s}$

energy of photon = h × frequency = $(6.625 \times 10^{-34}\text{J-s})(1.00 \times 10^{15}/\text{s})$

= 6.63×10^{-19} J

(*Chemistry* 7th ed. page 278 / 8th ed. page 287-289)

15. **E** ℓ cannot be larger than n-1. (*Chemistry* 7th ed. pages 293-294 / 8th ed. pages 303-304)

FREE RESPONSE

1.
   (a)   Frequency = speed of light / wavelength

$$= \frac{2.998 \times 10^8 \text{ m/s}}{505 \text{ nm} \times 10^{-9} \text{ m/nm}}$$

$$= 5.94 \times 10^{14} \text{ s}^{-1}$$

$$= 5.94 \times 10^{14} \text{ Hz.}$$

   (b) Energy = Planck's constant × frequency

$$= 6.626 \times 10^{-34} \text{ J . s} \times 5.90 \times 10^{14} \text{ s}^{-1}$$

$$= 3.93 \times 10^{-19} \text{J}$$

   (c) From red → violet
   i.   frequency increases
   ii. wavelength decreases
   iii. energy increases

(*Chemistry* 7th ed. pages 278–281 / 8th ed. pages 288–291).

2. The Bohr model of the atom involves a specific circular orbital for the electron, whereas the quantum mechanical model is an electron density plot or probability model of orbitals based on an understanding of standing waves. According to Bohr's model, as the electron revolves around the nucleus, it should emit light, lose energy, and be drawn into the nucleus. None of these things happen. (You might also note the differences in the terms "orbit" and "orbital." In an orbit, the electron is always found at the given distance from the nucleus; the orbital describes the most probable distance at which the electron is found (*Chemistry* 7th ed. pages 285–293 / 8th ed. pages 295–303).

3. The energy of the electron is restricted only to certain values; it is quantized. This means that only a certain few changes in the energy of the electron can occur and not all possible changes in

energy. If this were so then the spectrum that you see would be a continuous spectrum made of all the colors in the visible region (*Chemistry* 7th ed. pages 284–285 / 8th ed. pages 294–295).

8

BONDING, STRUCTURE, AND ORGANIC CHEMISTRY

All chemical reactions involve making and breaking chemical bonds. In this chapter, you will learn why different types of bonds form, the nature of those bonds, and you will study a model to predict the three-dimensional structure of molecules formed from covalent bonds.

Bond formation always releases energy and bond breaking always requires energy. The net enthalpy for a chemical reaction will depend on whether it takes more energy to break the old bonds in the reactants than is released when new bonds form in the products. The change in enthalpy for reactions will be calculated from bond energies.

Intermolecular forces and their effect on properties of molecules will also be reviewed.

An overview of organic chemistry will cover basic nomenclature, isomerism, and functional groups.

You should know:
- The characteristics of ionic and covalent bonding.
- The relative sizes of ions.
- The effect of lattice energy on melting points of ionic compounds.
- How to use electronegativity to predict the polarity of covalent bonds.
- How to draw Lewis symbols for atoms and Lewis structures for molecular compounds and polyatomic ions.
- How to draw resonance structures.
- How to assign molecular shapes using the VSEPR Theory for molecules, polyatomic ions, and multi-centered

molecules and polyatomic ions, through the octahedral shape.

- How to assign hybrid orbitals to central atoms of molecules and polyatomic ions.
- How to calculate the enthalpy of a reaction using bond energies.
- How to compare bond lengths in molecules.
- How to write formulas, names, and structures for organic compounds including alkanes, alkenes, alkynes, cyclic compounds and compounds with functional groups.
- How to draw and name isomers of alkanes and alkenes.
- How to identify the intermolecular forces in a substance.

IONIC BONDING

(*Chemistry* 7th ed. pages 330–331 / 7th ed. pages 341–342)

Ionic bonding is the result of electrostatic attraction of oppositely charged ions. Ionic bonds form when an atom which loses electrons easily reacts with an atom that has a high affinity for electrons. Most of the time, an ionic compound is formed when a metal reacts with a nonmetal.

COULOMB'S LAW

(*Chemistry* 7th ed. pages 330–331 / 8th ed. pages 341–342)

Coulomb's law expresses the energy of interaction between a pair of ions. This is directly proportional to the charge on each ion and inversely proportional to the distance between the ions. The greater the charge on the ions, the stronger the ionic bond, and if the charges are equal, then we expect a smaller ion to have a stronger ionic bond.

$$E = (2.31 \times 10^{-19} \text{ J} \bullet \text{nm})\left(\frac{Q_1 Q_2}{r}\right).$$

E is the energy of interaction between a pair of ions, in joules.

r is the distance in nanometers between ion centers.

Q_1 and Q_2 are the charges of the ions.

IONS: ELECTRON CONFIGURATION AND SIZES

(*Chemistry* 7th ed. pages 338–342 / 8th ed. pages 350–353)

PREDICTING FORMULAS OF IONIC COMPOUNDS; FORMATION OF IONIC COMPOUNDS

When a metal and a nonmetal react, ions form so that the valence electron configuration of the nonmetal achieves the electron configuration of the next noble gas atom by gaining electrons and the main group metals lose all of their valence electrons. Usually both ions achieve noble gas configurations.

EXAMPLE: Explain how atoms of sodium and oxygen form an ionic compound.

Sodium and chlorine have the following valence electron configurations:

Na: $[Ne]3s^1$

O: $[He]2s^22p^4$

The electronegativity of oxygen is much greater than that of sodium, so electrons are transferred from sodium to oxygen. Oxygen needs two electrons to fill its 2p valence orbitals and to achieve the configuration of neon. Sodium loses one electron to achieve the electron configuration of neon. Two sodium atoms give one oxygen atom the two electrons necessary to fill its valence orbitals. The oxide ion has an electron configuration like that of neon.

$$2\,Na \;\rightarrow\; 2\,Na^+ + 2\,e^-$$
$$O \;+\; 2\,e^- \rightarrow O^{2-}$$

Using electron configurations:

$$Na \;\rightarrow\; Na^+ + e^- \qquad O + 2\,e^- \rightarrow O^{2-}$$

$$[Ne]3s^1 \rightarrow [Ne] + e^- \qquad [He]2s^22p^4 + 2\,e^- \rightarrow [He]2s^22p^6 = [Ne]$$

SIZES OF IONS

(*Chemistry* 7th ed. pages 340–341 / 8th ed. pages 351–352)

Cations are smaller than their parent atoms and anions are larger than their parent atoms. The size of an ion depends on its nuclear charge, the number of electrons it contains, and the outer energy level of the atom. Positive ions are always smaller than neutral atoms because the nuclear charge is the same but the number of electrons has decreased increasing the effective nuclear charge felt by each remaining electron. When all the valence electrons are lost, then the change in volume is greater because as n decreases, the radial probability is smaller so the radius is smaller.

Anions are larger than their parent atoms because the electrons gained cause the proton to electron ratio to decrease, and the effective nuclear charge felt by each electron decreases. Electron–electron repulsions also increase. The electrons spread out more to accommodate the additional electrons.

In an isoelectronic series, ions contain the same number of electrons.

EXAMPLE: Arrange the following ions in order of increasing size: Ca^{2+}, S^{2-}, K^+, Cl^-.

All of the ions mentioned above are isoelectronic, containing 18 electrons each. In order of increasing size $Ca^{2+} < K^+ < Cl^- < S^{2-}$. Ca^{2+} is the smallest because it contains the greatest number of protons. The greater the positive charge on the nucleus, the stronger the attraction for electrons.

LATTICE ENERGY

(*Chemistry* 7th ed. pages 344–346 / 8th ed. pages 355–357)

Lattice energy is the energy release that occurs when separated gaseous ions are packed together to form an ionic solid.

$$X^{x+}(g) + Y^{y-}(g) \rightarrow X_yY_x\,(s) + energy$$

$$Lattice\ energy = k\left(\frac{Q_1Q_2}{r}\right).$$

k is a proportionality constant dependent on the structure of the solid and the electron configuration of the ions.

r is the distance in nanometers between the ion centers.

Q_1 and Q_2 are the charges of the ions.

> EXAMPLE: Which has the most exothermic lattice energy, NaCl or KCl?
>
> NaCl has the most exothermic lattice energy since all of the ions in both compounds have either +1 or –1 charges (the distance between the charges needs to be considered). Na^+ is smaller than K^+. So the distance between the centers of the sodium and chloride ions is less than the distance between the potassium and chloride ions.

COVALENT BONDING

(*Chemistry* 7th ed. page 331 / 8th ed. page 342)

Covalent bonding involves sharing electrons between nuclei with similar electronegativities. A covalent or molecular compound is usually formed by bonding two nonmetals. There are some metallic compounds that are covalent, such as beryllium chloride, $BeCl_2$.

ELECTRONEGATIVITY

(*Chemistry* 7th ed. pages 333–335 / 8th ed. pages 344–346)

Electronegativity measures the ability of an atom in a molecule to attract shared electrons to itself. In general, the trend in electronegativity follows the trend in atomic size. The smaller the atom, the greater is its electronegativity. Electronegativity increases across a period with increasing atomic number and decreases going down a group.

EXAMPLE: Place the following atoms in order of increasing electronegativity: Cs, Rb, S, Al, Sr, O.

In order of increasing electronegativity: Cs < Rb< Sr < Al< S < O.

Electronegativity increases from the lower left to the upper right of the periodic table.

POLAR BOND

(*Chemistry* 7th ed. pages 332–338 / 8th ed. pages 343–349)

The type of bond formed between two atoms is related to the difference in electronegativity between the two atoms. For identical atoms, such as a molecule of hydrogen, the electronegativity difference is zero. There is an equal sharing of electrons between the atoms, and the bond formed is a nonpolar covalent bond. When the difference in electronegativities is very small, the bond is described as nonpolar. An important nonpolar covalent bond is the C–H bond in organic compounds.

When two atoms have a very large difference in electronegativity, an ionic bond forms. For example, atoms of lithium and chlorine have a large difference in electronegativity and form an ionic compound.

A polar covalent bond results when there is an unequal sharing of electrons. The difference in the electronegativities of atoms in the polar covalent bond is between that of the ionic and nonpolar bonds. An example of a polar covalent bond is HF. HF has the following charge distribution.

$$H–F$$
$$\delta + \ \delta-$$

EXAMPLE: Which of the following three bonds will be most polar? O–F, N–F, C–F.

The most polar bond will be the bond in which the difference in electronegativities is the greatest. The C–F bond has the greatest electronegativity difference and will be the most polar bond in the example. C is the least electronegative among C, N, and O.

BOND POLARITY AND DIPOLE MOMENTS

Fluorine is the most electronegative element in the periodic table. When HF is placed in an electric field, the molecules tend to orient themselves with the fluoride end closest to the positive pole and the hydrogen end closest to the negative pole. HF is said to have a dipole moment, having two poles. An arrow points to the negative charge center and the tail of the arrow indicates the positive center of charge.

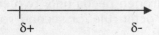

$$\delta+ \qquad\qquad \delta-$$

A diatomic molecule with a polar bond has a dipole moment. Molecules with more than two atoms can also have dipolar behavior.

This is addressed in the section on molecular shapes where we look at the distribution of charge in the molecule.

LOCALIZED ELECTRON BONDING MODEL

(*Chemistry* 7th ed. page 353 / 8th ed. page 364)

The localized electron model assumes that a molecule is composed of atoms that are bound together by sharing pairs of electrons using the atomic orbitals of the bound atoms. Lone pairs are pairs of electrons which are localized in an atom. Bonding pairs occupy the space between atoms.

LEWIS STRUCTURES

(*Chemistry* 7th ed. pages 354–361 / 8th ed. pages 365–372)

The Lewis structure shows how the valence electrons are arranged among atoms in a molecule.

Steps for writing Lewis structures:
1. Sum the total valence electrons from each atom. For a polyatomic ion, add one electron for each negative charge on the anion, or take away one electron for each positive charge on the cation.

2. Place the least electronegative element in the middle. (Hydrogen can never be in the middle because it can form only one bond.) Draw the outer atoms around the central atom. Chemical formulas are often written in the order in which they are connected, such as HCN. The central atom is sometimes written first in a formula, when a central atom has a group of other atoms bonded to it, as in SO_4^{2-}.

3. Connect the central atom to the outer atoms using a line to indicate each pair of bonding electrons.

4. Count the valence electrons used in the bonds in step 3 and subtract these from the total valence electrons in step 1 to determine how many valence electrons remain.

   Valence electrons left =
   Total valence electrons (step 1) – Valence electrons used (step 3).

5. Arrange the remaining electrons around the atoms to follow the octet rule for families 4A–7A outer atoms to satisfy the duet rule for hydrogen (hydrogen needs only two electrons to have a filled orbital) and the octet rule for second-row elements and the periods following.

6. Determine the number of valence electrons remaining as in step 4. Place the remaining electrons on the central atom to satisfy the octet rule for elements in the second-row and after.

a. If there are no electrons remaining to satisfy the central atom, then add a double bond for every pair of electrons you are "short" and readjust the number of electrons on each surrounding atom.

b. Elements in periods higher than the second can have more than eight electrons. If $n > 2$, then d orbitals exist and the number of electrons that can surround the element can exceed eight.

7. Check that the valence electrons drawn = the total valence electrons in step 1.

EXAMPLE: Draw the Lewis Structure for NF$_3$

Step 1: Total valence electrons = [1 × N(5e$^-$)+ 3 × F(7e$^-$)] =

5 e$^-$ + 21e$^-$ = 26 valence electrons.

Steps 2 and 3: Place nitrogen in the middle and connect the fluorine atoms to nitrogen using a line to represent a single bond which is a pair of electrons.

$$
\begin{array}{c}
\text{F} \\
| \\
\text{F}\!-\!\!-\!\text{N}\!-\!\!-\!\text{F}
\end{array}
$$

Step 4: Count the valence electrons used. Count two electrons for each single bond (2 × 3 single bonds = 6 electrons used). Count the electrons left. 26 – 6 = 20 electrons left.

Step 5: Satisfy the octet rule for each of the outer fluorine, F, atoms. Each F atom needs six more to complete the octet.

$$
\begin{array}{c}
:\!\ddot{\text{F}}\!: \\
| \\
:\!\ddot{\text{F}}\!\!-\!\!\text{N}\!-\!\!\ddot{\text{F}}\!:
\end{array}
$$

Step 6: Count electrons remaining after satisfying the outer atoms. 20 – 6 (3) used = 2 electrons left. Place remaining electrons on central atom, nitrogen, and its octet is completed.

$$
\begin{array}{c}
:\!\ddot{\text{F}}\!: \\
| \\
:\!\ddot{\text{F}}\!\!-\!\!\ddot{\text{N}}\!-\!\!\ddot{\text{F}}\!:
\end{array}
$$

Step 7: Check that electrons drawn, 26, equals the total valence electrons in step 1.

RESONANCE

(*Chemistry* 7th ed. pages 362–363 / 8th ed. pages 373–374)

Resonance occurs when more than one Lewis structure can be drawn for a molecule or ion.

EXAMPLE: Draw the Lewis structure for the nitrate ion, NO_3^-.

Step 1: Total valence electrons $= 5 + 3(6) + 1 = 24$ valence electrons (5 from N and 6 from each O and 1 from the negative charge on the ion).

Step 2: Place nitrogen in the middle and connect the oxygen atoms to nitrogen.

Step 4: Count valence electrons used, 2 for each single bond (2 × 3 single bonds = 6 electrons used). Count the electrons left. 24 – 6 = 18 electrons left.

Step 5: Satisfy the octet rule for each of the outer atoms of oxygen. Each O atom needs six more to complete the octet.

Step 6: Count electrons remaining after satisfying the outer atoms. 18 – 6(3) used = 0 electrons left. The central atom, N, still needs two more electrons to satisfy the octet rule. You can add a double bond for every pair of electrons you are "short."

One double bond forms between the N and the O. A double bond is made of two pairs of electrons. However, the nitrate ion does not have one double bond and two single bonds. All of the three bonds are the same length, somewhere intermediate between a single bond and a double bond. The nitrate ion exists as an average of three resonance structures as pictured below.

Step 7: Check that electrons drawn, 26, equals the total valence electrons in step 1.

EXCEPTIONS TO THE OCTET RULE

Boron tends to form compounds in which boron can have less than eight electrons, such as boron trifluoride, BF_3. Boron can have six electrons around it.

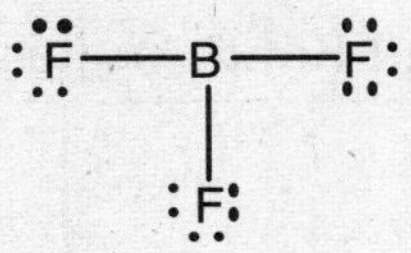

$$:\ddot{F}—B—\ddot{F}:$$
$$|$$
$$:\ddot{F}:$$

Some elements in period 3 of the periodic table and beyond can exceed the octet rule. For example, ClF_3, has 10 electrons around the central atom.

$$:\ddot{F}—\ddot{Cl}—\ddot{F}:$$
$$|$$
$$:\ddot{F}:$$

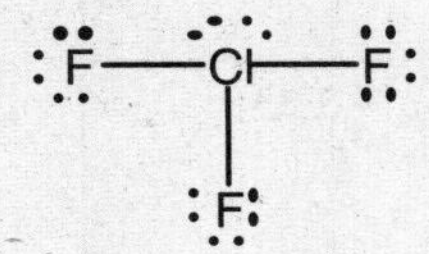

Elements in the third period have 3s, 3p, and 3d valence orbitals. The valence electrons in chlorine occupy the 3s and 3p orbitals. The 3d orbital is empty and can hold extra electrons. Elements in the second period have only 2s and 2p orbitals. These elements cannot have more than eight electrons in their valence orbitals because they do not have d orbitals available.

MOLECULAR STRUCTURE: THE VSEPR MODEL

(*Chemistry* 7th ed. pages 367–379 / 8th ed. pages 378–389)

The valence shell electron-pair repulsion (VSEPR) model predicts the geometries of molecules and polyatomic ions. The structure around the central atom is determined by minimizing electron-pair repulsions. Bonding and nonbonding pairs of electron are positioned as far apart as possible. Nonbonding or lone pairs of electrons require more room than bonding pairs and tend to compress the angle between bonding pairs.

Steps to apply the VSEPR model:

1. Draw the Lewis structure of the molecule.

2. Count the bonding and nonbonding electron pairs around the central atom. (For structures with multiple bonds, count the multiple bond as one effective pair of electrons.)

3. Count the number of atoms attached to the central atom. Look in the chart on the next page to find the arrangement that minimizes that repulsion, placing the electrons as far apart as possible. For example, BeH_2, another exception to the octet rule, has two electron pairs and two atoms attached to the central atom. If the two bonded pairs of electrons are placed 180^0 apart, the resulting linear geometry has minimal repulsion.

Molecular Shapes and Bond Angles

No. of Electron Pairs	Electron Pair Geometry (Bond Angle)	No. of Bonding Electron Pairs	No. of Lone Electron Pairs	Molecular Geometry	Formula	2D Structure	Hybrid Orbital
2	Linear (180°)	2	0	Linear	BeH_2		sp
3	Trigonal planar (120°)	3	0	Trigonal planar	CO_3^{2-}		sp²
3	Trigonal planar (120°)	3	1	Bent	NO_2^-		sp²
4	Tetrahedral (109.5°)	4	0	Tetrahedral	CH_4		sp³

No. of Electron Pairs	Electron Pair Geometry (Bond Angle)	No. of Bonding Electron Pairs	No. of Lone Electron Pairs	Molecular Geometry	Formula	2D Structure	Hybrid Orbital
4	Tetrahedral (>109.5°)	3	1	Trigonal pyramidal	NH_3		sp^3
4	Tetrahedral (>>109.5°)	2	2	Bent	H_2O		sp^3
5	Trigonal bipyramidal (90°, 120°)	5	0	Trigonal bipyramidal	PCl_5		sp^3d
5	Trigonal bipyramidal (90° and >120°)	4	1	Seesaw unsymmetrical tetrahedron	SF_4		sp^3d

No. of Electron Pairs	Electron Pair Geometry (Bond Angle)	No. of Bonding Electron Pairs	No. of Lone Electron Pairs	Molecular Geometry	Formula	2D Structure	Hybrid Orbital
5	Trigonal bipyra-midal (90° and 180°)	3	2	T-shaped	BrF₃		sp³d
5	Trigonal bipyra-midal (180°)	2	3	Linear	ICl₂⁻		sp³d
6	Octahedral (90°)	6	6	Octahe-dral	SF₆		sp³d²
6	Octahedral (90°)	5	1	Square pyramidal	BrF₅		sp³d²

No. of Electron Pairs	Electron Pair Geometry (Bond Angle)	No. of Bonding Electron Pairs	No. of Lone Electron Pairs	Molecular Geometry	Formula	2D Structure	
6	Octahedral (90°)	4	2	Square planar	ICl_4^-		sp^3d^2

EXAMPLE: Predict the molecular structure of the carbon dioxide molecule. Is this molecule expected to have a dipole moment?

First we must draw the Lewis structure for the CO_2 molecule.

$$\ddot{O}=C=\ddot{O}$$

In this structure for CO_2, there are two effective pairs around the central atom (each double bond is counted as one effective pair). There are two atoms attached to the central atom.

According to the table, a linear arrangement is required.

Each C–O bond is polar, but, since the molecule is linear, the dipoles cancel out and the molecule is nonpolar.

EXAMPLE: Draw the Lewis structure for PF_5 and identify the molecular geometry.

PF_5 has 40 valence electrons. Connecting 5 F atoms to P, the central atom, uses 10 electrons in five single bonds. The 30 remaining electrons can be used to satisfy the octet rule for each F atom. There are 5 electron pairs around P and 5 atoms attached to P, resulting in a trigonal bipyramidal shape.

EXAMPLE: Draw the Lewis structure for CH_3OH.

This molecule has more than one central atom, the carbon and the oxygen. Hydrogen cannot be in the middle since it can form only one bond. Draw the correct Lewis structure and then using the rules for VSEPR, assign the molecular geometry around each central atom. The carbon atom has four electron pairs around it and four atoms attached resulting in tetrahedral geometry. Oxygen also has four electron pairs around it, but only has two atoms attached. This leads to bent geometry around oxygen.

EXAMPLE: Arrange the following molecules in order of increasing bond angles: H_2S, CCl_4, NF_3.

First, draw the correct Lewis structure for each molecule and then assign an approximate bond angle for each.

Each of the four molecules has four electron pairs around the central atom, resulting in a tetrahedral arrangement of the electron pairs about the central atom and approximately a 109.5° bond angle. H_2S has two lone pairs on the central atom; NF_3 has one lone pair on the central atom. These lone pairs repel more than the bonded electron pairs and will cause the bond angle to be smaller than the expected 109.5°. The more lone pairs, the smaller the bond angle.

In order of increasing bond angles, the answer is H_2S < NF_3 < CCl_4.

HYBRIDIZATION

Hybridization describes the mixing of atomic orbitals to form special orbitals that share electrons during bond formation. For example, in methane (CH_4), the 2s orbital and three 2p orbitals of carbon mix together to form four sp^3 hybrid orbitals. Each sp^3 orbital of carbon overlaps the 1s orbital of hydrogen forming a sigma bond. A sigma bond, σ, is formed from an overlap of orbitals. Single bonds are sigma bonds.

In general, the sum of the superscripts on the hybrid orbitals equals the number of electron groups (multiple bonds count as one group) around the central atom. The carbon atom in CH_4 has 4 electron groups around the central atom. The sum of the superscripts in sp^3 equals 4.

The carbon atom of the double bond in *cis*-2-pentene on page 158 of this guide is sp^2 hybridized. There are 3 electron groups around the C atom of the double bond. The sum of the superscripts in sp^2 is 3. The 2s orbital and two of the 2p orbitals of carbon mix to form sp^2, leaving one 2p orbital unused. Each carbon atom of the double bond has an unhybridized p orbital perpendicular to the sp^2 orbital. These parallel p orbitals can share a pair of electrons above and below the carbon-carbon bond, forming a pi (π) bond.

A double bond, such as the one in *cis*-2-pentene, consists of one sigma bond because of the head-on overlap of sp^2 orbitals of each carbon atom and one pi bond due to the parallel p orbitals.

The hybrid orbitals are oriented in space according to the VSEPR Theory. For example, the four sp^3 orbitals have a tetrahedral arrangement (there are 4 electron groups around the central atom).

BOND ENERGIES AND ENTHALPY

(*Chemistry* 7th ed. pages 350–353 / 8th ed. pages 361–364)

Estimate the change in enthalpy, ΔH, for the following reaction using the table of bond energies on page 372 of the text:

$$N_2 + 3\,H_2 \rightarrow 2NH_3$$

First, draw the Lewis structures for the reactants and products.

In the reactants, one triple bond between the atoms of nitrogen and one single bond between atoms of hydrogen are broken. Bond breaking is an endothermic process that has a positive value for enthalpy. Energy must be added to break the bonds.

Three single bonds between hydrogen and nitrogen are formed in the products. Bond formation is exothermic having a negative value for enthalpy.

$\Delta H = \Sigma$ energies to break bonds $+ \Sigma$ energies released to form new bonds.

REACTANT BONDS BROKEN

N_2: 1 mol $N\equiv N$	1 mol $\times$ 941 kJ/mol	= 941 kJ
H_2: 3 mol $H-H$	3 mol $\times$ 432 kJ/mol	= 1296 kJ
	Total energy required	= 2237 kJ

PRODUCT BONDS FORMED

NH_3: 3 $N-H$	6 mol $\times$ –391 kJ/mol	= –2346 kJ

$\Delta H = 2237$ kJ $+ -2346$ kJ $= -109$ kJ

BOND LENGTH

(*Chemistry* 7th ed. page 352 / 8th ed. page 363)

As the number of bonds between two atoms increases, the bond grows shorter and stronger.

EXAMPLE: Arrange the following molecules in order of decreasing C–C bond length: C_2H_4, C_2H_2, C_2H_6.

In order of decreasing C–C bond length: C_2H_6, C_2H_4, C_2H_2.

First, draw the Lewis structure for each molecule. The C–C triple bond is the shortest, followed by the double bond, and the single bond.

ORGANIC CHEMISTRY

In this section, the structure of carbon-containing compounds and their properties will be reviewed. You should know how to name the compounds, draw their structures inclusive of isomers, and identify trends in physical properties of the compounds based on their intermolecular forces.

HYDROCARBONS

(*Chemistry* 7th ed. pages 997–998 / 8th ed. pages 1006–1007)

Hydrocarbons contain hydrogen and carbon. Saturated hydrocarbons have all carbon–carbon single bonds. Unsaturated hydrocarbons contain carbon–carbon multiple bonds.

ALKANES

(*Chemistry* 7th ed. pages 997–1001 / 8th ed. pages 1006–1011)

Each alkane has the general formula C_nH_{2n+2}; n is the number of carbon atoms. This is a saturated hydrocarbon. In straight-chain hydrocarbons, also called unbranched or normal, all of the carbon atoms are connected in a long chain. The name of each alkane ends in –ane. The names of the first ten members of the alkane series appear in the following table.

Alkanes: Names and Formulas

Name	Formula	Name	Formula
Methane	CH_4	Hexane	C_6H_{14}
Ethane	C_2H_6	Heptane	C_7H_{16}
Propane	C_3H_8	Octane	C_8H_{18}
Butane	C_4H_{10}	Nonane	C_9H_{20}
Pentane	C_5H_{12}	Decane	$C_{10}H_{22}$

ISOMERISM IN ALKANES
(*Chemistry* 7th ed. pages 1001–1003 / 8th ed. pages 1011–1012)

Structural isomers occur when two molecules have the same chemical formulas but different structural arrangements of atoms. These structural differences result in differences in melting point, viscosity, and other properties that we examine later in this chapter.

EXAMPLE: Draw all the isomers of hexane and name each one.
Step 1: The first isomer is the straight chain alkane, C_6H_{14}.

$CH_3\text{-}CH_2\text{-}CH_2\text{-}CH_2\text{-}CH_2\text{-}CH_3$

This is the condensed structural formula. Each C has 4 bonds.

Step 2: Take one C atom off of the end of the chain and place it anywhere along the chain except on the end carbon which will result in the same straight chain isomer.

$CH_3\text{-}CH\text{-}CH_2\text{-}CH_2\text{-}CH_3$
       |
      CH_3

Step 3: To name this isomer, circle the longest continuous chain that determines the root or parent name. In this case the longest chain is pentane. This is the last part of the alkane name,

$$\underset{\text{CH}_3\text{-CH-CH}_2\text{-CH}_2\text{-CH}_3}{\overset{1\quad 2\quad 3\quad 4\quad 5}{}}$$

$$\underset{\text{CH}_3}{|}$$

Next, name the alkyl groups which are substituted on the parent chain. Count the number of carbons in the alkyl group. Using the root of name for the number of carbon atoms (meth, eth, and so on) drop the -ane from the hydrocarbon and change it to –yl. In the example here the branch or substituent group is called the methyl group.

Step 4: Specify the location of the substituent groups by numbering the longest chain of carbon atoms sequentially, starting at the end closest to the branching. Note, numbering left to right allows the methyl group to be at position number 2. If the longest chain is numbered from right to left, the methyl group would be on position number 4. The substituent groups should always have the lowest number possible, so the chain must be numbered from left to right.

The name of this branched alkane is 2-methylpentane.
A hyphen separates the number from the name.

Step 5: Continue to place the substituted group, methyl in this case, in as many places as you can along the chain such that you arrive at a different name. If the name of the branched alkane is the same as the ones you've already found, then a new isomer has not been found. The methyl group in this example can also be placed on carbon #3 of the parent chain. The name of the resulting alkane is 3-methyl pentane.

   Note that placing the methyl group on carbon #4 yields the same name as 2-methylpentane which has already been found. Remember, the parent chain must be numbered to give the lowest number for the substituents.

Step 6: When you have identified all possibilities in step 5, remove two C atoms off of the straight chain hydrocarbon in step 1. Place these C atoms along the remaining parent chain, naming each one. A new correct name, means a new isomer.

$$\text{CH}_3\text{-CH - CH -CH}_3 \quad \text{2,3 – dimethylbutane.}$$
$$\underset{\text{CH}_3 \ \ \text{CH}_3}{|\qquad |}$$

The longest chain is 4 C atoms. Numbers are separated by commas. Prefixes di-, tri-, and so on, are used to indicate multiple identical substituents.

Another isomer

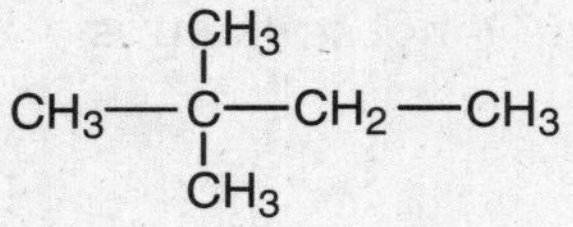
CH3
CH3—C—CH2—CH3
CH3

2,2–dimethylbutane.

All isomers based on the parent chain butane have been identified. If you try to remove an additional atom, for a total of three, from the straight chain hexane and place the remaining three on the parent chain propane, you end up duplicating isomers which have already been identified. For example, if you draw 2-ethyl-2-methylpropane and then name it, circling the longest parent chain, the correct name is 2,2–dimethylbutane.

To summarize the naming of branched alkanes: The location and number of each substituent are followed by the root alkane name, the substituents are listed in alphabetical order, and the prefixes di-, tri-, and so on, are used to indicate multiple, identical substituents.

ALKENES

(*Chemistry* 7th ed. pages 1005–1006 / 8th ed. pages 1014–1015)

Alkenes are hydrocarbons that contain at least one double bond. Alkenes have the general formula, C_nH_{2n}.

NOMENCLATURE OF ALKENES The root ending for alkenes is –ene instead of –ane.

For alkenes which have more than three carbon atoms, the lowest numbered carbon atom is the location of the double bond.

Isomers of 2-pentene appear on the next page. The longest continuous chain of carbon atoms is five. So the parent name of the alkene will be pentene. There are no substituted groups on the pentene chain. Numbering the chain from left to right will yield a parent name of 2-pentene. The double bond must have the lowest number of carbon atoms involved in the bond.

Alkenes exhibit *cis–trans* isomerism because the rotation of the double bond in alkenes is restricted. Look at each C atom of the double bond. The heavier of the two groups (or atoms) attached to each carbon is circled in the structure below. So on C atom #2, CH_3 is heavier than H. In the *cis* isomer, the heavier groups are on the same side of the double bond. In the *trans* isomer, the heavier groups are on opposite sides of the double bond.

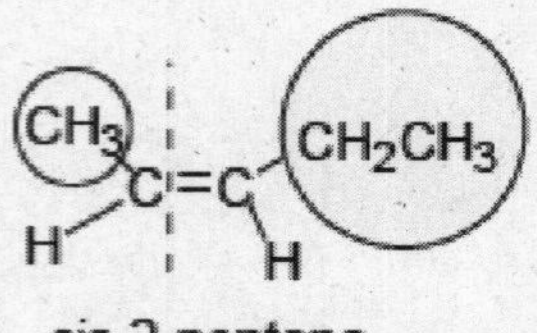
CH3
CH2CH3
C=C
H
H

***cis*-2-pentene**

The CH_3 and CH_2CH_3 groups which are in the longest chain containing the C=C lie on the same side of the C=C.

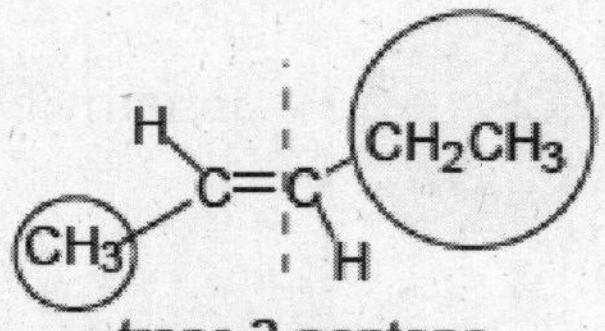
H
CH2CH3
C=C
CH3
H

***trans*-2-pentene**

The CH_3 and CH_2CH_3 groups which are in the longest chain containing the C=C lie across the C=C.

ALKYNES

(*Chemistry* 7th ed. page 1006 / 8th ed. page 1015)

Alkynes are hydrocarbons that contain at least one triple bond. An example of an alkyne is ethyne, C_2H_2, more commonly called acetylene. Alkynes have the general formula C_nH_{2n-2} and are named similarly to the alkenes; the –yne ending is substituted for the –ane on the longest alkane chain.

CYCLIC HYDROCARBONS

(*Chemistry* 7th ed. pages 1004–1005, 1006–1007 / 8th ed. pages 1013–1014, 1015–1016)

Carbon atoms also form ring-shaped structures.
Cyclic alkanes and alkenes are represented below.
A carbon atom is located at the vertex of each polygon.
For each cyclic alkane, only two hydrogen atoms can attach to each carbon atom.

Cyclopropane Cyclobutane Cyclopentane Cyclohexene
C_3H_6 C_4H_8 C_5H_{10} C_6H_{10}

Substitution can also occur in cyclic hydrocarbons.
methylcyclobutane

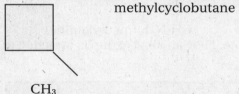

CH$_3$

AROMATIC HYDROCARBONS

(*Chemistry* 7th ed. pages 1007–1008 / 8th ed. pages 1015–1017)

Aromatic compounds are a special class of cyclic unsaturated hydrocarbons. The simplest structure is benzene, C_6H_6, which appears below. Benzene has alternating single and double bonds. Benzene exhibits resonance. More complex aromatic hydrocarbons can be thought of as consisting of a number of "fused" benzene rings.

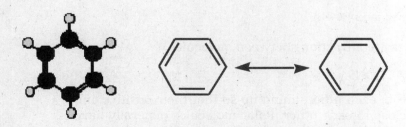

FUNCTIONAL GROUPS

(Chemistry 7th ed. pages 1010–1015 / 8th ed. pages 1019–1025)

Many organic molecules contain elements in addition to hydrogen and carbon. Functional groups are atoms or groups of atoms attached to the hydrocarbon which have characteristic chemistry.

Functional Group	Name of Group	Example of IUPAC name, Common Name
–X	Halide	CH_3CH_2Br (bromoethane)
–OH	Alcohol	CH_3OH (methanol)
–O–	Ether	CH_3OCH_3 (dimethyl ether)
$\overset{O}{\underset{}{-C-H}}$	Aldehyde	CH_3CHO (ethanal, acetaldehyde)
$\overset{O}{\underset{}{-C-}}$	Ketone	CH_3COCH_3 (propanone, acetone)
$\overset{O}{\underset{}{-C-OH}}$	Carboxylic acid	CH_3CO_2H (ethanoic acid, acetic acid)
$\overset{O}{\underset{}{-C-O-}}$	Ester	$CH_3CO_2CH_3$ (methyl ethanoate, methyl acetate)
$\overset{O}{\underset{}{-C-NH_2}}$	Amide	CH_3CONH_2 (acetamide)
$-\overset{\vert}{N}-$	Amine	CH_3NH_2 (methylamine) Note: H or an alkyl group is attached to N.

AP tip

You should know how to name compounds containing functional groups, draw their structures, identify bond angles, determine changes in enthalpy from bond energy values and predict trends in properties based on intermolecular forces. On the AP exam organic chemistry is often included in other types of problems such as thermochemistry and stoichiometry of combustion reactions.

INTERMOLECULAR FORCES

(Chemistry 7th ed. pages 426–429 / 8th ed. pages 440–443)

Intermolecular forces are attractions between molecules.

DIPOLE–DIPOLE ATTRACTIONS

Polar molecules attract each other, lining up so that their positive and negative poles are close to each other. Polar molecules generally have higher boiling points than nonpolar molecules of similar molar mass because they have dipole–dipole attractions in addition to London forces. An example of a polar molecule is SO_2.

HYDROGEN BONDNG

Hydrogen bonding is an unusually strong dipole–dipole force among molecules in which hydrogen is bound to a highly electronegative atom such as nitrogen, oxygen, or fluorine. The hydrogen bond occurs between the hydrogen on one atom and the O, F, or N on another molecule. This is the strongest type of intermolecular force. Examples of molecules which exhibit H-bonding include H_2O, NH_3, and HF.

LONDON DISPERSION FORCES

London dispersion forces exist between all molecules and account for the boiling points of the noble gases. If there were no attractions between molecules, then they would never liquefy. We believe that the electron cloud can experience temporary shifts that result in one side of the molecule becoming more negative than the other, causing a temporary dipole. This instantaneous dipole can induce a similar dipole on a neighboring atom.

We'd expect that the greater the charge on the nucleus and the larger the number of electrons in the molecule, the greater the induced dipole. As the atomic number increases down a group, atoms become more polarizable, that is, the electron cloud can become polarized due to the instantaneous dipole. For example, the halogens all experience London dispersion forces, but the force becomes stronger toward the bottom of the group. F_2 and Cl_2 are found as gases in nature, Br_2 is a liquid, and I_2 is a solid due to increasing London forces as the atomic number and size of the electron cloud increases in the group.

> **EXAMPLE:** Arrange C_2H_6, CH_4, C_4H_{10} in order of increasing boiling point.
>
> In order of increasing boiling point $CH_4 < C_2H_6 < C_4H_{10}$. All of these molecules exhibit London dispersion forces. The higher the molar mass, the more polarizable the molecule becomes. C_4H_{10} has the greatest molar mass, followed by C_2H_6, and has the strongest London dispersion forces.

MULTIPLE-CHOICE QUESTIONS

No calculators are to be used in this section.

1. The compound most likely to be ionic is
    (A) KF.
    (B) CCl_4.
    (C) CO_2.
    (D) ICl.
    (E) CS_2.

2. Ranking the ions S^{2-}, Ca^{2+}, K^+, Cl^-, from smallest to largest gives the order as
   (A) S^{2-}, Cl^-, K^+, Ca^{2+}.
   (B) Ca^{2+}, K^+, Cl^-, S^{2-}.
   (C) K^+, Ca^{2+}, Cl^-, S^{2-}.
   (D) Cl^-, S^{2-}, K^+, Ca^{2+}.
   (E) Ca^{2+}, K^+, Cl^-, S^{2-}.

3. The type of bonding within a water molecule is
   (A) ionic bonding.
   (B) polar covalent bonding.
   (C) nonpolar covalent bonding.
   (D) metallic bonding.
   (E) hydrogen 'bonding.'

4. Dinitrogen oxide, N_2O, has two double bonds. The general structure is N=N=O. The formal charge on the oxygen atom in this molecule is
   (A) zero.
   (B) positive one (+1).
   (C) positive two (+2).
   (D) negative one (-1).
   (E) negative two (-2).

5. The Lewis structure of which molecule requires resonance structures?
   (A) $MgCl_2$
   (B) PCl_5
   (C) SiO_2
   (D) SO_2
   (E) OCl_2

6. The predicted geometry (shape) of PH_3, according to the VESPR theory, is
   (A) linear.
   (B) bent or angular.
   (C) trigonal pyramidal.
   (D) tetrahedral.
   (E) trigonal planar.

7. Rank the following from lowest to highest boiling temperature:
   I C_2H_6 II C_2H_5OH III C_2H_5Cl
   (A) I<II<III
   (B) II<III<I
   (C) III<II<I
   (D) III<I<II
   (E) I<III<II

8. The total number of lone pairs in PCl_3 is
   (A) 1.
   (B) 8.
   (C) 10.
   (D) 12.
   (E) 14.

9. The Lewis structure for carbon dioxide is
   (A) :C=O-O:
   (B) :C̈=O-O:
   (C) :C̈-Ö-Ö
   (D) :Ö=C=Ö:
   (E) :Ö=C=Ö:

10. Which of the following molecules has a net dipole (dipole moment)?
    I. CO_2 II. SO_3 III. CCl_4 IV. PF_3.
    (A) I and II only
    (B) II and III only
    (C) III only
    (D) IV only
    (E) II, III, and IV

11. A triple bond is comprised of:
    (A) three σ (sigma) bonds
    (B) two σ (sigma) bonds and one π (pi) bond
    (C) one σ (sigma) bond and two π (pi) bonds
    (D) three π (pi) bonds
    (E) one σ (sigma) bond and two δ (delta) bonds

12. A substance with strong intermolecular forces of attraction would
    be expected to have
    (A) a low boiling point
    (B) a high vapor pressure
    (C) a high heat of vaporization
    (D) a low melting point
    (E) a low viscosity

13. What is the ELECTRON PAIR (or ELECTRON REGION)
    arrangement around the central atom in the molecule IF_5?
    (A) trigonal pyramidal
    (B) square planar
    (C) tetrahedral
    (D) octahedral
    (E) square pyramidal

14. This compound is used in the manufacture of antiseptics. To which family of compounds does it belong?

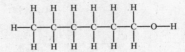

(A) alkanes
(B) alkenes
(C) alcohols
(D) aldehydes
(E) ketones

15. In which bond does the oxygen atom possess a partial positive charge?
(A) O-H
(B) O-F
(C) N-O
(D) O-C
(E) P-O

FREE-RESPONSE QUESTIONS

1. Is it possible to have
   (a) a nonpolar molecule that is made of polar bonds?

   Support your answer with discussion.

   (b) Sodium iodide boils at 1300°C. Sulfur hexafluoride sublimes at –64°C. Yet the difference in electronegativity between sodium and iodine (1.6) is almost exactly the same as that between sulfur and fluorine (1.5). Explain why there are such differences in physical properties between NaI and SF_6.

2. The combinations of nitrogen and oxygen forming different molecules are many and interesting. Their properties differ greatly. For this question consider nitrogen dioxide.
   (a) Draw the Lewis structure for NO_2, which is an odd electron compound.
   (b) What kind of hybridization does the nitrogen display in NO_2? Explain how you arrived at this answer. Compare this with the hybridization in the linear NO_2^+ ion.
   (c) Estimate the O–N–O bond angle in NO_2. What is this geometric shape called? Explain how you arrived at this answer.
   (d) Discuss the reactivity of the NO_2 molecule based on its structure. Explain how you arrived at this answer.

3. (a) Show the complete equation for the combustion of the flammable gas butene, C_4H_8.

(b) Given the following table of bond energies, estimate the enthalpy change, *H*, for the reaction noted in 3(a).

(c) Often the heat of reaction (enthalpy) calculated from bond energies differs by 10–20% or more from the laboratory determined values. Suggest why this is so.

(d) Show two of the several isomers of butene.

(e) Draw a Lewis structure diagram of butene. Indicate the kinds of hybridization of carbon in this molecule. Indicate all bond angles.

Answers

MULTIPLE CHOICE

1. **A** Bonds formed from elements with greater differences in electronegativity are most likely ionic in nature. (Electronegativity increases for elements as you move up and to the right on the Periodic Table.) (*Chemistry* 7th ed. pages 339–340 / 8th ed. pages 350–351)

2. **A** Note that all ions in this isoelectronic series have an Ar electron configuration; therefore the nuclear charge determines the size (*Chemistry* 7th ed. pages 340–342 / 8th ed. pages 351–353).

3. **B** This question is asking you to consider the nature of the bonding between hydrogen and oxygen in water (note the term *within* in the question!). With an electronegativity difference (H = 2.1 and O = 3.5) of 1.4, this suggests a bond of a very polar covalent nature due to the uneven sharing of the pair of electrons between hydrogen and oxygen. (Note: The force <u>between</u> one water molecule and another water molecule is known as a hydrogen 'bond.') (*Chemistry* 7th ed. pages 335–339 / 8th ed. pages 346–350).

4. **A** To determine the formal charge on an atom you need to determine the difference in the number of electrons assigned to the atom in the molecule and the number of outermost (valence) electrons on the free atom. In this case, the double bond between nitrogen and oxygen allows oxygen to "own" two (one-half of the four) electrons in the bond, plus totally own the other four electrons. (If this is not clear to you, draw the Lewis structure for this molecule.) Since the number of electrons assigned to oxygen in the molecule and as a free atom is the same, the formal charge is zero (*Chemistry* 7th ed. pages 363–367 / 8th ed. pages 374–378).

5. **D** Watch for *resonance* when it is possible to draw more than one valid Lewis structure. This tends to occur when the same kind of atoms are bonded once as a single bond and again with a multiple bond holding the same kind of atoms together. In this case there is a single bond between sulfur and one of the oxygens, and a double bond between sulfur and the other oxygen (*Chemistry* 7th ed. pages 362–363 / 8th ed. pages 373–374).

6. **C** It is the position of the nuclei upon which we base the shape of a molecule. From the Lewis structure, you can see an unshared pair of electrons on the P; it is this unshared pair which repels the bonded pairs (between P and H) down toward the corners of a tetrahedron. However, the top pair does not contribute to the shape; hence the geometry is a trigonal pyramid (*Chemistry* 7th ed. pages 367–379 / 8th ed. pages 378–389).

7. **E** Consider the IMF (intermolecular forces) between molecules. C_2H_6, with the lowest boiling temperature is nonpolar and hence has a low IMF. C_2H_5OH has hydrogen 'bonding' between molecules. Watch for these high IMF when hydrogen is bonded to O, N, or F. C_2H_5Cl is polar covalent and has intermediate strength IMF between molecules. (Measured boiling temperatures are as follows: for C_2H_6 it is -88.3°C, for C_2H_5OH it is +78.5°C, and for C_2H_5Cl it is +12.3°C.) (*Chemistry* 7th ed. pages 429–429 / 8th ed. page 443)

8. **C** A sketch of the Lewis structure will show three lone pairs on each of the chlorine atoms and one more lone pair on the phosphorus atom, for a total of 10 lone pairs (*Chemistry* 7th ed. pages 338–340 / 8th ed. pages 349–351).

9. **D** First count the number of total outermost (valence) electrons. Each oxygen has six and the carbon has four, for a total of sixteen. Only response D, with two double bonds, has sixteen valence electrons (*Chemistry* 7th ed. pages 353–361 / 8th ed. pages 364–372).

10. **D** In order to have a dipole moment there must be polar covalent bonds present and an assymetrical distribution of charge. In order to determine this, you must consider both the nature of the bonding (extent of polarity) and the arrangement of these bonds (the molecular shape). If all the vectors, resulting from polar bonds, cancel each other, then there is zero dipole moment. In the case of PF_3, the shape is that of a trigonal pyramid, the vectors do not cancel each other, and the molecule is polar (*Chemistry* 7th ed. pages 335–338 / 8th ed. pages 346–350).

11. **C**(*Chemistry* 7th ed. pages 394-397 / 8th ed. pages 408-410)

12. **C** A great deal of energy would be required to overcome strong intermolecular forces between the molecules within the substance in order for them to move from liquid to gaseous state (*Chemistry* 7th ed. pages 426-430 / 8th ed. pages 440-444).

13. **D** Iodine has 7 valence electrons and each fluorine atom has 7 valence electrons; so 7 + 5(7) = 42 electrons. Thus there are 5 bonding pairs and one lone pair around the central atom. The six pairs or regions of electrons assume an octahedral arrangement (*Chemistry* 7th ed. pages 367-379 / 8th ed. pages 378-387).

14. **C** (*Chemistry* 7th ed. pages 997-1013 / 8th ed. pages 1006, 1014, 1017, 1019-1020)

15. **B** Since fluorine is the most electronegative element in the Periodic Table, the oxygen atom, when covalently bonded to the fluorine atom, will have a partial positive charge (*Chemistry* 7th ed. pages 333-335 / 8th ed. pages 344-349).

FREE RESPONSE

1. (a) Yes, it is possible to have a nonpolar molecule that contains polar bonds. This is possible if the polar bonds are arranged so that their charge distributions cancel each other, thereby having no separation between the centers of positive and negative charge.

   (b) NaI is ionic; in crystal form, each positive Na ion is surrounded by six negative I ions, and each I^- is surrounded by six Na^+. Powerful ionic bonds lead to a high boiling temperature for NaI. SF_6 is a covalent compound. In this compound, six fluorine atoms surround one sulfur atom to form a neutral molecule. The symmetry of SF_6 (an octahedral molecule) and the high electron density of fluorine lead to little attraction, hence low IMF, and a lower boiling temperature (*Chemistry* 7th ed. pages 332–338, 342–348, 426–429, 454–459, and 462 / 8th ed. pages 343–349, 353-359, 440-443, 466-471, and 474).

2. (a) :Ö=N̈-Ö:

   (b) Nitrogen displays sp^2 hybridization (leaving a 'p' orbital unhybridized for the pi bonding). The hybridization of nitrogen in the ion (:O=N⁺=O:) is sp, allowing for formation of the double bonds between nitrogen and each of the oxygens.

   (c) This would usually suggest a bond angle of 120°. However, because the lower repulsion between the lone nonbonded electron and the bonded pairs is lower than the repulsion between the bonded pairs on the oxygen, the O-N-O angle should be greater than 120° (Linus Pauling, in his book *Nature of the Chemical Bond* gives 140°). The shape is 'bent' (angular).

   (d) The properties of this molecule would certainly be related to that lone electron on the nitrogen. For example, this material is very reactive and dimerizes (reacts with itself to form N_2O_4). (*Chemistry* 7th ed. pages 354–362, 363–367, 369–376, and 397–398 / 8th ed. pages 365–373, 374-378, 381-387, 410-411).

3. (a) $C_4H_8(g) + 6O_2(g) \rightarrow 4CO_2(g) + 4H_2O(g)$

   (*Chemistry* 6th ed. pages 102–106)

   (b) Average Bond energies (kJ/mol)

C-H	413
C-C	347 (single)
C=C	614 (double)
C ≡ C	839 (triple)

C-O   358

C=O   799

H-O   467

H-H   432

O=O   495

From $CH_2CHCH_2CH_3 + 6O\text{-}O \rightarrow 4O\text{=}C\text{=}O + 4H\text{-}O$

$\backslash$

H

Bonds broken (reactants):

1 C=C   1 mol × 614 kJ/mol = 614 kJ

2 C-C   2   × 347   = 694

8 C-H   8   × 413   = 3304

6 O-O   6   × 495   = 2970

Total energy required to break bonds = +7582 kJ.

Bonds formed (products):

8 C=O   8 mol × 799 kJ/mol = 6392 kJ

8 H–O   8   × 467   = 3736

Total energy released as bonds form = –10128 kJ.

Note that it is a negative value, denoting an exothermic reaction.

H = –2546 kJ/mol C_4H_8 reacting.

(*Chemistry* 7th ed. pages 351–353 / 8th ed. pages 362–364)

(c) The surrounding bonds often affect the strength of a given bond. For example, the average C=O bond energy is 745 kJ/mol, but the C=O bond energy in CO_2 is 799 kJ/mol (the value given for your use in the table). Examining the structure of butane, you might guess that the C-H bond energies of bonds near a C=C double bond would be somewhat different from those with a C-C single bond attached to the same carbon. The environment of the bond does make a difference (*Chemistry* 7th ed. pages 350–351 / 8th ed. pages 361–362).

(d) 1-butene is $CH_3\text{-}CH_2\text{-}CH\text{=}CH_2$

2-butene is $CH_3\text{-}CH\text{=}CH\text{-}CH_3$

this form might also be shown as both *cis*- and *trans*-isomers)

2-methylpropene
$$CH_2\text{=}C\text{-}CH_3$$
$$|$$
$$CH_3$$

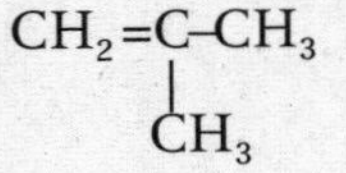

(*Chemistry* 7th ed. pages 960–965, 1006 / 8th ed. pages 969–974, 1015)

(e)
```
    H      H  H  H
     \    /   |  |
      C = C - C - C - H
     /        |  |
    H         H  H
```

The first and second carbons display sp² hybridization with bond angles of 120°.

The other two carbons are sp³ hybridized with bond angles of 109° (*Chemistry* 7th ed. pages 1005–1006 / 8th ed. pages 1014–1015).

9

LIQUIDS AND SOLIDS

The forces that hold together solids and liquids, the condensed states of matter, are similar. The effect of these forces on properties such as surface tension and vapor pressure will be reviewed. The bonding models, structure, and properties of liquids and solids will be discussed. Changes in state from solid to liquid to gas will also be considered.

You should be able to

- ■ Explain how intermolecular forces such as surface tension, capillary action, viscosity, vapor pressure, and boiling point affect the properties of liquids.
- ■ Rank substances in order of decreasing or increasing properties such as boiling points.
- ■ Understand the different types of solids and know examples of each.
- ■ Know characteristic points of heating curves and phase diagrams.

AP tip

Although the topics in this chapter can appear in the multiple-choice section of the exam, it is often the case that many of these topics such as the effect of intermolecular forces on properties of liquids and the structures and properties of solids appear in the essay section.

INTERMOLECULAR FORCES

(*Chemistry* 7th ed. pages 426–429 / 8th ed. pages 440–443)

Intermolecular forces were introduced in the last chapter. In this section, you will see the effect of intermolecular forces on properties of liquids such as boiling point, surface tension, and vapor pressure.

When molecular compounds such as water undergo changes in state, a disruption of forces between molecules is involved. The covalent bonds in the molecules are not broken. When ionic compounds such as sodium chloride change state, ionic bonds are broken.

PROPERTIES OF LIQUIDS

SURFACE TENSION

(*Chemistry* 7th ed. page 429 / 8th ed. page 443)

Molecules at the surface of a liquid are pulled inward by the molecules beneath them. The surface tension of a liquid is the resistance of a liquid to increasing its surface area. Liquids with strong intermolecular forces tend to have high surface tensions.

EXAMPLE: Equal-sized drops of Hg, Br_2, or H_2O are placed onto the same glass surface. Which of the three liquids will have the most spherical shape?

ANSWER: Hg will have the most spherical shape due to the metallic bonding which occurs between mercury atoms. Metallic bonding is stronger than both the hydrogen bonding in water and the dispersion forces of bromine.

CAPILLARY ACTION

(*Chemistry* 7th ed. page 429 / 8th ed. page 443)

The rise of liquids up very narrow tubes is called capillary action. Water and dissolved nutrients move upward through plants via capillary action.

Adhesive forces, the attractions between a liquid and its container, cause the liquid to creep up the walls of the container. Adhesive forces are opposed by cohesive forces, the intermolecular forces in the liquid that try to decrease the liquid's surface area.

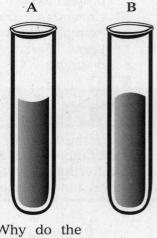

A   B

EXAMPLE: Two different liquids are placed in separate glass tubes. Liquid A forms a concave meniscus. Liquid B forms a convex meniscus. Why do the meniscuses of liquids A and B have two different shapes? Suggest a possible identity for each liquid.

ANSWER: Liquid A's adhesive forces are stronger than its cohesive forces, as in water. Liquid B's cohesive forces are stronger than its adhesive forces, as in mercury.

VISCOSITY

(Chemistry 7th ed. pages 429–430 / 8th ed. pages 443–444)

Viscosity is a measure of a liquid's resistance to flow. The greater its resistance, the more slowly it will flow. Molasses, motor oil, and glycerol are viscous liquids. Highly viscous liquids have large intermolecular forces or are very complex molecules with large molar masses such as polymers. For example, grease is viscous because it contains carbon chains 20 to 25 carbons long which can get tangled. Gasoline consists of shorter carbon chains three to eight carbons long and is therefore a nonviscous liquid. Viscosity decreases with temperature because more molecules will have greater average kinetic energy to overcome the intermolecular forces of the liquid.

VAPOR PRESSURE

(Chemistry 7th ed. pages 459–463 / 8th ed. pages 471–475)

In the process of vaporization (evaporation), liquid molecules leave the surface and form a vapor. The heat of vaporization (enthalpy of vaporization, ΔH_{vap}) is the energy required to vaporize 1 mole of a liquid at its normal boiling point (1 atm). This quantity is endothermic because energy is required to overcome the intermolecular forces. In the reverse process, condensation, energy is released as the IMFs form between the particles when the gas molecules become closer together in the liquid.

In a closed container, a liquid is at equilibrium when the rate of evaporation equals the rate of condensation. The equilibrium vapor pressure is the pressure above the liquid from the molecules of the liquid in the gas phase.

Volatile liquids are liquids with high vapor pressures. The vapor pressure at a given temperature is determined by the intermolecular forces of the liquid. Strong intermolecular forces result in lower vapor pressure. Weak intermolecular forces result in higher vapor pressure because liquid molecules are not held as strongly by intermolecular forces within the liquid and can readily escape from the surface.

The vapor pressure of a liquid increases with increasing temperature. At higher temperatures, more molecules will have sufficient kinetic energy to overcome intermolecular forces and break free from the surface of the liquid.

The dependence of vapor pressure on temperature is represented in the figure on the next page.

> **EXAMPLE:** Explain why diethyl ether, $CH_3CH_2OCH_2CH_3$, is more volatile than ethanol, CH_3CH_2OH, at 25°C.
>
> **ANSWER:** The dipole forces in diethyl ether are weaker than the hydrogen bonding in ethanol. The weaker the intermolecular forces, the higher the liquid's vapor pressure and volatility.

BOILING

(Chemistry 7th ed. page 466 / 8th ed. page 478)

The boiling of a liquid occurs in a container open to the atmosphere when the vapor pressure of the liquid equals atmospheric pressure. Bubbles containing the vapor form within the liquid and rise to the surface due to their reduced density. During boiling, the temperature remains constant until all of the liquid boils away. The normal boiling point of a liquid is the temperature at which the vapor pressure of a liquid is equal to standard atmospheric pressure (1 atm = 760 mm Hg). To determine the normal boiling point of a liquid, interpolate the vapor pressure curve for the liquid at 1 atmosphere.

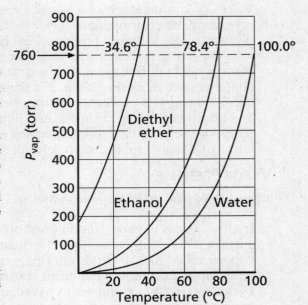

EXAMPLE: What are the normal boiling points of diethyl ether and ethanol?

ANSWER: The normal boiling points for diethyl ether and ethanol are approximately 34.6°C and 78.4°C, respectively. The normal boiling point occurs at 1 atm and can be found by interpolating the vapor pressure curves above.

EXAMPLE: At a location 19,340 ft above sea level, the atmospheric pressure is 350 torr. At what temperature will water boil? Will it take more or less time to cook an egg at this location than at sea level or 1 atm?

ANSWER: Using the vapor pressure curve above, at 350 torr, water will boil at about 80°C. It will take more time to hard-boil an egg at this elevation because water boils at a temperature lower than that at sea level.

EXAMPLE: Pressure cookers heat a small amount of water under high constant pressure. What effect does using a pressure cooker have on cooking time?

ANSWER: Food in a pressure cooker will take less time to prepare because it will cook at a higher temperature. When the pressure is increased, the boiling point will increase until the vapor pressure of the liquid equals the pressure above the liquid. Once the boiling point is reached, the temperature will not change as long as there is liquid present. Vapor pressure is directly proportional to temperature.

USING VAPOR PRESSURE DATA

(*Chemistry* 7th ed. pages 461–463 / 8th ed. pages 473–475)

The Clausius-Clapeyron equation can be used to determine the vapor pressure of a liquid at a different temperature or to find the heat of vaporization of a substance.

$$\ln\left(\frac{P_{vap,T_1}}{P_{vap,T_2}}\right) = \frac{\Delta H_{vap}}{R}\left(\frac{1}{T_2} - \frac{1}{T_1}\right).$$

In an experiment, plotting $\ln P$ vs. $1/T$ (K^{-1}) will result in a line whose slope is equal to $-\Delta H_{vap}/R$, where $R = 8.3145$ J mol^{-1}K^{-1}.

> **EXAMPLE:** Determine the boiling point of water in Breckenridge, Colorado, where the atmospheric pressure is 520 torr. For water, $\Delta H_{vap} = 40.7$ kJ/mol.
>
> **ANSWER:** 362 K or 89°C.
>
> At 100°C, 373 K, the vapor pressure of water is 1.00 atm.
>
> $$\ln\left(\frac{520 \text{ torr}}{760 \text{ torr}}\right) = \left[\frac{(4.07 \times 10^4 \text{ J/mol})}{(8.3145 \text{ J/K} \times \text{mol})}\right] \times \left(\frac{1}{373 \text{ K}} - \frac{1}{T_2}\right)$$
> $$T_2 = 362 \text{ K or } 89°C.$$

STRUCTURE AND TYPES OF SOLIDS

(*Chemistry* 7th ed. pages 430–432, 435–436 / 8th ed. pages 444–446, 449–451)

CRYSTALLINE SOLIDS

Crystalline solids are classified according to the type of particle that occupies the lattice points: atoms, molecules, or ions. The four types of crystalline solids are molecular, ionic, metallic, and network covalent. The table below describes each of these solids.

Type of Solid	Molecular	Ionic	Atomic	Metallic	Network Covalent
Examples	$C_6H_{12}O_6$	NaCl	Ar	Cu	C (diamond)
Particles occupying lattice points	Molecule	Ions	Atoms	Atoms	Atoms
Bonds or forces between lattice points	Dispersion forces	Ionic	Dispersion forces	Delocalized nondirectional metallic	Covalent
Melting points	Low	High	Low	High (1083°C)	Very high (3500°C)
Electrical conductivity as solid	None	None	None	Yes	None

Type of Solid	Molecular	Ionic	Atomic	Metallic	Network Covalent
Electrical conductivity when melted	None	Yes	None	Yes	None
Other properties		brittle		Malleable and ductile	Insulator very hard

A lattice is a three-dimensional system of points designating the points of the components (atoms, molecules, or ions) that make up the substance. A unit cell is the smallest repeating unit of a lattice.

AMORPHOUS SOLIDS

Amorphous solids, such as rubber and glass, have particles with no orderly structure.

STRUCTURE AND BONDING IN METALS

It is said that the bonding is nondirectional; the metallic atoms can easily slide over each other because the environment will be the same. Metals are malleable, can be formed into thin sheets, are ductile, and can be pulled into wires.

ALLOYS

Alloys are mixtures of metallic elements and have metallic properties. Substitutional alloys, like brass and sterling silver, have host metal atoms replaced by atoms of similar size. Brass is made of copper and zinc. Interstitial alloys have smaller atoms occupying the holes in the metal structure. The presence of the smaller atoms changes the properties of the host making it harder and stronger. Steel contains iron and carbon.

CHANGES IN STATE

HEATING CURVES

(*Chemistry* 7th ed. pages 463–466 / 8th ed. pages 473–478)

The chapter on thermochemistry discussed the quantitative aspects of the heating curve. Energy is added at a constant rate to a solid and a plot of temperature vs. time is drawn. The first plateau on the heating curve represents the melting point of the substance. The second plateau on the heating curve represents the boiling point of the substance. The temperature does not change during melting or boiling because the energy added is used to overcome intermolecular forces.

Usually, the plateau for the boiling point is longer in duration than the melting plateau because it takes more energy to separate the molecules from a liquid into a gaseous state than from a solid into a liquid state. This greater energy is related to the greater forces that must be overcome.

PHASE DIAGRAMS

(*Chemistry* 7th ed. pages 467–471 / 8th ed. pages 479–483)

Phase diagrams graphically summarize the conditions under which equilibria exist between different states of matter of a substance. The state of matter, which is stable at a given set of temperature and pressure conditions, can be determined from the phase diagram.

You should be able to interpret and sketch the phase diagram of a substance.

> **EXAMPLE:** Consider the phase diagram for water. At 1.0 atmosphere, describe the phases of matter present as the temperature is increased from very low to high temperatures in experiment 1.
>
> **ANSWER:** At low temperatures, solid water (ice) is present. As the temperature is increased to 0°C, water reaches its normal melting point at 1 atm. Water in its solid and liquid state is present along the solid/liquid line from point A to point B. The line from point A through point B represents the change in melting point with pressure.

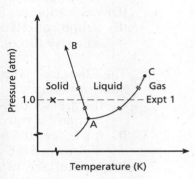

Line AB, the solid–liquid equilibrium line for water, has a negative slope. Ice is less dense than liquid water. Carbon dioxide, on the contrary, has a solid–liquid line with a positive slope; solid CO_2 is denser than liquid CO_2. The triple point of water is lower than that of CO_2. This means that water exists as a solid, liquid, and a gas at about 1 atm. Because the triple point of solid CO_2 is at 5.11 atm, it does not melt at pressures less than 5.11 atm; it sublimes and becomes a gas.

Line AC represents the vapor pressure curve of the liquid. Point C is the critical point of the liquid. It corresponds to the critical temperature and the critical pressure. The critical temperature is the highest temperature at which a liquid can exist. Beyond the critical temperature, no amount of pressure will liquefy a gas. The critical pressure is the pressure required to liquefy a vapor at the critical temperature. The liquid vapor line on the phase diagram ends at this point.

At point A, the triple point, all three states of matter are present. Under these conditions, the three states of the substance will coexist in a closed system.

MULTIPLE-CHOICE QUESTIONS

No calculators are to be used in this section.

1. Water has a higher capillary action than mercury due to
   (A) higher dipole-dipole forces between the water molecules.
   (B) strong cohesive forces within water.
   (C) very significant induced intermolecular attractions.
   (D) weak adhesive forces in water.
   (E) strong cohesive forces in water which work with strong adhesive forces.

2. The Clausius-Clapeyron equation indicates that
   I. the greater the vapor pressure, the higher the entropy of vaporization.
   II. the higher the heat of vaporization, the steeper the slope of $\ln P$ vs. $1/T$.
   III. the enthalpy of vaporization is equal to the slope of the vapor pressure vs. temperature.

   (A) I only
   (B) II only
   (C) III only
   (D) I and II only
   (E) I, II, and III

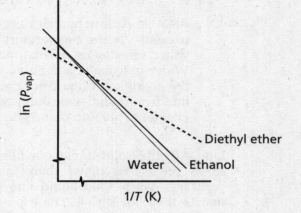

3. Small drops of water tend to bead up because of
   (A) high capillary action.
   (B) the shape of the meniscus.
   (C) the resistance to increased surface area.
   (D) low London dispersion forces.
   (E) weak covalent bonds.

4. The vapor pressure increases in a predictable order as shown as
   (A) $CH_4 < C_2H_5OH < C_2H_5–O–C_2H_5 < Ne$.
   (B) $Ne < CH_4 < C_2H_5–O–C_2H_5 < C_2H_5–OH$.
   (C) $C_2H_5–OH < C_2H_5–O–C_2H_5 < CH_4 < Ne$.
   (D) $C_2H_5–O–C_2H_5 < C_2H_5–OH < CH_4 < Ne$.
   (E) $C_2H_5–O–C_2H_5 < C_2H_5–OH < Ne < CH_4$.

5. Several liquids are compared by adding them to a series of 50-mL graduated cylinders, then dropping a steel ball of uniform size and mass into each. The time required for the ball to reach the bottom of the cylinder is noted. This is a method used to compare the differences in a property of liquids known as
   (A) surface tension.
   (B) buoyancy.
   (C) capillary action.
   (D) viscosity.
   (E) surface contraction.

6. The properties of solids vary with their bonding. An example of this is shown by
   (A) ionic solids with strong electrostatic attractions called ionic bonds, which have high melting temperatures.
   (B) molecular solids with high intermolecular forces which have high melting temperatures.
   (C) ionic solids with highly mobile ions which have high conductance.
   (D) amorphous solids with strong London dispersion forces and high vapor pressure.
   (E) network solids with low melting temperatures due to strong covalent bonds.

7. As you go down the noble gas family on the periodic table, the boiling temperature increases. This trend is due mainly to
   (A) an increase in hydrogen "bonding."
   (B) a decrease in dipole–dipole forces.
   (C) the lower atomic masses as you go down the family.
   (D) an increase in London dispersion forces.
   (E) an increase in lattice energy.

8. Carbon dioxide and silicon dioxide have properties which differ dramatically. This is due to
   (A) the hardness of silica compounds.
   (B) the size difference of carbon and silicon, resulting in less effective overlap with the smaller orbits of oxygen.
   (C) shared CO_2 and CO_4 tetrahedra.
   (D) the formation of a glass when silicates are heated.
   (E) the repeating sigma bonds formed in carbon dioxide.

9. This question refers to the phase diagram shown. Select the true statement.
   (A) Point E represents both critical temperature and critical pressure.
   (B) Moving from point A to point F is sublimation.
   (C) Point C represents both solid and vapor states.
   (D) Point B represents both solid and vapor states.
   (E) Point D represents a point where, no matter how much pressure is exerted, a liquid cannot form.

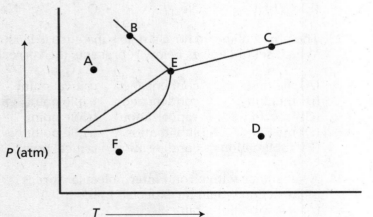

10. The densities of certain compounds are greater as liquids than as solids. This means that increasing the pressure will result in
    (A) a solid becoming a liquid.
    (B) a liquid becoming a solid.
    (C) lowering of the critical temperature.
    (D) elevation of the freezing temperature.
    (E) lowering of the triple point.

11. Which of the following properties of water is/are due to the intermolecular hydrogen bonding between the water molecules?
    1. Water has a high heat of vaporization.
    2. Ice floats on water.
    3. The boiling point of water is much higher than the boiling points of molecules of comparable size.
    4. The specific heat of water (4.184 J/°C-g) is much higher than that of lead (0.128 J/°C-g).

    (A) 1 and 2
    (B) 1 and 3
    (C) 1 and 4
    (D) 1, 2 and 3
    (E) All of them

12. The substances neon, sodium fluoride, carbon monoxide, and methylamine all have low molar masses. Order these substances in increasing boiling points or melting points.
    (A) Ne   < CO   < CH₃NH₂ < NaF
    (B) NaF   < CO   < CH₃NH₂ < Ne
    (C) Ne   < CH₃NH₂ < CO   < NaF
    (D) CH₃NH₂ < CO   < NaF   < Ne
    (E) CH₃NH₂ < Ne   < CO   < NaF

13. The phase diagram for a single substance is illustrated on the right. Which set of choices below represents the correct descriptions?

	A	B	C
(A)	melting	condensation	critical point
(B)	melting	condensation	triple point
(C)	freezing	vaporization	triple point
(D)	melting	sublimation	triple point
(E)	sublimation	condensation	critical point

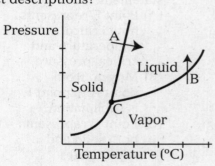

14. A substance with strong intermolecular forces of attraction would be expected to have
    (A) a low boiling point
    (B) a high vapor pressure
    (C) a high heat of vaporization
    (D) a low melting point
    (E) a high solubility in water

15. Consider a closed container containing a liquid and its vapor. Which statement is **incorrect**?
    (A) The vapor exerts a pressure called the vapor pressure.
    (B) Increasing the temperature of the liquid would lead to a greater vapor pressure.
    (C) Evaporation and condensation will eventually cease after a constant pressure has been attained.
    (D) Decreasing the volume of the container at constant temperature would cause increased condensation until the pressure of the vapor was once again the same as it had been.
    (E) The rate of evaporation is equal to the rate of condensation.

FREE-RESPONSE QUESTION

Calculators and Equation Tables may be used.

1. a. Compare the structural models used to describe solids, liquids, and gases. Include (1) distance between particles, (2) relative strength of intermolecular forces, and (3) the extent of entropy.
   b. (i) Is vaporization an endothermic or exothermic process? Support your answer with an explanation.
      (ii) In what ways is the high heat of vaporization of water important to each of us personally?

2. The vapor pressure of water at 25°C is 23.8 torr; ΔH_{vapor} = 43.9 kJ / mol.
   a. Determine the vapor pressure of water at 70°C.
   b. Determine the percentage error if your experimental results for vapor pressure at 70°C give a result of 225 torr.
   c. Give structural reasons that support the fact that the boiling temperature of C_2H_5OH is 78°C and water is 100°C, (both at 760 torr).

Answers

MULTIPLE CHOICE

1. **(E)** The strong adhesive forces lead to a creeping effect as water moves up the narrow tubing and the strong cohesive forces attempt to minimize the surface area (*Chemistry* 7th ed. pages 429–430 / 8th ed. pages 443–444).

2. **(B)** The Clausius-Clapeyron equation relates the ln vapor pressure to $-\Delta H_{vapor}$. It is most often used to determine the value of ΔH_{vapor} by measuring changes in vapor pressure with changes in temperature (see figure at right). You should be able to use this important equation (*Chemistry* 7th ed. pages 459–463 / 7th ed. pages 471–474).

3. **(C)** This is a description of surface tension, which is a result of high dipole–dipole forces between water molecules. These intermolecular forces are also called hydrogen "bonds" (*Chemistry* 7th ed. pages 429–430 / 8th ed. pages 443–444).

4. **(C)** Examining the IMF will suggest that only the alcohol (C_2H_5–OH) has an exposed –OH, suggesting strong hydrogen "bonding." The other three are essentially controlled by weaker London dispersion forces, which are greater in the larger, more massive compound. Realize also that the higher the IMF, the lower the vapor pressure (*Chemistry* 7th ed. pages 459–461 / 8th ed. pages 471–473).

5. **(D)** The resistance to flow of any fluid is called viscosity. As you would predict, liquids with high viscosity (e.g., maple syrup) have large intermolecular forces (*Chemistry* 7th ed. pages 429–430 / 8th ed. pages 443–444).

6. **(A)** Ionic bonds are unusually strong, requiring high temperatures to melt ionic substances. Table salt, for example, which is Na^+Cl^-, melts at 804°C (*Chemistry* 7th ed. pages 454–459, esp. Table 10.7 / 8th ed. pages 466–471, esp. Table 10.7).

7. **(D)** These very symmetrical atoms are nonpolar; except for the induced dispersion forces, they would all boil at zero Kelvin. As it is, He boils at 4K and Ne at 25K (*Chemistry* 7th ed. pages 426–429 / 8th ed. pages 440–443).

8. **(B)** Carbon dioxide acts as individual molecules and has a very low boiling temperature; SiO_2 is a network solid with chains of SiO_4 tetrahedra sharing oxygen atoms; it has both a very high melting and boiling temperature (*Chemistry* 7th ed. pages 446–448, esp. Figure 10.26 / 8th ed. pages 459–461, esp. Figure 10.26).

9. **(B)** Be sure you know the details of common pressure vs. temperature phase diagrams, which sections represent each physical state, and what process the movement from one section to another represents. In this case, moving in a line from point A to point F represents going from the solid to vapor (gaseous) state (*Chemistry* 7th ed. pages 467–471 / 8th ed. pages 479–483).

10. **(A)** Since the density of the solid is less than that of the liquid, the slope of the solid/liquid line must be negative (like water, unlike carbon dioxide and most other substances). If the pressure is increased, then you move from the solid section of the phase diagram into the liquid section (*Chemistry* 7th ed. pages 467–471 / 8th ed. pages 479–483).

11. **(E)** All of the anomalous properties of water are due to the exceptionally effective intermolecular hydrogen bonding (*Chemistry* 7th ed. pages 425–430, 459 / 8th ed. pages 439–444, 450, 467–468).

12. **(A)** The intermolecular forces in order are London dispersion forces, the weakest, then dipole-dipole, then hydrogen bonding and lastly the strongest being ionic forces holding the lattice of NaF together (*Chemistry* 7th ed. pages 426–429 / 8th ed. pages 440–443).

13. **(B)** Melting is the changing of a solid into a liquid. Condensation is the changing of vapor back into its liquid state. The triple point is that set of temperature and pressure conditions at which all three phases of matter coexist simultaneously (*Chemistry* 7th ed. pages 467–471 / 8th ed. pages 479–481).

14. **(C)** A high heat of vaporization would be expected of a substance with strong intermolecular forces because a significant amount of energy would be required to liberate the molecules from the liquid to gaseous state. Low boiling point, low melting point, high solubility in water, and a high vapor pressure would indicate that the forces holding the molecules together are relatively weak; therefore, the molecules are able to be separated more easily by added energy or outside forces of attraction, as is the case with the solvent in its ability to act on the solute (*Chemistry* 7th ed. pages 426–429 / 8th ed. pages 440–443).

15. **(C)** Once an equilibrium vapor pressure has been reached at a given temperature within the container, the rate of condensation and rate of evaporation become equal but neither one ceases (*Chemistry* 7th ed. pages 459–466 / 8th ed. pages 471–478).

FREE RESPONSE

1. a

	Distance	**Forces**	**Entropy**
Solid	Shortest	Strongest	Lowest
Liquid	Short	Strong	Low
Gas	Greatest	Weakest	Highest

b. Because vaporization is a bond breaking process (overcoming intermolecular forces) and thereby forms a higher potential energy state, it must be endothermic (ΔH = +). The high value of ΔH_{vapor} for water becomes important in many ways, from removing heat from our bodies (perspiration) to moderating temperature changes in areas near large bodies of water (*Chemistry* 7th ed. pages 430, 459–460 / 8th ed. pages 444, 471–472).

2 a.

$$\ln\left(\frac{P_{v1}}{P_{v2}}\right) = \frac{\Delta H_{vapor}}{R}\left(\frac{1}{T_2} - \frac{1}{T_1}\right)$$

$$\ln\left(\frac{23.8}{X}\right) = \frac{43900 \text{ J/mol}}{8.314 \text{ J/mol} \bullet \text{K}}\left(\frac{1}{343} - \frac{1}{298}\right) = -2.32$$

$$\frac{23.8}{X} = 0.0983$$

$$X = 242 \text{ torr.}$$

(Note: Be sure to change the kJ/mol for ΔH_{vapor} to J.)

b.

$$\% \text{ error} = \left(\frac{\text{your experimental value} - \text{accepted value}}{\text{accepted value}} \right) \times 100\%$$

$$= \left(\frac{225 - 242}{242} \right) \times 100\% = -7.0\%.$$

c. The molecules of water are smaller and each has two –OH portions (twice as many as the alcohol), giving the opportunity for stronger hydrogen "bonding" (*Chemistry* 7th ed. pages 460–463, 426–428 / 8th ed. pages 472–474, 440-442).

10

SOLUTIONS

In this section, the properties of liquid solutions will be reviewed. Solution composition, factors affecting solubility, and colligative properties such as freezing point depression, boiling point elevation, and osmotic pressure will be covered.

You should be able to
- Perform calculations with different solution concentrations such as molarity, mass percent, molality, and mole fraction.
- Discuss the effects of temperature, pressure, and structure on solubility.
- Perform calculations with Raoult's law.
- Understand colligative properties such as boiling point elevation, freezing point depression, and osmotic pressure.
- Use colligative properties to determine the molar mass of a solute.

AP tips

Questions from this section may appear on both the free-response and multiple-choice sections of the AP Exam. Questions on the solubility or miscibility of substances in one another may appear in the essay section. Questions involving the determination of molar mass by freezing point depression could also appear in the lab essay.

197

SOLUTIONS AND THEIR COMPOSITIONS

(*Chemistry* 7th ed. pages 485–487 / 8th ed. pages 498–500)

TYPES OF SOLUTIONS

(*Chemistry* 7th ed. page 485 / 7th ed. page 498)

A solution is a homogeneous mixture. The table below summarizes the different types of solutions that can exist. In this chapter, the focus will be on liquid solutions.

Various Types of Solutions

Example	State of Solution	State of Solute	State of Solvent
Air, natural gas	Gas	Gas	Gas
Vodka in water, antifreeze	Liquid	Liquid	Liquid
Brass	Solid	Solid	Solid
Carbonated water (soda)	Liquid	Gas	Liquid
Seawater, sugar solution	Liquid	Solid	Liquid
Hydrogen in platinum	Solid	Gas	Solid

A solute, the substance being dissolved, is added to a solvent, which is present in the largest amount when referring to a liquid–liquid or gas–gas solution.

COMPOSITION

(*Chemistry* 7th ed. pages 485–487 / 8th ed. pages 498–500)

MOLARITY   The molarity, M, of a solution is the number of moles of solute per liter of solution.

MASS PERCENT   Mass percent, also called weight percent, is the percent by mass of solute in a solution.

MOLE FRACTION   Mole Fraction, χ, is the ratio of moles of a given component to the total number of moles of solution. For a two-component solution, where n_A and n_B are the moles of the two components:

$$\text{Mole fraction of component A} = \chi_A = \frac{n_A}{n_A + n_B}$$

MOLALITY Molality, *m*, is the number of moles of solute per kilogram of solvent.

$$\text{Molality} = \frac{\text{moles solute}}{\text{kilogram of solvent}}$$

EXAMPLE: A solution is prepared by mixing 30.0 mL of butane (C_4H_{10}, $d = 0.600$ g/mL) with 65.0 mL of octane (C_8H_{18}, $d = 0.700$ g/mL). Assuming that the volumes add in mixing, calculate the following for butane:

(a) molarity

(b) mass percent

(c) mole fraction

(d) molality

SOLUTION:

(a) The molarity of butane is 3.26 M.

$$30.0 \text{ mL} \times \frac{0.600 \text{ g}}{1 \text{ mL}} \times \frac{1 \text{ mol}}{58.1 \text{ g}} = \frac{0.310 \text{ mol } C_4H_{10}}{0.0950 \text{ L}} = 3.26 \text{ M}$$

(b) The mass percent is 28.3% butane.

Mass solution = mass solute + mass of solvent

$$= \left(30.0 \text{ mL} \times \frac{0.600 \text{ g}}{1 \text{ mL}} \right) + \left(65.0 \text{ mL} \times \frac{0.700 \text{ g}}{1 \text{ mL}} \right) = 63.5 \text{ g}$$

$$\text{Mass solute} = 30.0 \text{ mL} \times \frac{0.600 \text{ g}}{1 \text{ mL}} = 18.0 \text{ g}$$

$$\text{Mass percent} = \frac{18.0 \text{ g}}{63.5 \text{ g}} \times 100\% = 28.3\% \text{ butane}$$

(c) The mole fraction of butane is 0.437.

$$\text{Moles of octane} = 65.0 \text{ mL} \times \frac{0.700 \text{ g}}{1 \text{ mL}} \times \frac{1 \text{ mol}}{114 \text{ g}} = 0.399 \text{ mol octane}$$

$$\text{Mole fraction of butane} = \frac{0.310 \text{ mol butane}}{(0.310 \text{ mol butane} + 0.399 \text{ mol octane})} = 0.437$$

(d) 0.310 moles butane/0.0455 kg = 6.81 *m*

FACTORS AFFECTING SOLUBILITY

(Chemistry 7th ed. pages 492–496 / 8th ed. pages 504–507)

The formation of a liquid solution begins by separating the solute into its individual components. Next, the solvent's intermolecular forces must be overcome to make room for the solute. The solute and solvent then interact to form the solution.

STRUCTURAL EFFECTS

The phrase "like dissolves like" means that solubility is favored if the solute and solvent have similar polarities, as determined by their structure. The table below summarizes the solubilities of different types of solutes in different types of solvents.

Type of Solute	Type of Solvent	Solubility	Example
Ionic	Polar	Usually soluble	LiCl in H_2O
Polar	Polar	Soluble (miscible)	CH_3OH in H_2O
Nonpolar	Polar	Immiscible	C_6H_{14} in H_2O
Nonpolar	Nonpolar	Miscible	C_6H_{14} in CCl_4

EXAMPLE: Discuss the solubility of each of the following solutes in carbon tetrachloride: ammonium nitrate, 1-pentanol, and pentane. Explain why each solute will or will not dissolve.

SOLUTION: Carbon tetrachloride is nonpolar and nonpolar solutes will dissolve in it. Ammonium nitrate is ionic and will not dissolve in a nonpolar solvent. It will dissolve in a polar solvent like water. 1-Propanol is polar and will not be soluble in carbon tetrachloride. Pentane is nonpolar and will be miscible (will form a homogeneous mixture) with carbon tetrachloride.

PRESSURE EFFECTS

Pressure has little effect on the solubilities of liquids and solids. Gases become more soluble in liquids when the pressure of the gas above the liquid increases.

TEMPERATURE EFFECTS

The solubility of most solids increases with temperature. However, some solids decrease solubility with temperature. The energy needed to separate the solute–solute particles is usually greater than the energy released from the solute–solvent attractions. It is difficult to predict the temperature dependence of the solubilities of solids. Gases, on the other hand, always decrease in solubility with increasing temperature.

THE VAPOR PRESSURES OF SOLUTIONS

(*Chemistry* 7th ed. pages 497–504 / 8th ed. pages 509–516)

A nonvolatile solute lowers the vapor pressure of the solvent. The nonvolatile solute decreases the escaping tendency of the solvent molecules.

RAOULT'S LAW

Raoult's law states that the vapor pressure of a solution is directly proportional to the mole fraction of the solvent present.

$$P_{soln} = \chi_{solvent} P^o_{solvent}$$

P_{soln} is the observed vapor pressure of the solution.

$\chi_{solvent}$ is the mole fraction of the solvent.

$P^o_{solvent}$ is the vapor pressure of the pure solvent.

For an ideal solution, a plot of P_{soln} vs. $\chi_{solvent}$ at constant temperature gives a straight line with a slope equal to $P^o_{solvent}$.

EXAMPLE: The vapor pressure of a solution containing 53.6 g of glycerin, $C_3H_8O_3$, in 133.7 g of ethanol, C_2H_5OH is 113 torr at 40°C. Calculate the vapor pressure of the pure ethanol at 40°C assuming that glycerin is a nonvolatile, nonelectrolyte solute in ethanol.

SOLUTION: The vapor pressure of pure ethanol at 40°C is 136 torr.

$$P_{ethanol} = \chi_{ethanol} P^o_{ethanol}$$

$$\chi_{ethanol} = \frac{\text{mol ethanol}}{\text{total mol in solution}}$$

$$\frac{133.7 \text{ g } C_2H_5OH \times 1 \text{ mol}}{46.07 \text{ g}} = 2.90 \text{ mol } C_2H_5OH$$

$$\frac{53.6 \text{ g } C_3H_8O_3 \times 1 \text{ mol}}{92.09 \text{ g}} = 0.582 \text{ mol } C_3H_8O_3$$

Total mol = 0.582 mol + 2.90 mol = 3.48 mol

$$P^o_{ethanol} = \frac{P_{ethanol}}{\chi_{ethanol}} = \frac{113 \text{ torr}}{(2.90 \text{mol} / 3.48 \text{ mol})} = 136 \text{ torr}$$

REAL VS. IDEAL SOLUTIONS Ideal solutions obey Raoult's law. Real solutions best approximate the behavior of ideal solutions when the solute concentration is low and the solute and solvent have similar types of intermolecular forces and molecular sizes.

Deviations from Raoult's law occur when the interactions between the solute and solvent are extremely strong, as in hydrogen bonding, and the vapor pressure of the solution is lower than predicted. Also, the vapor pressure of a real solution is greater than predicted when the intermolecular forces between the solute and solvent are weaker than the intermolecular forces between the solute–solute or solvent–solvent.

COLLIGATIVE PROPERTIES

(*Chemistry* 7th ed. pages 504–513 / 8th ed. pages 516–525)

Colligative properties depend on the number, not the identity, of the solute particles in an ideal solution.

BOILING POINT ELEVATION

A nonvolatile solute elevates the boiling point of a solvent. Because a nonvolatile solute lowers the vapor pressure of a solution, it must be heated to a temperature higher than the boiling point of the pure solvent. The normal boiling point is the point where the vapor pressure of liquid equals 1 atmosphere. The change in boiling point due to the presence of the nonvolatile solute is represented by

$$\Delta T_b = K_b m.$$

ΔT is the difference between the boiling point of the solution and that of the pure solvent.

K_b is the molal boiling point elevation constant of the solvent.

m is the molality of the solute in solution.

> **EXAMPLE:** 2.00g of a large biomolecule was dissolved in 15.0 g of carbon tetrachloride. The boiling point of this solution was 77.85°C. Calculate the molar mass of the biomolecule. For carbon tetrachloride, the boiling point constant is 5.03°C kg/mol, and the boiling point of pure carbon tetrachloride is 76.50°C.
>
> **SOLUTION:** The molar mass is 498 g/mol.
>
> $$\Delta T_b = 77.85°C - 76.50°C = 1.35°C.$$
>
> $$m = \frac{1.35°C}{(5.03°C \text{ kg/mol})} = 0.268 \text{ mol/ kg.}$$
>
> Mol of biomolecule = 0.0150 kg solvent × 0.268 mol hydrocarbon/kg solvent
>
> $$= 4.02 \times 10^{-3} \text{ mol.}$$
>
> $$\text{Molar mass} = \frac{2.00 \text{ g}}{4.02 \times 10^{-3} \text{ mol}} = 498 \text{ g/mol.}$$

FREEZING POINT DEPRESSION

A nonvolatile solute decreases the freezing point of a solvent.

$$\Delta T_f = K_f m.$$

ΔT is the difference between the boiling point of the solution and that of the pure solvent.

K_f is the molal freezing point depression constant of the solvent.

> **EXAMPLE:** Which of the following solutions would have the largest freezing point depression? Explain your answer.
>
> (a) 0.010 m NaCl
>
> (b) 0.010 m Na_3PO_4
>
> (c) 0.010 m $Al_2(SO_4)_3$
>
> (d) 0.010 m $C_6H_{12}O_6$

SOLUTION: The solution with the lowest freezing point depression is 0.010 m $Al_2(SO_4)_3$, the solution which yields the highest molality (or the most) particles when dissolved. 0.010 m $Al_2(SO_4)_3$ yields five particles (ions) when dissolved or the total concentration of solute particles is 0.050 m. Two particles (or 0.020 m) are present in a solution of 0.010 m NaCl. Four particles (or 0.040 m) are present in a solution of 0.010 m Na_3PO_4. One particle (or 0.010 m) is present in a solution of 0.010 m $C_6H_{12}O_6$.

OSMOTIC PRESSURE

Osmosis is the flow of a pure solvent into a solution through a semipermeable membrane. Osmotic pressure is the pressure that must be applied to a solution to stop osmosis.

$$\pi = MRT.$$

π is the osmotic pressure in atmospheres.

M is the molarity of the solution.

R is the gas law constant, 0.08206 L atm/mol K.

T is temperature in Kelvin.

> **EXAMPLE:** An aqueous solution of 10.00 g of catalase, an enzyme found in the liver, has a volume of 1.00 L at 27°C. The solution's osmotic pressure at 27°C is 0.745 torr. Calculate the molar mass of the catalase.

SOLUTION: The molar mass of catalase is 2.51×10^5 g/mol.

$$M = \frac{\pi}{RT} = \frac{0.745 \text{ torr} \times \left(\dfrac{1 \text{ atm}}{760 \text{ torr}}\right)}{\left[0.08206 \text{L} \times \dfrac{\text{atm}}{(\text{mol} \times \text{K})}\right] 300\text{K}} = 3.98 \times 10^{-5} \text{ M}$$

$$1.00 \text{ L} \times 3.98 \times 10^{-5} \text{ mol/L} = 3.98 \times 10^{-5} \text{ mol}$$

$$\text{Molar mass} = \frac{10.00 \text{ g}}{3.98 \times 10^{-5} \text{ mol}} = 2.51 \times 10^5 \text{ g/mol}$$

MULTIPLE-CHOICE QUESTIONS

No calculators may be used on this part of the exam.

1. As temperature decreases
   (A) solutes always become less soluble in solvents.
   (B) polar solutes always become less soluble in polar solvents.
   (C) solid solutes always become less soluble in liquid solvents.
   (D) gaseous solutes always become more soluble in liquid solvents.
   (E) there is little effect on solubility.

2. The aqueous solution with the lowest freezing temperature
   (greatest freezing temperature depression) would be
   (A) 0.010 M NaCl.
   (B) 0.020 M NaCl.
   (C) 0.030 M $C_6H_{12}O_6$.
   (D) 0.030 M NaNO$_3$.
   (E) 0.030 M HC$_2$H$_3$O$_2$.

3. The expected value for i, the van't Hoff factor, is often only a first
   approximation due to
   (A) ion pairing.
   (B) moles of undissolved solute.
   (C) an exothermic heat of solution.
   (D) a variation in the freezing or boiling temperature constant.
   (E) complete dissociation into component ions.

4. The number of moles of Al(NO$_3$)$_3$ which must be added to water to
   form 2.00 L of 0.30 M NO$_3^-$ ions is
   (A) 0.30 mol.
   (B) 0.60 mol.
   (C) 0.20 mol.
   (D) 2.4 mol.
   (E) 8.0 mol.

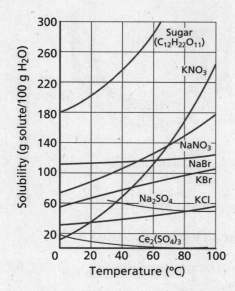

5. Refer to the solubility graph above. When the temperature of a saturated solution of potassium nitrate is lowered from 80°C to 40°C, the grams of KNO₃ lost from 2.00 g of water would be about
   (A) 2 g.
   (B) 7 g.
   (C) 17 g.
   (D) 100 g.
   (E) 200 g.

6. Not all liquid–liquid solutions are ideal. Those that are usually have
   (A) solute–solvent interactions that are weaker than the interactions in pure liquid.
   (B) solute–solvent interactions that are stronger than the interactions in pure liquid.
   (C) similar liquids with an enthalpy of solution near zero.
   (D) an observed vapor pressure lower than that predicted by Raoult's law.
   (E) high dipole–dipole interactions between solute and solvent molecules.

7. Mole fraction is a method of expressing concentration that is needed in several kinds of problems, such as those dealing with vapor pressure of a solution. To calculate the mole fraction, it is necessary to
   (A) divide the total number of moles of solution by 2.
   (B) divide the moles of solute by moles of solvent.
   (C) divide the moles of one component by the sum of the moles of all components.
   (D) divide the mass of the solute by the mass of the solution.
   (E) divide the mass of the solute by 1000 g of solvent.

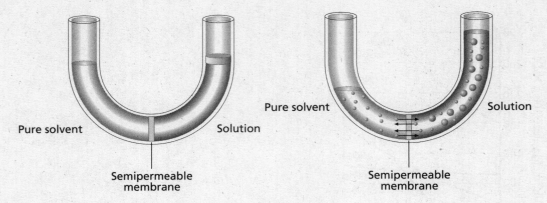

Pure solvent   Solution

Semipermeable
membrane

Pure solvent   Solution

Semipermeable
membrane

8. Refer to the two diagrams above. Osmotic pressure may be defined
    as
    I.   the excess pressure shown by different liquid levels on the two
         sides.
    II. the pressure which must be applied to stop osmosis.
    III. flow through a semipermeable membrane.
    (A) I only
    (B) II only
    (C) III only
    (D) I and II only
    (E) I, II, and III

9. In the following chart, which compares a KCl aqueous solution with
    pure water, the row which correctly identifies the salt solution's
    characteristics is

Row	Boiling Temp.	Freezing Temp.	Vapor Pressure
(A)	Lower	Lower	Lower
(B)	Lower	Lower	Higher
(C)	Lower	Higher	Lower
(D)	Higher	Lower	Higher
(E)	Higher	Lower	Lower

10. Toluene has half the molecular mass (g/mol) of glucose. If 100. g of
    each of these two substances are mixed and all of the glucose
    dissolves in the toluene, the mole fraction of toluene in this solution
    would be
    (A) 1/3.
    (B) 1/2.
    (C) 2/3.
    (D) 3/2.
    (E) 5/6.

11. What is the molality of a 1.00 L of a 50.0% by mass aqueous solution
    of acetic acid, CH_3CO_2H?
    (A) 1.7 m
    (B) 6.4 m
    (C) 8.3 m
    (D) 13.9 m
    (E) 16.7 m

12. The vapor pressure of pure benzene (C_6H_6) and toluene (C_7H_8) at 25 °C are 95.1 and 28.4 mm Hg, respectively. A solution is prepared with a mole fraction of toluene of 0.750. Assume the solution to be ideal and determine the total vapor pressure above the solution, in mm Hg.
   (A) 61.8
   (B) 66.7
   (C) 123
   (D) 45.1
   (E) 78.4

13. When an ionic salt dissolves in water, the solute-solvent interaction is
   (A) hydrogen bonding
   (B) London forces
   (C) ion-ion forces
   (D) ion-dipole forces
   (E) dipole-dipole forces.

14. Concerning vapor pressures, which of the following is **not** a correct statement?
   (A) The vapor pressure of a solvent is proportional to the mole fraction of solvent in solution.
   (B) Raoult's Law is applicable to ideal solutions, but often works for dilute nonideal solutions.
   (C) The total vapor pressure of a system is equal to the product of the vapor pressures of the individual components.
   (D) The vapor pressure above a solution containing two volatile components is enriched in the lower boiling-point component.
   (E) The addition of a nonvolatile solute to a solvent lowers the vapor pressure of the solvent.

15. Which statement is **not** true regarding colligative properties?
   (A) The magnitude depends on the concentration.
   (B) The magnitude depends on whether the solute is an electrolyte or not.
   (C) The magnitude depends on the identity of the solute.
   (D) Raoult's Law describes the vapor pressure above a solution.
   (E) Since the vapor pressure of the solvent is lowered by a nonvolatile solute, the boiling point of the solution is higher.

FREE-RESPONSE QUESTIONS

1. (a) Molarity is a temperature dependent quantity, whereas molality is not. Explain why this is so.
   (b) When the concentration of very dilute aqueous solutions are calculated, the values of molarity and molality are the same. Explain.
   (c) The heat of solution of salts may be either endothermic (which most are) or exothermic. Explain what factors determine whether $\Delta H_{sol'n}$ is endothermic or exothermic.
   (d) As a solution freezes, the freezing temperature continues to decrease. Why is this so?

2. You are to determine the formula and the molecular mass for an organic compound from the following information: 5.00 g of the organic solid was dissolved in 100.0 g of benzene. The boiling temperature of this solution was 82.3°C. The organic compound is 46.7% nitrogen, 6.67% hydrogen, 26.7% oxygen, and the remainder is carbon. The boiling temperature of pure benzene is 80.2°C; K_b = 2.53 °C kg / mol.
   (a) Determine the molecular mass of the organic solid.
   (b) Determine the molecular formula of the solid.
   (c) Determine the mole fraction of the organic solid in the solution.
   (d) If the density of this solution is 0.8989 g / mL, calculate the molarity of the solution.

Answers

MULTIPLE CHOICE

1. **D** Only in the case of a gas dissolved in a liquid can we always be certain that as the temperature decreases, the gas will become more soluble (*Chemistry* 7th ed. pages 495–496 / 8th ed. pages 506–507).

2. **D** Freezing temperature depression is a colligative property. The greater the concentration of particles in solution, the greater the freezing temperature lowering. With response (A) 0.010 M NaCl, you expect 0.0100 M Na^+ ions and 0.010 M Cl^- ions, for a total concentration of 0.020 M ions. In response (D), we would have the greatest expected concentration of ions, 0.030 M Na^+ and 0.030 M NO_3^- (*Chemistry* 7th ed. pages 506–507 / 8th ed. pages 518–519).

3. **A** Ion pairing suggests that a few of the ionic solute ions stay together, acting as one unit rather than two (or more) particles. This is an important effect in more concentrated solutions (*Chemistry* 7th ed. pages 512–513 / 8th ed. pages 524–525).

4. **C** Since every $Al(NO_3)_3$ unit contains 3 units of NO_3^- ions, you need 1/3 as many $Al(NO_3)_3$ units as NO_3^- units. Needed are 0.30 mol/L of $NO_3^- \times 2.00$ L = 0.60 mol NO_3^-. Hence $1/3 \times 0.60$ mol = 0.20 mol

$Al(NO_3)_3$ required (*Chemistry* 7th ed. pages 512–513 / 8th ed. pages 524–525).

5. **A** The solubility at 80°C is about 170 g KNO_3/100 g H_2O, dropping to 70 g KNO_3/100 g H_2O at 40°C, for a loss of 100 g KNO_3. If only 2/100 as much water were used, then only 2/100 of the 100 g of KNO_3 would come out of solution (*Chemistry* 7th ed. pages 493–496 / 8th ed. pages 505–508).

6. **C** Liquid–liquid solutions that obey Raoult's law are said to be ideal solutions. This is most likely when the solute–solute, solvent–solvent, and solute–solvent interactions are of about the same magnitude. This means that the tendency of solvent molecules to escape, thereby producing a vapor pressure, is about what is expected (*Chemistry* 7th ed. pages 501–504 / 8th ed. pages 513–516).

7. **C** Mole fraction is the ratio of the number of moles of a given component to the total number of moles of solute plus solvent; it is most often applied to two-component solutions of liquids in liquids at this level of the study of chemistry (*Chemistry* 7th ed. pages 485–487 / 8th ed. pages 498–500).

8. **D** Selection III defines osmosis, not osmotic pressure. Both of the other two definitions are valid (*Chemistry* 7th ed. page 508–509 / 8th ed. pages 520–521).

9. **E** Boiling temperature is always elevated, freezing temperature is always depressed, and vapor pressure is lower for an ionic aqueous solution than for the pure solvent (*Chemistry* 7th ed. pages 501–507 / 8th ed. pages 513–519).

10. **C** Assume that you have a mass (grams) of one mol of glucose. That same mass (in grams) would be two moles of toluene. Of the total 3 moles, 2 moles would be toluene, hence $2/(1 + 2) = 2/3$ for the mol fraction (*Chemistry* 7th ed. pages 485–486 / 8th ed. pages 498–499).

11. **E** Assume 100 grams of the solution. Then the mass of acetic acid is 50% of 100 grams or 50 grams. Moles of acetic acid = 50 grams/ 60.05 g/mol or 0.833 moles acetic acid. Mass of water is 50g = 0.050 kg. The definition of molality is moles of solute/kg of solvent so: 0.833 mol/0.0500kg = 16.7 molal (*Chemistry* 7th ed. page 486 / 8th ed. page 500).

12. **D** The key idea is that Raoult's law is not limited to the solvent if a solution also contains volatile solutes. A Raoult's law expression can be written for any volatile component in a solution, whether it is the solvent or solute, as long as the solution is nearly idea. Since we know the mole fractions of each component, we can calculate the partial pressure of each component and sum to get the total pressure.

Since $\qquad$ $X_{benzene} + X_{toluene} = 1$ and $X_{toluene} = 0.750$,

$X_{benzene} = 1.0 - 0.750 = 0.250$

$p_{benzene} = X_{benzene} P^o_{benzene} = (0.250)(95.1 \text{ mm Hg}) = 23.8 \text{ mm Hg}$

$p_{toluene} = X_{toluene} P^o_{toluene} = (0.750)(28.4 \text{ mm Hg}) = 21.3 \text{ mm Hg}$

$p_{benzene} + p_{toluene} = P_T = 45.1 \text{ mm Hg}$

(*Chemistry* 7th ed. pages 498–503 / 8th ed. pages 510–514)

13. D Water has a dipole which can either be attracted to the negative anion within the crystal lattice or the positive cation (*Chemistry* 7th ed. pages 127–129, 488–492 / 8th ed. pages 130–132, 501–505).

14. C The total vapor pressure of the system is actually the sum of the vapor pressures of the individual components taking into consideration the respective mole fractions (*Chemistry* 7th ed. pages 497–504 / 8th ed. pages 509–515).

15. C The magnitude has nothing to do with the identity, but rather the concentration and if it is an electrolyte, how many ions are produced upon dissolution (*Chemistry* 7th ed. pages 504–513 / 8th ed. pages 516–523).

FREE RESPONSE

1. (a) Molarity depends upon the volume of the solution and this property changes (generally rather slightly with liquid solvents) with temperature. Molality depends on mass which is not temperature dependent (*Chemistry* 7th ed. pages 485–486 / 8th ed. pages 498–499).

   (b) For very dilute aqueous solutions, the solvent (water) mass is 1000 g, which is a volume of 1 liter; because the solution is "very dilute," the solution is almost totally solvent (*Chemistry* 7th ed. pages 485–486 / 8th ed. pages 498–499).

   (c) The heat of solution, $\Delta H_{sol'n}$ may be considered as the sum of the energy required to separate the ions in the salt (plus, perhaps, the energy to separate the particles of the liquid) and the energy given off when the ions become "bonded" to liquid particles. In this expression, $\Delta H_{sol'n} = \Delta H_{c.e.} + \Delta H_{hyd}$, the crystal energy (or lattice energy) is always positive (the energy to separate charged particles must be added, as in "bond breaking"), and energy is always given off as charged particles move nearer to each other ("bond forming"). In summary, $\Delta H_{c.e.}$ is always (+) and ΔH_{hyd} is always (–); whichever of these dominates determines the sign of $\Delta H_{sol'n}$ (*Chemistry* 7th ed. pages 488–492 / 8th ed. pages 501–504).

(d) As a solution freezes, the solvent molecules are removed from the solution which causes an increase in the concentration. This causes further freezing temperature depression (*Chemistry* 7th ed. pages 504–505 / 8th ed. pages 516–517).

2. (a) The molecular mass of the organic solid:

$$\Delta T_b = 82.3°C - 80.2°C = 2.1°C.$$

$$\text{Mass of solvent} = \frac{100.0 \text{ g} \times 1 \text{ kg}}{1000 \text{ g}} = 0.1000 \text{ kg of solvent.}$$

$$m = \frac{\Delta T}{k} = \frac{2.1°C}{2.53°C \times \text{kg/mol}} = 0.83 \ m = 0.83 \text{ mol/kg.}$$

$$\frac{5.00 \text{ g}}{0.1000 \text{ kg}} = 50.0 \text{ g/kg} = 0.83 \text{ mol/kg}$$

Hence, $\dfrac{50.0 \text{ g}}{0.83 \text{ mol}} = 60. \text{ g/mol.}$

(*Chemistry* 7th ed. pages 504–505 / 8th ed. pages 516–517)

(b) The formula:

$$\text{N} \quad \frac{46.7 \text{ g}}{14.0 \text{ g/mol}} = 3.34 \text{ mol N.}$$

$$\text{H} \quad \frac{6.67 \text{ g}}{1.00 \text{ g/mol}} = 6.67 \text{ mol H.}$$

$$\text{O} \quad \frac{26.7 \text{ g}}{16.0 \text{ g/mol}} = 1.67 \text{ mol O.}$$

$$\text{C} \quad \frac{19.9 \text{ g}}{12.0 \text{ g/mol}} = 1.66 \text{ mol C.}$$

$$\frac{3.34}{1.67} = 2 \text{ mol N.}$$

$$\frac{6.67}{1.67} = 4 \text{ mol H.}$$

$$\frac{1.67}{1.67} = 1 \text{ mol O and 1 mol C.}$$

This gives a formula of CON_2H_4, which has a molecular mass [see part (a)] of 60.g/mol.

This means that this is both the empirical and the molecular formula (*Chemistry* 7th ed. pages 93–95 / 8th ed. pages 94–96).

(c) Mole fraction:

$$0.83 \ m = \frac{0.83 \text{ mol of organic solid}}{1000 \text{ g of benzene}} ; \quad \frac{1000 \text{ g}}{78.0 \text{ g/mol}} = 12.8 \text{ mol benzene.}$$

$$\frac{0.83 \text{ mol}}{(0.83 + 12.8 \text{ mol})} = \frac{0.83}{13.6} = 0.061.$$

(*Chemistry* 7th ed. pages 487–488 / 8th ed. pages 500–501).

(d) Molarity:

Mass of solution = 5.00 g solute + 100.0 g benzene = 105.0 g of solution.

$$105.0 \text{ g} \times \frac{1 \text{ mL}}{0.8989 \text{ g}} = 116.8 \text{ mL of solution.}$$

$$\frac{0.083 \text{ mol}}{\left(116.8 \text{ mL} \times \dfrac{1 \text{ L}}{1000 \text{ mL}}\right)} = \frac{0.083 \text{ mol}}{0.1168 \text{ L}} = 0.71 \text{ M.}$$

(*Chemistry* 7th ed. pages 485–487 / 8th ed. pages 498–499).

11

KINETICS

Chemical kinetics is the study of the rates of reactions and the mechanisms, the series of steps by which each reaction occurs. You will review the factors which affect the rate of a reaction and the determination of the rate law and mechanism for a reaction.

In this chapter, you should be able to:

- Identify factors which affect reaction rates.
- Calculate the rate of production of a product or consumption of a reactant using mole ratios and the given rate.
- Determine the rate law for a reaction from given data, overall order, and value of the rate constant, inclusive of units.
- Determine the instantaneous rate of a reaction.
- Use integrated rate laws to determine concentrations at a certain time, t, and create graphs to determine the order of a reaction. Also, determine the half-life of a reaction.
- Write the rate law from a given mechanism given the speeds of each elementary step.
- Write the overall reaction for a mechanism and identify catalysts and intermediates present.
- Determine the activation energy for the reaction using the Arrhenius equation.
- Graphically determine the activation energy using the Arrhenius equation.

213

FACTORS AFFECTING REACTION RATES

Increasing the concentrations of reactants increases the chances for more molecular collisions which make products. Usually the increase in concentrations of the reactants will increase the rate of a reaction.

Increasing the surface area of the reactants will also increase the number of collisions. For example, granular zinc will react more quickly with hydrochloric acid than a strip of zinc.

Increasing the temperature of a reaction will also speed it up. Increasing the temperature increases the rate of both forward and reverse reactions, increasing the number of collisions over a certain period of time.

In the next section, the method by which a catalyst speeds up a reaction will be discussed.

COLLISON THEORY

(*Chemistry* 7th ed. pages 552–553 / 8th ed. pages 565–566)

The collision theory model accounts for the observed characteristics of reaction rates. For a reaction to occur, the molecules must collide in the correct orientation with sufficient energy. The theory assumes that all but the simplest reactions take place in a series of two-particle collisions. This sequence of two-particle collisions is called the reaction mechanism.

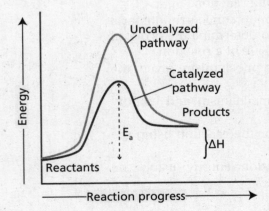

The potential energy diagram for a reaction, the reaction progress, is shown in the graph above. At the top of the "hill," or barrier, is the activated complex, or transition state. Once the energy in the transition state, the activation energy, is overcome, the reactants can become products. Activation energy is the amount of energy required to form an activated complex.

Because products have a higher potential energy than reactants, the reaction represented in the above diagram is endothermic. Recall that $\Delta H_{rxn} = \Delta H_{products} - \Delta H_{reactants}$. The difference between the potential energies of the products and the reactants gives the enthalpy of the reaction.

If asked to draw the reaction progress for an exothermic reaction, the energy of the products would be lower than that of the reactants.

The diagram shows the pathway for both catalyzed and uncatalyzed reactions. A catalyst speeds up a reaction by lowering the activation energy of the reaction without being consumed itself.

REACTION RATES

(*Chemistry* 7th ed. pages 527–532 / 8th ed. pages 540–545)

The reaction rate is the change in the concentration of a reactant or product per unit time.

Consider the reaction, A ⟶ B.

$$Rate = -\frac{\Delta[A]}{\Delta t}$$

The concentration of A in moles per liter is represented by [A]. The change in time is represented by Δt. The quantity, $-\Delta[A]/\Delta t$, is negative because the reactants are disappearing and forming products.

EXAMPLE:

Consider the reaction $4PH_3(g) \longrightarrow P_4(g) + 6 H_2(g)$.

If 0.0048 mol of PH_3 is consumed in a 2.0-L container during each second of the reaction, what are the rates of production for P_4 and H_2?

The rate at which PH_3 is being consumed is $-\frac{\Delta[PH_3]}{\Delta t}$.

0.0048 mol PH_3 / (2.0 L × s) = 0.0024 mol L^{-1} s^{-1} PH_3

The rate at which P_4 and H_2 are being produced can be determined by using mole ratios:

$$\frac{0.0024 \text{ mol } PH_3}{L \times s} \times \frac{1 \text{ mol } P_4}{4 \text{ mol } PH_3} = 0.0060 \text{ mol } L^{-1} s^{-1} P_4$$

$$\frac{PH_3}{L \times s} \times \frac{6 \text{ mol } H_2}{4 \text{ mol } PH_3} = 0.0036 \text{ mol } L^{-1} s^{-1} H_2$$

RATE LAWS

(Chemistry 7th ed. pages 532–547 / 8th ed. pages 545–559)

Rate laws show how the rate of a reaction depends on the concentrations of reactants. Except for an elementary reaction (one in which the balanced equation represents the mechanism), the rate law cannot be determined from the balanced equation.

For the reaction

$$2NO_2(g) \longrightarrow 2NO(g) + O_2(g),$$

the rate law can be written as

$$\text{Rate} = k[NO_2]^n.$$

The proportionality constant, k, is called the rate constant and is determined by experiment. For a given reaction at a given temperature, this value is constant. Its units depend on the order of the reactants.

The order, n, of the reactant must also be determined by experiment. It is the power to which the reactant concentration much be raised in the rate law. For example, if the reaction A → B is first order, then the rate law is Rate = $k[A]$; doubling the concentration of the reactant doubles the rate of the reaction. If the reaction is second order, then Rate= $k[A]^2$; doubling the concentration of the reactant will result in the rate quadrupling. If the reactant concentration is changed and the rate is not affected, the order of the reactant is zero. The rate law would be Rate = $k[A]^0$.

INSTANTANEOUS RATES

(Chemistry 7th ed. pages 529–530 / 8th ed. pages 542–543)

One way to determine the rate of a reaction at a particular time, the instantaneous rate, is to plot the reactant concentration versus time and take the slope of the tangent to the curve at time t.

If the slopes of tangents to the curve at two different concentrations are calculated, the rate law of a reaction can be determined by comparing the changes in rate to the changes in concentration.

In the graph to the right, when the concentration of the reactant, N_2O_5, is halved, the rate is also halved. The reaction is first order. The rate law for the reaction is Rate = $k[N_2O_5]$.

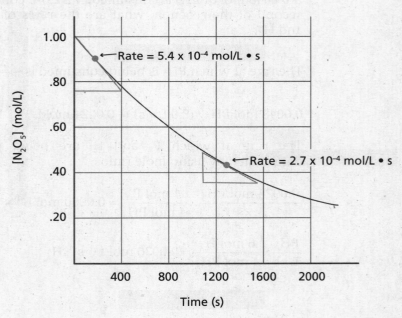

METHOD OF INITIAL RATES

(*Chemistry* 7th ed. pages 535–538 / 8th ed. pages 548–551)

The initial rate of a reaction is the instantaneous rate just after the reaction begins (just after $t = 0$ and before the initial concentrations of the reactants have changed.)

The rate law of a reaction can be determined by performing a few trials with different reaction concentrations and measuring the initial rate for each trial. To find the order of one reactant, change its concentration while holding the concentration of the other reactants constant.

EXAMPLE: Determine the rate law and the value of the rate constant for the reaction at –10°C:

$$2NO(g) + Cl_2(g) \longrightarrow 2NOCl(g)$$

Trial	$[NO]_o$ (mol/L)	$[Cl_2]_o$ (mol/L)	Initial rate (mol/L • s)
1	0.10	0.10	0.18
2	0.10	0.20	0.36
3	0.20	0.20	1.45

First, write the general form of the rate law: Rate = $k[NO]^x[Cl_2]^y$

Your goal is to determine x and y.

To find x, the order with respect to NO, pick two trials in which NO changes and Cl_2 remains the same.

Compare Trial 3 to Trial 2.

Write a ratio of the rate law of trial 3 to the rate of the rate law for trial 2. Substitute the values for the rates and known concentrations and solve for the order, x.

$$\frac{\text{Rate (Trial 3)}}{\text{Rate (Trial 2)}} = \frac{k[NO]^x \cancel{[Cl_2]^y}}{k[NO]^x \cancel{[Cl_2]^y}}; \quad \frac{1.45}{0.36} = \frac{(0.20)^x}{(0.10)^x}; \quad 4.0 = 2^x; x = 2.0$$

The order, x, with respect to NO is 2. This means that when the concentration of NO is doubled, the rate quadruples. Note that the rate constant and the concentration of Cl_2 cancel out because they are the same in Trials 2 and 3.

To find y, the order with respect to Cl_2, pick two trials in which Cl_2 changes and NO remains the same.

Compare Trial 2 to Trial 1.

$$\frac{\text{Rate (Trial 2)}}{\text{Rate (Trial 1)}} = \frac{\cancel{k[NO]^x}[Cl_2]^y}{\cancel{k[NO]^x}[Cl_2]^y}; \quad \frac{0.36}{0.18} = \frac{(0.20)^x}{(0.10)^x}; \quad 2.0 = 2^y; y = 1$$

The order, y, with respect to Cl_2, is 1. This means that when the concentration of Cl_2 is doubled, the rate also doubles. Note that the rate constant and the concentration of NO cancel out because they are the same in Trials 1 and 2.

The rate law for the reaction is Rate = $k[NO]^2[Cl_2]$.

The overall order of a reaction is the sum of the reaction orders. For this example, the overall order is $2 + 1 = 3$.

Note: The order with respect to each reactant will not always be the same as the coefficients in the balanced equation.

To determine the value of the rate constant, including its units, use the rate law and experimental data from any given trial.

k = Rate / ($[NO]^2[Cl_2]$)

Using values from Trial 1:

= $(0.18 \text{ M s}^{-1}) / (0.10\text{M})^2(0.10\text{M}) = 180 \text{ L}^2/(\text{mol}^2 \text{ s})$.

It helps to work out the units separately.

$$\left(\frac{\text{mol}}{\text{L}\bullet\text{s}}\right)\left(\frac{\text{L}^2}{\text{mol}^2}\right)\left(\frac{\text{L}}{\text{mol}}\right) = \frac{\text{L}^2}{\text{mol}^2 \bullet \text{s}}$$

If the rate constant is known and the order with respect to one of the reactants has been determined, the order with respect to the other can be calculated even if its concentration is not held constant between any of two trials.

INTEGRATED RATE LAWS

(*Chemistry* 7th ed. pages 538–547 / 8th ed. pages 551–560)

An integrated rate law, derived from the differential rate law, expresses the reactant concentration as a function of time. The table below summarizes the integrated rate laws for the reaction

$$A \longrightarrow \text{Products}.$$

	Order		
	Zero	*First*	*Second*
Rate Law:	Rate = k	Rate = $k[A]$	Rate = $k[A]^2$
Integrated Rate Law:	$[A] = -kt + [A]_0$	$\ln[A] = -kt + \ln[A]_0$	$\dfrac{1}{[A]} = kt + \dfrac{1}{[A]_0}$
Plot Needed to Give a Straight Line	$[A]$ versus t	$\ln[A]$ versus t	$\dfrac{1}{[A]}$ versus t
Relationship of Rate Constant to the Slope of Straight Line:	Slope = $-k$	Slope = $-k$	Slope = k
Half-life:	$t_{1/2} = \dfrac{[A]_0}{2k}$	$t_{1/2} = \dfrac{0.693}{k}$	$t_{1/2} = \dfrac{1}{k[A]_0}$

You can determine the order of the reactant graphically if you know the concentration of A at various times, t, during the reaction. For example, if you plot $\ln[A]$ vs. t and obtain a straight line, the reaction is first order in A. If the graph is not linear, then the reaction is not first order. The table above summarizes what is graphed to test for the order of the reactant.

The half-life of a reactant, $t_{1/2}$, is the time required for a reactant to reach half of its original concentration. The general equations for the half-lives for each reactant order appear in the table above.

EXAMPLE: The rate of the reaction

$$NO_2(g) + CO(g) \longrightarrow NO(g) + CO_2(g)$$

depends only on the concentration of nitrogen dioxide below 225°C. At a temperature below 225°C, the following data were collected.

Time (s)	$[NO_2]$ (M)	ln $[NO_2]$	$1/[NO_2]$ (M^{-1})
0	0.500	−0.693	2.00
1.20×10^3	0.444	−0.812	2.25
3.00×10^3	0.381	−0.965	2.62
4.50×10^3	0.340	−1.079	2.94
9.00×10^3	0.250	−1.386	4.00
1.80×10^4	0.174	−1.749	5.75

Assume that the data are first order and see if the plot of $\ln[NO_2]$ vs. time is linear. If this isn't linear, try the second-order plot of $1/[NO_2]$ vs. time. The data and plots follow.

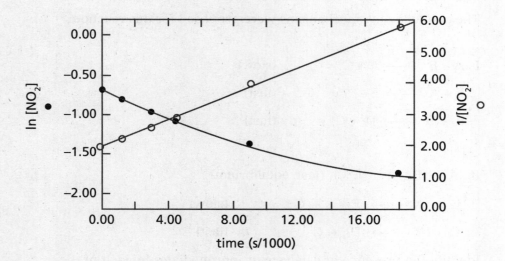

The plot of $1/[NO_2]$ vs. time is linear, so the slope of this line gives the value of k.

$$\text{Slope} = k = \frac{\Delta y}{\Delta x} = \frac{(5.75 - 2.00)\ M^{-1}}{(1.8 \times 10^4 - 0)\ s} = 2.08 \times 10^{-4}\ \text{L/mol s}$$

To determine $[NO_2]$ at 2.70×10^4 s, use the integrated rate law where $1/[NO_2]_o = 1/0.500\ M = 2.00\ M^{-1}$.

$$1/[NO_2] = kt + 1/[NO_2]_0$$

$$\frac{1}{[NO_2]} = \frac{2.08 \times 10^{-4} \text{ L}}{\text{mol s}} \times 2.70 \times 10^4 \text{ s} + 2.00 \text{ M}^{-1}$$

$$\frac{1}{[NO_2]} = 7.62, \ [NO_2] = 0.131 \text{ M}$$

MECHANISMS

(*Chemistry* 7th ed. pages 549–552 / 8th ed. pages 562–565)

A mechanism is a series of steps by which a reaction occurs. The elementary steps, each step in the mechanism, must add up to give the overall balanced equation for the mechanism. The slowest step in the mechanism must agree with the experimentally determined rate law.

Once the rate law has been determined experimentally, then chemists propose several mechanisms consistent with that rate law and design experiments to determine which matches the data. You will have to know how to identify catalysts and intermediates in a mechanism and write the overall equation for a mechanism. You will also need to be able to write a rate law from a given mechanism.

EXAMPLE: Consider the balanced equation for the hypothetical reaction

$$2A + 2B \longrightarrow 2C + D_2.$$

The experimentally determined rate law is Rate = $k[A]^2[B]$.

The two following mechanisms were proposed for this reaction:

I. $A + B \longrightarrow E + C$ (slow)

   $E + A \longrightarrow E_2$ (fast)

   $E_2 + B \longrightarrow D_2 + C$ (fast)

II. $A + B \rightleftharpoons E$ (fast, equilibrium)

   $E + A \longrightarrow E_2 + C$ (slow)

   $E_2 + B \longrightarrow D_2 + C$ (fast)

Identify the presence of catalysts or intermediates in each of the mechanisms.

A catalyst is present in the reactants and then appears in the products. Recall that the catalyst is not consumed in the reaction. The catalyst does not appear in the overall reaction. In the two mechanisms above, there are no catalysts. An example of a mechanism involving a catalyst involves the decomposition of ozone by atoms of chlorine.

$$Cl(g) + O_3(g) \longrightarrow ClO(g) + O_2(g)$$
$$O(g) + ClO(g) \longrightarrow Cl(g) + O_2(g)$$
$$\overline{O(g) + O_3(g) \longrightarrow 2O_2(g)}$$

An intermediate is produced in one elementary step and then consumed in the next. It also does not appear in the overall reaction for the mechanism because it cancels out. In both mechanisms, E is an intermediate. It is produced in the first step and consumed in the second.

Which of the following two mechanisms is consistent with the rate law? For the proposed mechanism to be correct, the overall reaction and the rate law for the mechanism must agree with the given reaction and the experimentally determined rate law.

Write the overall reaction for each mechanism. Add up the elementary steps. Cancel out the catalysts and intermediates that appear.

I. $A + B \longrightarrow E + C$ (slow)

   $E + A \longrightarrow E_2$ (fast)

   $\underline{E_2 + B \longrightarrow D_2 + C}$ (fast)

   $2A + 2B \longrightarrow 2C + D_2$ is the overall reaction.

II. $A + B \rightleftharpoons E$ (fast, with equal rates)

   $E + A \longrightarrow E_2 + C$ (slow)

   $\underline{E_2 + B \longrightarrow D_2 + C}$ (fast)

   $2A + 2B \longrightarrow 2C + D_2$ is the overall reaction.

The overall reaction for each mechanism matches the given reaction.

DETERMINE THE RATE LAW FROM THE MECHANISM

Begin by writing the rate law for the slow step of each mechanism.

The rate law for the slow step of mechanism I is Rate = $k[A][B]$ which is not consistent with the rate law determined by experiment. Mechanism I is not possible for this reaction.

The rate law for the slow step of mechanism II is Rate = $k[E][A]$.

[E] is an intermediate and may not be included in the rate law. Because the first step in the mechanism is reversible and fast, the rate of the forward reaction equals the rate of the reverse reaction.

Rate (forward) = $k_f[A][B]$

Rate (reverse) = $k_r[E]$

Rate (forward) = Rate(reverse), therefore $k_f[A][B] = k_r[E]$

$[E] = (k_f/k_r)\,[A][B] = k'\,[A][B]$

Substitute for [E] in the rate law for the slow step in mechanism II to get Rate = $kk'[A]^2[B]$ or $k''[A]^2[B]$ where k'' is the rate constant of the experimentally determined rate law.   Mechanism II is possible because its rate law agrees with the experimentally determined rate law.

DETERMINATION OF ACTIVATION ENERGY

(*Chemistry* 7th ed. pages 554–557 / 8th ed. pages 567–570)

Activation energy can be determined using the Arrhenius equation

$$\ln(k) = -\frac{E_a}{R}\left(\frac{1}{T}\right) + \ln(A).$$

The rate constant is represented by k.

E_a is the activation energy.

R is the gas constant, 8.3145 J/(molK).

T is the Kelvin temperature.

One way to determine the activation energy, E_a, is to measure the rate constant, k, at several different temperatures, and then graph $\ln(k)$ vs. $1/T$ which gives a straight line with the slope equal to $-E_a/R$.

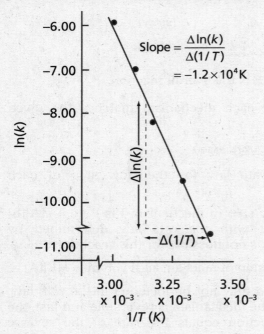

The activation energy can also be calculated from the values of k at only two temperatures using the equation:

$$\ln\frac{k_2}{k_1} = \frac{E_a}{R}\left(\frac{1}{T_1} - \frac{1}{T_2}\right)$$

EXAMPLE: A first order reaction has constants of 4.6×10^{-2} s^{-1} and 8.1×10^{-2} s^{-1} at 0°C and 20.°C, respectively. Calculate the value of the activation energy.

$$\ln\left(\frac{8.1 \times 10^{-2}\,\text{s}^{-1}}{4.6 \times 10^{-2}\,\text{s}^{-1}}\right) = \frac{E_a}{8.3145\ \text{J/mol}\bullet\text{K}}\left(\frac{1}{273\ \text{K}} - \frac{1}{293\ \text{K}}\right)$$

$0.57 = E_a/8.3145\ (2.5 \times 10^{-4})$

$E_a = 1.9 \times 10^4$ J/mol $= 19$ kJ/mol

MULTIPLE-CHOICE QUESTIONS

No calculators are to be used in this section.

Each of the following questions or incomplete statements has five suggested responses. Select the one which best answers the question or incomplete statement.

1. The reaction $H_2(g) + Cl_2(g) \longrightarrow 2HCl(g)$
   (A) has a high positive ΔG value and is therefore a very fast reaction.
   (B) has a high negative ΔG value and is therefore a very fast reaction.
   (C) has ΔG = zero indicating the rate of reaction is not great.
   (D) has a ΔG value which does not relate to the reaction rate.
   (E) has a positive ΔG value indicating a spontaneous reaction.

2. The rate constant
   (A) always shows an exponential increase with the Kelvin or absolute temperature.
   (B) increases with increasing concentration.
   (C) usually increases with increased pressure for gases.
   (D) never changes (it is a constant).
   (E) is the same for a given reaction at the same Kelvin or absolute temperature.

3. A reaction mechanism
   (A) is the sum of all steps in a reaction except the rate determining step.
   (B) has a H equal to the H for the most energy demanding step.
   (C) always has a rate determining step.
   (D) may be absolutely proven from the rate law.
   (E) is determined from the balanced expression only.

4. If both the H and E_a for this forward reaction are known, the reverse reaction would have an E_a
    (A) of $(-H \longrightarrow) + E_{a\rightarrow}$.
    (B) of $H \longrightarrow + E_{a\rightarrow}$.
    (C) equal to E_a for the forward reaction.
    (D) equal to $(-E_a)$ for the forward reaction.
    (E) but none of the above describe the value of E_a.

5. The rate of a chemical reaction
    (A) is always dependent on the concentration of all reactants.
    (B) is always increased with increasing temperatures.
    (C) is directly proportional to the value of E.
    (D) is greater with higher activation energies.
    (E) is independent of surface area.

6. The values for the change in enthalpy, ΔH, and the activation energy, E_a, for a given reaction are known. The value of E_a for the reverse reaction equals
    (A) E_a for the forward reaction.
    (B) $-(E_a)$ for the forward reaction.
    (C) the sum of $-\Delta H$ and E_a.
    (D) the sum of ΔH and E_a.
    (E) the difference of ΔH and E_a.

7. For two first-order reactions of different substances A and X

    $$A \longrightarrow B \quad t_{1/2} = 30.0 \text{ min}$$

    $$X \longrightarrow Y \quad t_{1/2} = 60.0 \text{ min}$$

    This means that
    (A) doubling the concentration of A will have 1/2 the effect on half-life that doubling the concentration of B will have on its half-life.
    (B) a certain number of grams of A will react twice as fast as the same number of grams of X.
    (C) a certain number of grams of X will react twice as fast as the same number of grams of A.
    (D) the rate constant for A $\longrightarrow$ B is lower than the rate constant of X $\longrightarrow$ Y.
    (E) 3 moles of A will react more rapidly than 3 moles of X.

8. A reaction is first order with respect to [X] and second order with respect to [Y]. When [X] is 0.20 M and [Y] = 0.20 M the rate is 8.00×10^{-3} M/min. The value of the rate constant, including correct units, is
    (A) 1.00 M min^{-1}.
    (B) 1.00 M^{-2} min^{-1}.
    (C) 2.00 M^{-1} min^{-1}.
    (D) 2.0 M^{-2} min^{-1}.
    (E) 8.00×10^{-3} min^{-3}.

9. The activation energy for this reaction, X + 2Y ⟶ 3Z, shown in the potential energy diagram, could be
   (A) increased by increasing [X].
   (B) increased by increasing [X] and [Y].
   (C) increased by increasing the temperature.
   (D) decreased by removing Z from the system as it forms.
   (E) decreased by adding a suitable catalyst.

10. For all zero-order reactions
    (A) a plot of time vs. concentration squared is linear.
    (B) E_a is very low.
    (C) the concentration of reactants is constant.
    (D) the rate is independent of time.
    (E) the rate constant is zero.

Use the following information to answer questions 11 through 13.

Carbon monoxide reacts with oxygen: $CO(g) + \frac{1}{2} O_2(g) \rightarrow CO_2(g)$

The rate law for the reaction is: Rate $= k[CO]^m[O_2]^n$

Experimental information for the reaction is given in the following table:

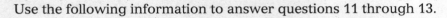

	[CO] mol/L	[O₂] mol/L	Initial rate, mol/L/min
Experiment 1	0.020	0.020	3.68×10^{-5}
Experiment 2	0.040	0.020	1.47×10^{-4}
Experiment 3	0.020	0.040	7.36×10^{-5}

11. Which values correspond to the reaction orders m and n for CO and O₂?
    (A) $m = 4, n = 2$
    (B) $m = 1, n = \frac{1}{2}$
    (C) $m = 1, n = 1$
    (D) $m = 2, n = 1$
    (E) $m = 2, n = 2$

12. What is the overall reaction order?
    (A) 0
    (B) 1
    (C) 2
    (D) 3
    (E) −1

13. What is the value for the rate constant k for this reaction?
    (A) 3.68×10^{-5} mol/L·min
    (B) 4.60 mol^{-2}L^{2}min^{-1}
    (C) 2.93×10^{-25} mol/min
    (D) 1.54×10^4 min^{-1}
    (E) 9.20×10^{-5} mol^{-2}L^2/min

14. Which of the following statements would be correct regarding the following reaction?
    $$2 H_2 (g) + O_2 (g) \rightarrow 2 H_2O (g)$$
    (A) The rate of O_2 disappearance is twice the rate of H_2 disappearance.
    (B) The rate of H_2 disappearance is twice the rate of O_2 disappearance.
    (C) The rate of H_2O disappearance is twice the rate of O_2 disappearance.
    (D) The rate of H_2O appearance is equal to the rate of O_2 disappearance.
    (E) The rate of H_2 disappearance is equal to the rate of O_2 disappearance.

15. According to the collision theory of kinetics, which statement best describes the rate of a chemical reaction?
    (A) All collisions result in a chemical reaction.
    (B) All collisions between molecules with at least a minimum kinetic energy result in reaction.
    (C) All collisions between molecules with at least a minimum kinetic energy and the proper orientation result in reaction.
    (D) The greater the difference in energy between the reactants and the products, the faster is the reaction.
    (E) The greater the difference in energy between the reactants and the transition state, the faster is the reaction.

FREE-RESPONSE QUESTIONS

Calculators may be used for this section.

1. The data below are for the reaction $2X + 2Y \longrightarrow Q_2 + 2Z$. The rate has been determined in terms of $[Q_2]$ formed for various concentrations of the two reactants. Temperature is held constant.

Exp.	Initial [X]	[Y]	Rate $\Delta[Q_2]/\Delta t$
1	0.20 M	0.10 M	0.048 M s^{-1}
2	0.40 M	0.10 M	0.19 M s^{-1}
3	0.40 M	0.20 M	0.38 M s^{-1}
4	0.60 M	0.60 M	?

(a) Determine the rate law for X + 2Y ⟶ Q$_2$ + 2Z based on the above information.
(b) Indicate the order of this reaction.
(c) Determine the value of the rate constant.
(d) Calculate the rate for Exp. 4.
(e) Determine the rate for Exp. 3 in terms of Δ[X]/Δt.

2. Three major methods used to increase the rate of a reaction are
(a) adding a catalyst,
(b) increasing the temperature, and
(c) increasing the concentration of a reactant.

From the perspective of collision theory, explain how each of these methods increases the reaction rate.

(d) For the same reaction run at two different temperatures, the rate constant is shown to change as:

Temp.	k
25°C	4.2×10^{-5}
37°C	2.6×10^{-4}

Determine the activation energy for this reaction from these data.

Answers

MULTIPLE CHOICE

1. **D** While it may seem counterintuitive, spontaneity [ΔG = (–)] refers to the tendency to react and not to the rate of reaction (*Chemistry* 7th ed. page 527 / 8th ed. page 540).

2. **E** In this question be sure to note that it is the rate constant that is involved. Refer to the Arrhenius equation for the specific relationship (*Chemistry* 7th ed. pages 552–555 / 8th ed. pages 565–568).

3. **C** In all series of steps (the mechanism for the overall reaction), there must always be a slowest step. The rate determining step is the slowest step (*Chemistry* 7th ed. page 550–551 / 8th ed. pages 563–564).

4. **A** This activation energy for the reverse reaction is difference between the potential energy of the products and the (top of the curve) energy of the activated complex, hence the sum of the reverse of the enthalpy change of the forward reaction plus the activation energy of the forward reaction. This can be understood by starting with the energy of the products and tracing the curve and

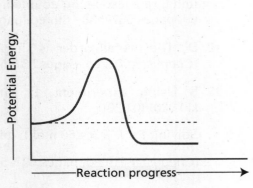

resulting energy change to the top of the curve (*Chemistry* 7th ed. page 553 / 8th ed. page 566).

5. **B** The collision model indicates that molecular collision is necessary for reaction. Because an increase in temperature raises molecular velocities and the percentage of effective collisions, the reaction rate increases (*Chemistry* 7th ed. pages 552–555 / 8th ed. pages 565–568).

6. **C** Examine the diagram that accompanies Question 4 in this chapter for a clear understanding (*Chemistry* 7th ed. pages 552–555 / 8th ed. pages 565–568).

7. **E** It takes half as much time for A to form B as for X to form Y, as seen by the smaller half-life. Note that option "B" would be incorrect as the grams of A and the grams of X are not the same number of moles (*Chemistry* 7th ed. pages 541–543 / 8th ed. pages 555–556).

8. **B** From these data, it follows that the rate law is Rate = $k[X][Y]^2$. Solving for the rate constant and substituting data for this reaction:
k = Rate / $[X][Y]^2$
= 8.00×10^{-3}M/min/$(0.200$M$)(0.200$M$)^2$
= 0.008 M/min/0.008 M^3
= 1.00 M^{-2}min^{-1}.
(*Chemistry* 7th ed. pages 535–541, 548 / 8th ed. pages 548–554, 562)

9. **E** Adding a catalyst suitable for this reaction will lower the energy barrier (activation energy) by forming a different activated complex which has a lower potential energy (*Chemistry* 7th ed. pages 552–559 / 8th ed. pages 565–572).

10. **D** For zero-order reactions, Rate = $k[X]^0$. Because anything raised to the zero power is equal to one, Rate = k. This is another way of saying that the rates of zero-order reactions do not change; they do not speed up and they do not slow down, they either take place or they do not (*Chemistry* 7th ed. page 546 / 8th ed. page 559).

11. **D** Compare experiments 1 and 2, doubling the concentration of CO has caused the initial rate to increase by a factor of 4 or 2^2. Therefore m = 2. And comparing experiments 1 and 3, doubling the concentration of O_2, keeping the [CO] the same, resulted in the initial rate also being doubled, or 2^1. Therefore n = 1 (*Chemistry* 7th ed. pages 537–538 / 8th ed. pages 549–551).

12. **D** The overall order is simply the sum of $m + n$ or 2 + 1 = 3 (*Chemistry* 7th ed. pages 537–538 / 8th ed. pages 549–551).

13. **B** Using experiment 1, rate = $k[CO]^2[O_2]$; 3.68×10^{-5} = $k(0.020)^2(0.020)$

Solving for k, k = 4.60 mol^{2}L^{2}min^{-1}

(*Chemistry* 7th ed. pages 537–538 / 8th ed. pages 549–551)

14. **B** According to the balanced equation 2 moles of H_2 are used up for every mole of O_2, so during the same time period, the rate of disappearance of the H_2 would be twice as great (*Chemistry* 7th ed. 527–532 / 8th ed. pages 540–545).

15. **C** Collision theory states that the molecules colliding must not only have the minimum combined activation energy required for reaction, they must also have the proper spatial orientation when they collide (*Chemistry* 7th ed. pages 552–555 / 8th ed. pages 565–568).

FREE RESPONSE

1. (a) To determine how each of the concentrations of the reactants is related to the reaction rate, only the one reactant concentration may change. For example, as we go from Exp. 1 to Exp. 2, only [X] is changing. As it doubles, the reaction rate becomes four times greater, suggesting that rate is directly related to the square of the concentration of X. In like manner, compare the concentration changes from Exp. 2 to 3: [X] is held constant, [Y] is doubled, and the rate doubles. This suggests that rate is proportional to [Y]. Therefore, the rate law is Rate = $k[X]^2[Y]$.

   (b) From the rate law, we can say that this reaction is second order with respect to X, first order with respect to Y, and third order (2+1) overall.

   (c) Solving the rate law for k yields:

   k = Rate / ([X]^2[Y])

   = 0.048 M s^{-1} / (0.20M)^2(0.10M)

   = 12 M^{-2} s^{-1}.

   (d) Rate = 12 M^{-2} s^{-1} (0.60 M)^2 (0.60M) = 2.6 M s^{-1}.

   (e) Note that every time a mole of Q is formed, two moles of X disappear, so
   0.38 M s^{-1} × 2 = 0.76 M s^{-1}.

   (*Chemistry* 7th ed. pages 537–546 / 8th ed. pages 550–559)

2. (a) Adding a suitable catalyst lowers the energy demands of the reaction (the activation energy) by providing a less demanding pathway. It does this by forming a different activated complex with a lower activation energy.

   (b) Increasing the temperature increases the average kinetic energy of the molecules. However, what is most important here is that the fraction of molecules that have the energy necessary to react (the activation energy) is increased.

   (c) Increasing the concentration of a reactant increases the probability of collisions, and therefore increases the possibility of reaction, for the particles must collide to react. Note that not all collisions result in successful reactions.

(d) The activation from k values taken at different temperatures:

$$\ln\frac{k_2}{k_1} = \frac{E_a}{R}\left[\frac{(T_2 - T)_1}{(T_2 T_1)}\right];$$

the form of the equation given on the formula sheet for the AP exam is

$$\ln k = -\frac{E_a}{R}\left(\frac{1}{T}\right) + \ln A$$

$$\ln\left(\frac{2.6\times10^{-4}}{4.2\times10^{-5}}\right) = \frac{E_a}{8.314 \text{ J/molK}}\left(\frac{310 - 298}{310 \times 298}\right)$$

$$E_a = \frac{8.314 \times 1.82}{1.30\times10^{-4}} = 120{,}000 \text{ J} = 120 \text{ kJ}$$

Note: Because the two-point form of this equation is not given on the equation sheet, it is more likely that you will be given a graph and asked what the slope represents or asked to label the axes (*Chemistry* 7th ed. pages 552–557 / 8th ed. pages 565–572).

12

EQUILIBRIUM

In this chapter, you will review the characteristics of equilibrium and calculations involving the concentrations of reactants and products for a given system at equilibrium. Your knowledge of the basic equilibrium problems in this chapter will enable you to solve for the pH of acids, bases, salts, and buffers as well as the solubility of ionic compounds.

AP tip

In Section II of the AP Exam, the first question will always be an equilibrium problem, worth 20% of the free-response score. It can be one of the types of problems found in any of the three equilibrium chapters in this guide.

You should be able to

- Write equilibrium expressions for a given reaction.
- Calculate Q and compare it to K to determine if a reaction is at equilibrium.
- Manipulate K if a reaction is reversed or multiplied by a coefficient.
- Calculate K from given equilibrium concentrations, or if given K and all except one equilibrium concentration, solve for the missing value.
- Calculate equilibrium concentrations (or one of the missing variables) if given any two of the following values: K, the initial concentrations, one equilibrium concentration.
- Do calculations involving gaseous equilibria, partial pressures, and K_p.
- Use Le Châtelier's principle to determine in what direction the position of equilibrium will shift when a change is imposed.

■ Calculate the value of K from thermodynamic values such as $\Delta G°$.

EQUILIBRIUM CONDITION

Many reactions are reversible. When equilibrium is reached, the rate of the forward reaction equals the rate of the reverse reaction. The ratio of the product concentrations to the reactant concentrations is constant at equilibrium.

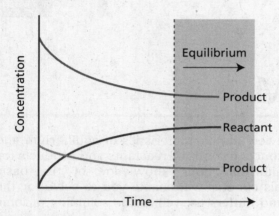

THE EQUILIBRIUM EXPRESSION

(*Chemistry* 7th ed. pages 582–583 / 8th ed. pages 597–598)

The equilibrium constant, K, measures the ratio of product concentrations to reactant concentrations.

For a hypothetical reaction $a\text{A} + b\text{B} \rightleftharpoons c\text{C} + d\text{D}$

$$K = \frac{[\text{A}]^a [\text{B}]^b}{[\text{C}]^c [\text{D}]^d}.$$

K is the equilibrium constant, usually given without units. [ ] represents the concentration of the reactants and products at equilibrium in moles/liter. The concentrations of products and reactants are raised to the powers of their respective coefficients in the balanced chemical equation.

You will be expected to write the equilibrium expression for a given reaction. Remember, the concentrations of solids and liquids are not included.

EXAMPLE: Write the equilibrium expression for the reaction

$$\text{P}_4(s) + 5\text{O}_2(g) \rightleftharpoons \text{P}_4\text{O}_{10}(s).$$

SOLUTION: The equilibrium expression for the reaction is

$$K = \frac{1}{[\text{O}_2]^5}.$$

The concentrations of the solids are not included in the expression because the concentration of a solid or liquid is a constant, so it is included in the value of K.

The value of the equilibrium constant measures the extent to which a reaction occurs.

$K > 1$: The concentrations of the products are greater than the concentrations of the reactants.

$K < 1$: The concentrations of the reactants are greater than the concentrations of the products. The reaction hardly proceeds toward completion.

REACTION QUOTIENT

(*Chemistry* 7th ed. page 593 / 8th ed. pages 608–609)

When reactants and products are mixed together, they may not be at equilibrium. The reaction quotient, Q, compared to the equilibrium constant, K, will determine which way a system will shift to reach equilibrium. If a system is not at equilibrium, it will move in a direction to reach equilibrium.

If a reactant or product in a reaction is not present and is at zero concentration, the reaction will move in the direction that produces the missing component.

If all reactants and products are present and have an initial concentration, you must determine the value of Q, the reaction quotient.

To determine if the reaction is at equilibrium or the direction it will shift to attain equilibrium, plug all of the initial concentrations into the reaction quotient, which is the same as the equilibrium expression (also called the law of mass action), and compare the value of Q to K.

Comparison of Q to K

If $Q = K$, the reaction is at equilibrium.

If $Q > K$, the reaction will shift to the left. A shift toward the reactants will consume products.

If $Q < K$, the reaction will shift to the right. A shift to produce more products will consume reactants.

EXAMPLE: For the reaction $2NO(g) \rightleftharpoons N_2(g) + O_2(g)$ $K = 2.4 \times 10^3$ at a particular temperature.

(a)   If the initial concentrations are 0.024 M NO, 2.0 M N_2, and 2.6 mol O_2, is the system at equilibrium?

(b)   If it is not at equilibrium, in which direction will the reaction shift?

SOLUTION: The reaction is not at equilibrium, $Q > K$, so it will shift to the left.

$$Q = \frac{[N_2][O_2]}{[NO]^2} = \frac{(2.0)(2.6)}{(0.024)^2} = 9.0 \times 10^3$$

TYPES OF EQUILIBRIUM PROBLEMS

(*Chemistry* 7th ed. pages 583, 596–604 / 8th ed. pages 598, 612–620)

The types of equilibrium problems you can expect to solve in this chapter and the two chapters in this book which follow include:
1. Manipulation of the equilibrium constant, K
    a. If the reaction is reversed, the equilibrium expression is the reciprocal of the expression for the forward reaction.
    EXAMPLE:

$$A \rightleftharpoons B: K = \frac{[B]}{[A]}$$

$$B \rightleftharpoons A: K' = \frac{1}{K} = \frac{[A]}{[B]}$$

    b. If the coefficients in a balanced equation are multiplied by a number, n, the equilibrium constant is raised to the power n.
    EXAMPLE:

$$A \rightleftharpoons B: K = \frac{[B]}{[A]}$$

$$2A \rightleftharpoons 2B: K'' = K^2 = \frac{[B]^2}{[A]^2}$$

2. If given all equilibrium concentrations, calculate the value of the constant, K.
3. If given the value of the equilibrium constant, K, and all but one of the equilibrium concentrations, solve for the missing concentration.
4. If given the value of the initial concentrations of the reactants and one of the equilibrium concentrations of either the reactants or products, solve for all equilibrium concentrations and the value of K.
5. If given the initial concentrations and the value of K, solve by approximation for the equilibrium concentrations.

The first four types of problems listed above can be solved using basic algebra. The fifth type of problem can be solved using the method of approximation.

EXAMPLE: (Type 1) For the reaction

$N_2(g) + 3H_2(g) \longrightarrow 2NH_3(g)$, $K = 1.3 \times 10^{-2}$ at a certain temperature.

Calculate the value of K, called K', for the reaction

$NH_3(g) \longrightarrow \frac{1}{2} N_2(g) + \frac{3}{2} H_2(g)$.

SOLUTION: The reaction is the reverse and one half of the one which is given.

$$K' = [N_2]^{1/2}[H_2]^{3/2} / [NH_3] = (1/K)^{1/2}; \left(\frac{1}{1.3 \times 10^{-2}}\right)^{\frac{1}{2}} = 8.8.$$

EXAMPLE: (Type 4) At a particular temperature, $K = 1.00 \times 10^2$ for the reaction

$$H_2(g) + I_2(g) \rightleftharpoons 2HI(g).$$

In an experiment, 1.00 mol H_2, 1.00 mol I_2, and 1.00 mol HI are introduced into a 1.00-L container. Calculate the equilibrium concentrations of all reactions and products.

SOLUTION: To begin, write the balanced equation for the reaction and the equilibrium expression, omitting pure solids and liquids.

	$H_2(g)$ +	$I_2(g)$	$\rightleftharpoons$	$2HI(g)$	$K = \dfrac{[HI]^2}{[H_2][I_2]}$
I:	1.00	1.00		1.00	
C:	$-x$	$-x$		$+2x$	
E:	$1.00 - x$	$1.00 - x$		$1.00 + 2x$	

Make an ICE chart under the balanced chemical equation.

I = initial concentration in mol/L (note units are omitted from chart)

C = the change to reach equilibrium represented by + or $-x$. A minus sign indicates a decrease in concentration, a plus sign indicates an increase. The coefficient in front of the reactant in the balanced equation is placed in front of the x in the change line.

In this example, $Q = 1$ which is less than K, so the reaction will shift to the right.

E = Equilibrium concentrations which are obtained by adding the I and C lines together.

Plug the equilibrium values into the expression and solve for x:

$$K = 100. = \frac{(1.00 + 2x)^2}{(1.00 - x)^2}.$$

Taking the square root of both sides:

$$10.0 = \frac{1.00 + 2x}{1.00 - x}, \ 10.0 - 10.0\,x = 1.00 + 2x, \ 12x = 9.0, \ x = 0.75 \text{ M}.$$

Use this value of x to solve for the equilibrium concentrations of all reactants and products.

$[H_2] = [I_2] = 1.00 - 0.75 = 0.25$ M; $[HI] = 1.00 + 2 (0.75) = 2.50$ M

Last, check your equilibrium concentrations by making sure that they equal the correct value of K.

EXAMPLE: (Type 5) For the reaction $N_2O_4(g) \rightleftharpoons 2NO_2(g)$,

$K = 4.0 \times 10^{-7}$ at a specific temperature.

In an experiment, 1.0 mol of N_2O_4 is placed in a 10.0-L vessel.

Calculate the equilibrium concentrations of NO_2 and N_2O_4.

SOLUTION: At equilibrium, $[N_2O_4] = 0.10$ M and $[NO_2] = 2.0 \times 10^{-4}$ M. To solve this problem, proceed as in the previous example.

	$N_2O_4(g)$	$\rightleftharpoons$	$2NO_2(g)$
I	0.10		0
C	$-x$		$+2x$
E	$0.10 - x$		$2x$

$$K = \frac{[NO_2]^2}{[N_2O_4]} = \frac{(2x)^2}{(0.10 - x)} = 4.0 \times 10^{-7}$$

Since the value of K is much smaller than 1, you can assume that the change from the initial concentration, x in $0.10 - x$, is so small that it is negligible, that is, $0.10 - x$ is about equal to 0.10.

This greatly simplifies the math to

$$K = \frac{(2x)^2}{(0.10)} = 4.0 \times 10^{-7}$$

$4x^2 = 4.0 \times 10^{-8}$, $x = 1.0 \times 10^{-4}$ M

It is okay to make this assumption if the change from the initial concentration, in this case, x is less than 5% of the initial concentration.

$$\frac{x}{0.10} \times 100\% = \frac{1.0 \times 10^{-4}}{0.10} \times 100\% = 0.10\%$$

This is less than 5%, so, it is okay to make the assumption.

The equilibrium concentrations are as follows:

$[N_2O_4] = 0.10 - x = 0.10 - 1.0 \times 10^{-4} = 0.10$ M

$[NO_2] = 2x = 2(1.0 \times 10^{-4}) = 2.0 \times 10^{-4}$ M

GASEOUS EQUILIBRIUM

(*Chemistry* 7th ed. pages 586–588, 594–596 / 8th ed. pages 601–604, 610–612)

For equilibrium in the gas phase, the equilibrium expression can be written in terms of the partial pressures of the gases.

For the reaction $AsH_3(g) \rightleftharpoons 2As(s) + 3 H_2(g)$

$$K_p = \frac{\left(P_{H_2}\right)^3}{P_{AsH_3}}$$

K_p is the equilibrium constant in terms of the partial pressures of the gases.

P represents the partial pressure of the gases raised to their coefficients in the balanced chemical equation.

EXAMPLE: Given the following reaction:

$2NO(g) + Br_2(g) \rightleftharpoons 2NOBr(g)$ $K_p = 109$ at 25°C.

The equilibrium partial pressure of $Br_2 = 0.0159$ atm and

NOBr = 0.0768 atm. Calculate the equilibrium partial pressure of NO.

$$Kp = \frac{\left(P_{NOBr}\right)^2}{\left(P_{NO}\right)^2 \left(P_{Br_2}\right)}$$

$$109 \text{ atm}^{-1} = \frac{(.0768 \text{ atm})^2}{\left(P_{NO}\right)^2 (0.0159 \text{ atm})}$$

$$P_{NO} = 0.0583 \text{ atm}$$

The relationship between K_p and K is given by $K_p = K(RT)^{\Delta n}$.

K_p is the equilibrium gas constant.

Δn is the sum of the coefficients of the gaseous products minus the sum of the coefficients of the gaseous reactants.

EXAMPLE: $2NO(g) + Cl_2(g) \rightleftharpoons 2NOCl(g)$ $K_p = 1.9 \times 10^3$ at 25°C.

Calculate the value for K_c at 25°C.

$\Delta n = 2 - (2+1) = -1$

$K_p = K(RT)^{-1} = K/RT$

$K = K_p RT = (1.9 \times 10^3)(0.08206)(298) = 4.6 \times 10^4$.

LECHÂTELIER'S PRINCIPLE

(*Chemistry* 7th ed. pages 604–610 / 8th ed. pages 620–625)

The value of K is a constant at a particular temperature. The only factor that changes the value of K is temperature. Pressure, a catalyst, and changes in concentration will not affect the value of K.

LeChâtelier's principle states that if a change is imposed in a system at equilibrium, the position of the equilibrium will shift in a direction that will counteract the change.

EXAMPLE: Consider the following changes on the system below at equilibrium.

$$2\ SO_3\ (g) \rightleftharpoons 2\ SO_2(g) +\ O_2(g)\quad \Delta H = 197\ kJ.$$

(a) Addition of O_2

The equilibrium will shift to the left to form more reactants. The addition of oxygen increases the rate of the reverse reaction so more reactants will form, until the rate of the forward reaction again equals the rate of the reverse reaction. Another way to explain the shift is that $Q > K$ so that must decrease the products and increase reactants for $Q = K$ again.

(b) Removal of SO_2

Removal of an equilibrium component such as SO_2 at constant pressure and temperature will cause the equilibrium to shift toward the removed component to increase its concentration. The reaction in this example will shift to the right.

(c) Increase in temperature

When you consider the effect of temperature change on the equilibrium constant, add the heat of the reaction to the reactant side in an endothermic reaction or to the product side when the reaction is exothermic. The direction of the shift can be predicted in the same way as the addition or removal of a reactant or product.

In the reaction for the decomposition of SO_3, the reaction is endothermic. Treating heat as a reactant, an increase in temperature will cause the reaction to shift right, producing more products, increasing the value of K. A decrease in temperature will cause a shift to the left, increasing the reactant concentrations, and lowering the value of K.

(d) An increase in pressure.

If the pressure is increased in an equilibrium system, the reaction will shift toward the side with fewer moles of gas. Decreasing the volume has the same effect because the only way the pressure can be increased without changing the temperature or number of moles is to decrease the volume. In this example, the reaction will shift to the left, toward SO_3, when the pressure is increased or the volume is decreased. If the moles of gas are the same on both sides of the reaction, no shift will occur.

e) Addition of a solid or inert gas, such as Ne.

If a solid or an inert gas (with no change in volume) is added to the reaction, there will be no shift in equilibrium. Neither the solid nor the inert gas is part of the equilibrium expression. The concentrations of all the components of the equilibrium expression remain unchanged.

f) An inert gas, such as Ne, is added at constant pressure.

There will be a shift in the equilibrium to the side of the equation with more moles of gas. To add Ne at constant pressure, the volume of the container must increase so the concentrations or partial pressures of all gases have decreased. If the moles of gas are greater in the reactants, then $Q < K$, so equilibrium can be reestablished only by increasing the products.

RELATIONSHIP OF K TO FREE ENERGY, ΔG^O

(*Chemistry* 7th ed. pages 774–778 / 8th ed. pages 798–802)

The equilibrium constant at standard temperature can be determined from the thermodynamic values of a reaction using the equation:

$\Delta G^0 = -RT \ln K$

$R = 8.314$ J/(mol•K)

EXAMPLE: Calculate the value of K at 25.0°C for the reaction $2NO_2(g) \rightleftharpoons N_2O_4(g)$

The values of $\Delta H°$ and $\Delta S°$ are –58.03 kJ/mol and –176.6 J/K•mol, respectively.

SOLUTION:

At 25°C, $\Delta G° = \Delta H° - T \Delta S°$

$\qquad = -58.03 \times 10^3$ J/mol $- (298$ K$)(-176.6$ J/(K•mol)$)$

$\qquad = -5.40 \times 10^3$ J/mol.

$\Delta G^0 = -RT \ln K;\ \ln K = -\Delta G/RT = 2.18\ ;\ K = 8.8.$

MULTIPLE-CHOICE QUESTIONS

No calculators are to be used in this section.

1. In the reaction $3W + X \rightleftharpoons 2Y + Z$, all substances are gases. The reaction is initiated by adding an equal number of moles of W and of X. When equilibrium is reached,
    (A) [Y] = [Z].
    (B) [X] = [Y].
    (C) [W] =[X].
    (D) [X] > [W].
    (E) [W] + [X] = [Y] + [Z].

2. Consider Rx.I   X $\rightleftharpoons$ Y $K = 1 \times 10^8$

    Rx.II Z $\rightleftharpoons$ Y $K = 1 \times 10^5$

    both at the same temperature:
    (A) I is 3 times faster than II.
    (B) I is 1000 times faster that II.
    (C) II is 3 times faster than I.
    (D) II is 1000 times faster than I.
    (E) The size of K and the time required to reach equilibrium are
        not directly related.

3. The reaction $3H_2(g) + N_2(g) \rightleftharpoons 2NH_3(g)$ has an enthalpy of

    change of –92 kJ. Increasing the temperature of this equilibrium
    system causes
    (A) an increase in $[NH_3]$.
    (B) an increase in $[N_2]$.
    (C) a decrease in $[H_2]$.
    (D) an increase in K.
    (E) a decrease in pressure at constant volume.

4. Consider $N_2(g) + O_2(g) \rightleftharpoons 2NO(g)$. The reaction was initiated

    by adding 15.0 moles of NO to a 1.0-L flask. At equilibrium, 3.0
    moles of oxygen are present in the 1.0-L flask. The value of K must
    be
    (A) 0.33.
    (B) 3.0.
    (C) 5.0.
    (D) 9.0.
    (E) 81.

5. At a certain temperature, the synthesis of ammonia gas from

    nitrogen and hydrogen gases, shown as $N_2 + 3 H_2 \rightleftharpoons 2 NH_3$, has

    a value for K of 3.0×10^{-2}. If $[H_2] = [N_2] = 0.10$ M and $[NH_3] = 0.20$ M,
    (A) the reaction would shift toward the ammonia.
    (B) the reaction would shift toward the N_2 and the H_2.
    (C) the system is at equilibrium, therefore no shifting will occur.
    (D) the reaction will shift toward a new equilibrium position, but
        the direction cannot be determined from these data.
    (E) the equilibrium may shift but it is not possible to calculate Q
        without knowing the temperature.

6. The equilibrium constant, K, may be used to determine the K of other reactions.

   I. When the equation is reversed the reciprocal of the original K becomes the value of K for the new equation.

   II. When the equation is doubled the square root of the original K becomes the value for the new equation.

   III. When the equation for a reaction is tripled the equilibrium expression for the new equation is triple the original K.

   Of the above three statements, those which are always true are
   (A) I only.
   (B) II only.
   (C) III only.
   (D) I and III.
   (E) I, II, and III.

7. At a certain temperature, it has been determined that $K = 8.0$ for $H_2O(g) + CO(g) \rightleftharpoons CO_2(g) + H_2(g)$. If we also determine that the equilibrium mixture contains 0.80 mol of $H_2O(g)$, 0.080 mole of $CO_2(g)$, and 0.080 mole of $CO(g)$, in an 8.0-L flask, what must be the number of moles of $H_2(g)$ at equilibrium?
   (A) 0.010 mole.
   (B) 0.80 mole.
   (C) 6.4 moles.
   (D) 8.0 moles.
   (E) 64 moles.

8. Which of the following systems at equilibrium are not affected by a change in pressure caused by changing the volume at constant temperature?
   (A) $H_2(g) + Cl_2(g) \rightleftharpoons 2HCl(g)$

   (B) $H_2(g) + I_2(s) \rightleftharpoons 2HI(g)$

   (C) $N_2(g) + 3H_2(g) \rightleftharpoons 2NH_3(g)$

   (D) $2NH_3(g) \rightleftharpoons N_2(g) + 3H_2(g)$

   (E) $3O_2(g) \rightleftharpoons 2O_3(g)$

9. The equilibrium $P_4(g) + 6Cl_2(g) \rightleftharpoons 4PCl_3(l)$ is established at $-10°C$.

   The equilibrium constant expression is

   (A) $K = \dfrac{[PCl_3]}{[P_4][Cl_2]}$.

   (B) $K = \dfrac{[PCl_3]^4}{[P_4][Cl_2]^6}$.

   (C) $K = \dfrac{[P_4][Cl_2]^6}{[PCl_3]^4}$.

   (D) $K = \dfrac{1}{[PCl_3]^4}$.

   (E) $K = \dfrac{1}{[P_4][Cl_2]^6}$.

10. Ammonium hydrogen sulfide will decompose into ammonia gas and hydrogen sulfide gas when heated. Consider the equilibrium system

$$NH_4HS(s) \rightleftharpoons NH_3(g) + H_2S(g),$$

   which is developed from 1.000 mole of NH_4HS in a 100-L cylinder. At equilibrium the total pressure is found to be 0.400 atm. K_p will be equal to
   (A) 2.00×10^{-1}.
   (B) 1.00×10^{-2}.
   (C) 4.00×10^{-2}.
   (D) 4.00.
   (E) a value impossible to calculate from only these data.

11. Which is **not** correct concerning equilibrium?
   (A) The concentration of reactants and products are no longer changing.
   (B) The rates of the forward and reverse reactions are the same.
   (C) Either the reactants or the products of a reaction could be used to attain equilibrium for a reversible reaction.
   (D) At equilibrium, the rates of the forward and reverse reaction become zero.
   (E) The ratio of the concentrations of products and reactants, raised to the appropriate powers, is a constant.

12. The position of equilibrium for the reaction

$$ZnO(s) + H_2(g) \rightleftharpoons Zn(s) + H_2O(g)$$

does **not** depend on which of the following:
   I. Concentration of $ZnO(s)$
   II. Concentration of $H_2(g)$
   III Concentration of $Zn(s)$
   IV. Concentration of $H_2O(g)$
   V. The value of K
(A) I, II, and V
(B) II, III and V
(C) I and III
(D) II and IV
(E) V only.

13. For which of the following values of K will the equilibrium mixture consist almost entirely of reactants:
(A) 0.030
(B) 1.00
(C) 1×10^{-10}
(D) 30
(E) 4×10^8

14. $K = 0.25$ for $2\,NOBr(g) \rightleftharpoons 2NO(g) + Br_2(g)$. At the same T and P, what is the K for $NO + \frac{1}{2}\,Br_2 \rightleftharpoons NOBr$?
(A) 2.0
(B) 4.0
(C) 0.50
(D) 0.63
(E) 1.0

15. Ammonia and oxygen react to establish the following equilibrium:

$$4NH_3(g) + 3O_2(g) \rightleftharpoons 2\,N_2(g) + 6\,H_2O(g)$$

If a one liter flask is filled with 4.0 mol of oxygen and 3.0 mol of ammonia and the system is allowed to come to equilibrium, the flask is found to contain 1.0 mol of nitrogen. How much oxygen is present at equilibrium?
(A) 0.50 mol
(B) 1.00 mol
(C) 1.50 mol
(D) 3.00 mol
(E) 2.50 mol

FREE-RESPONSE QUESTIONS

1. Consider the reaction $2HI(g) \rightleftharpoons H_2(g) + I_2(s)$ which is in equilibrium.
   (a). How will adding more hydrogen gas at constant volume affect the equilibrium?

    (b) Some of the iodine is removed. How will this affect the equilibrium?

    (c) The pressure is increased by pumping in pure neon gas. How will this affect the equilibrium? Explain.

    (d) The pressure is increased by decreasing the volume. How will this affect the equilibrium? Explain.

    (e) A suitable catalyst is added. How will this affect the equilibrium?

2. Refer to the system $PCl_3(g) + Cl_2(g) \rightleftharpoons PCl_5(g)$. To an empty 15.0-L cylinder, 0.500 moles of gaseous PCl_5 are added and allowed to reach equilibrium. The concentration of PCl_3 is found to be 0.0220M. Assume a temperature of 375 K.

    (a) How many moles of PCl_5 remain at equilibrium?

    (b) Write the equilibrium constant expression for the above reaction.

    (c) Determine the value of K.

    (d) Determine the value of K_p for this same system at the same temperature.

    (e) How would the value of K_p be effected by increasing the temperature of the system at equilibrium for this exothermic reaction?

Answers

MULTIPLE CHOICE

1. **D** Since every time a mole of X reacts, 3 moles of W must react, so the amount of W remaining must be less than the amount of X remaining (recall that you started with an equal number of moles of W and of X) (*Chemistry* 7th ed. pages 579–582 / 8th ed. pages 594–597).

2. **E** There is not a direct relationship between K and reaction rate (*Chemistry* 7th ed. page 593 / 8th ed. pages 608–609).

3. **B** Increasing the temperature causes the equilibrium to shift to the left. For an exothermic reaction, increasing the temperature increases the rate of both the forward and reverse reactions, but proportionally makes a greater increase in the reverse reaction since it has the higher activation energy, favoring the formation of more hydrogen gas and more nitrogen gas, and lowering the concentration of the ammonia. This forms more gaseous particles; therefore, the pressure increases at constant volume (*Chemistry* 7th ed. pages 609–610 / 8th ed. pages 624–626).

4. **D** If 3.0 moles of oxygen are formed, 6.0 moles of NO must have reacted, leaving 9.0 moles of NO at equilibrium (15.0 – 6.0 = 9.0 mol/L for [NO]). Each time 3.0 moles of oxygen form, the same number of moles of nitrogen are produced. Since the reaction (take

care here) is written showing NO as a product, the equilibrium constant expression is $K = \dfrac{[NO]^2}{[N_2][O_2]} = \dfrac{(9.0)^2}{(3.0)(3.0)} = \dfrac{81}{9.0} = 9.0$.

(*Chemistry* 7th ed. pages 596–601 / 8th ed. pages 612–617)

5. **B**

$$Q = \frac{[NH_3]^2}{[H_2]^3[N_2]} = \frac{(0.20)^2}{(0.10)^3(0.10)} = 400$$

Since Q is greater than K, the reaction will shift toward N_2 and H_2. Note that this setup for Q is easily handled if seen as

$$\frac{(2\times10^{-1})^2}{(1\times10^{-1})^3(10^{-1})} = \frac{4\times10^{-2}}{1\times10^{-4}} = 4\times10^2.$$

(*Chemistry* 7th ed. pages 593–594 / 8th ed. pages 608–610)

6. **A** Statement II is not valid; when the equation is doubled the K value becomes the square of the original K, ($K' = K^2$). When the equation is tripled the value for K for the new equation is the cube of the original K ($K' = K^3$) (*Chemistry* 7th ed. page 584 / 8th ed. page 599).

7. **C**

$$K = \frac{[CO_2][H_2]}{[H_2][CO]} = 8.0 = \frac{\left(\dfrac{0.080}{8.0}\right)(X)}{\left(\dfrac{0.80}{8.0}\right)\left(\dfrac{.080}{8.0}\right)}$$

$$\frac{X}{0.10} = 8.0$$

$$X = 0.80M$$

$$0.80\,\frac{mol}{L} \times 8.0L = 6.4 \text{ mole of } H_2$$

(*Chemistry* 7th ed. pages 596–599 / 8th ed. pages 612–615)

8. **A** Because you are seeking an equilibrium system which has not changed with a change in pressure due to a volume change, look for a system with an equal number of moles of both gaseous reactants and products (*Chemistry* 7th ed. pages 606–608 / 8th ed. pages 621–624).

9. **E** The equilibrium constant is a ratio of the concentration of products divided by the concentration of reactants, each taken to a power represented by their coefficients. Pure liquids and solids are not shown in the equilibrium constant expression (*Chemistry* 7th ed. pages 582–583, 588–590 / 8th ed. pages 597–599, 603-606).

10. **C** The two gases are formed in equal molar amounts (1:1); therefore half of the pressure is due to each gas

$\left(\dfrac{0.400 \text{ atm}}{2} = 0.200 \text{ atm}\right)$. $K_p = \left(P_{NH_3}\right)\left(P_{H_2S}\right) = (0.200) \times 0.200) =$ 0.0400 $= 4.00 \times 10^{-2}$ (*Chemistry* 7th ed. pages 587–590 / 8th ed. pages 602–606).

11. **D** The rate of forward and reverse reactions become equal, or what is termed, a dynamic equilibrium on the microscopic or molecular level. The rate of forward and reverse reactions becomes equal (*Chemistry* 7th ed. pages 579–582 / 8th ed. pages 594–597).

12. **C** Pure solids do not affect the position of an equilibrium (*Chemistry* 7th ed. pages 588–590 / 8th ed. pages 604–606).

13. **C** A small value for K indicates that in the ratio of products to reactants, there are considerably more reactants than products, resulting in a very small number significantly less than one (*Chemistry* 7th ed. pages 592–593 / 8th ed. page 608).

14. **A** First, recognize that the reaction requested is the reverse of the one for which the K is given. The value of K for the reversed equilibrium reaction is the reciprocal of K or $1/K$, which is $1/0.25 = 4$. The new reaction as written is then multiplied by a factor of ½, therefore the equilibrium expression for the new reaction is the original K raised to the ½ power or in this case, the square root of 4, which is 2.0 (*Chemistry* 7th ed. pages 582–585 / 8th ed. pages 598–600).

15. **E** Work the problem using the ICE method as follows:

	$4NH_3(g)$	$+ 3O_2(g)$	$\rightleftharpoons 2\,N_2(g)$	$+ 6\,H_2O(g)$
initial	3.0	4.0	0	0
change	$-1.00 \times (4/2)$	$-1 \times (3/2)$	$+1.0$	$+1.0 \times (6/2)$
equilibrium	1.0	2.50	1.0	3.0

(*Chemistry* 7th ed. pages 591–599 / 8th ed. pages 606–615)

FREE RESPONSE

1. (a) Adding hydrogen causes the equilibrium to shift toward the reactant forming more HI; this is due to the increase in [H_2], which means more collisions between H_2 and I_2, therefore more reactions forming HI. $Q > K$ will cause more reactants to form.

   (b) There is no effect on equilibrium since the concentration of iodine, a pure solid, is constant.

   (c) There is no effect. The neon gas will not react with any product or reactant in this system. Since the volume does not change, there is no change in the concentrations of the gases.

   (d) Note that there are two moles of gas forming one mole of gas. Increasing the pressure by decreasing the volume ($Q < K$)

causes the reaction to shift toward fewer moles of gas, (toward the products in this case), thereby opposing the pressure change by lowering this pressure somewhat.

(e) There is no effect from adding a catalyst once the system is in equilibrium. (The addition of a catalyst will may cause the system to reach equilibrium more rapidly, but the equilibrium concentrations will remain the same) (*Chemistry* 7th ed. pages 604–610 / 8th ed. pages 620–625).

2. (a) 0.0220 mol/L PCl_3 × 15.0 L = 0.330 mole PCl_3 = 0.330 mole PCl_5 that reacted.

   0.500 – 0.330 = 0.170 mole PCl_5 remains.

   (b) $K_c = \dfrac{[PCl_5]}{[PCl_3][Cl_2]}$

   (c) $\dfrac{\left(\dfrac{0.170}{15.0\ L}\right)}{(0.0220)(0.0220)} = 23.4$

   (d) $K_p = K_c\,(RT)^{\Delta n}$

   In this case, 1 + 1 mol of gas $\longrightarrow$ 1 mole of gas, so $n = -1$ moles

   $K_p = 23.4\,(0.08206 \times 375)^{-1} = 0.760$

   (e) Raising the temperature of an exothermic reaction opposes the forward reaction. Further, from $K_p = K_c\,(RT)^n$, if T increases, the value of K_p decreases (*Chemistry* 7th ed. pages 594–596, 587–588 / 8th ed. pages 609–612, 602-603).

13

ACIDS, BASES, AND SALTS

Concepts from the last chapter will be applied in this section covering the equilibria of weak acids, weak bases, and salts. In addition, acid base theories and properties of acids will be reviewed. Calculations involved in acid-base titrations, preparation of buffers, and salt hydrolysis will be demonstrated.

You should be able to

- Understand the acid-base theories of Arrhenius, Brønsted-Lowry, and Lewis.
- Identify strong acids and bases and calculate their pH's.
- Calculate the pH of a weak acid or base.
- Calculate the concentration of a strong or weak acid or base from its pH.
- Calculate the pH and ion concentrations in a polyprotic acid.
- Predict the pH of a salt from its formula and then calculate the pH of the salt.
- Identify the components of a buffer and perform calculations involving the preparation of a buffer and the addition of strong acid or strong base to a buffer.
- Perform calculations involving strong acid-strong base titrations as well as weak acid-strong base and weak base-strong acid calculations.
- Be familiar with titration curves and selection of an acid-base indicator.

AP tip

Be systematic. Acid-base equilibrium problems require a step-by-step process. Although the problems in this chapter have many similarities, you must make note of the differences.

ACID-BASE THEORIES

ARRHENIUS

(Chemistry 7th ed. page 623 / 8th ed. page 639)

The Arrhenius theory states that, in aqueous solution (water), acids produce hydrogen ions and bases produce hydroxide ions.

BRØNSTED-LOWRY

(Chemistry 7th ed. pages 623–626 / 8th ed. pages 639–642)

The Brønsted-Lowry theory says that an acid is a proton (H^+) donor and a base is a proton acceptor.

In the reaction below, HNO_3 transfers a proton to H_2O forming H_3O^+, the hydronium ion. H_3O^+ is the conjugate acid of H_2O and NO_3^- is the conjugate base of HNO_3. The formulas in a conjugate acid-base pair differ by one H^+.

$$HNO_3 \quad + \quad H_2O \longrightarrow \quad H_3O^+ \quad + \quad NO_3^-$$

Acid　　　　Base　　　　Conjugate acid　　Conjugate base

EXAMPLE: Give the formulas for the conjugate base of H_2SO_4 and the conjugate acid of CH_3NH_2.

ANSWER: HSO_4^- is the conjugate base of H_2SO_4.

$CH_3NH_3^+$ is the conjugate acid of CH_3NH_2.

Note that each conjugate acid-base pair differs by 1 H^+:

$HSO_4^- + 1H^+ = H_2SO_4$; $CH_3NH_3^+ = 1H^+ + CH_3NH_2$.

LEWIS ACID-BASE MODEL

(Chemistry 7th ed. pages 663–665 / 8th ed. pages 679–681)

According to this theory, a Lewis acid is an electron pair acceptor and a Lewis base is an electron pair donor. Remember that this is a much broader definition than the other two, in that the acid-base interaction does not need to involve a transfer of a proton H^+.

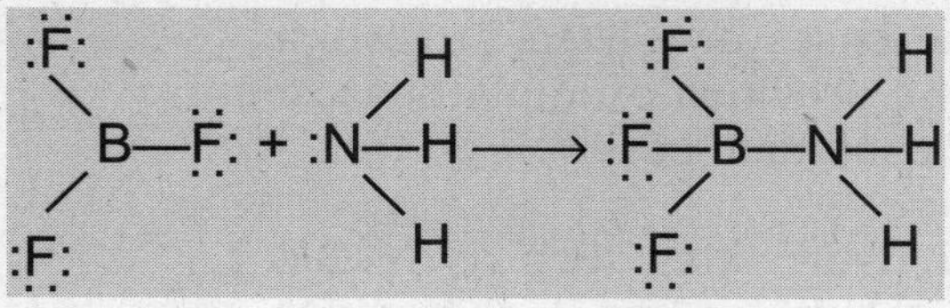

ACID AND BASE STRENGTH

(*Chemistry* 7th ed. pages 626–629, 644–645 / 8th ed. pages 642–645, 661–662)

The names and formulas of the six strong acids must be memorized. The six strong acids are HCl, HBr, HI, HNO_3, H_2SO_4, and $HClO_4$.

   If an acid is not one of the six in the list, then for purposes of the AP exam, you can assume it is a weak acid. The chart below compares the dissociation of strong acids to weak acids.

Comparison of Strong and Weak Acids			
Type of acid, HA	Reversibility of reaction	K_a value	Ions existing when acid, HA, dissociates in H_2O
Strong	Not reversible	K_a value very large	H^+ and A^-, only. No HA present.
Weak	reversible	K_a is small	H^+, A^-, and HA

Weak acids exist in equilibrium with their ions in aqueous solution. The acid dissociation constant, K_a, measures the extent to which the acid dissociates in water:

$$HA(aq) + H_2O\ (l) \rightleftharpoons H_3O^+(aq) + A^-(aq)$$

The equilibrium expression for the reaction is

$$K_a = \frac{\left[H_3O^+\right]\left[A^-\right]}{[HA]}$$

A table of K_a values for monoprotic acids, containing one acidic hydrogen, appears in the appendix of *Chemistry* and many other textbooks and is worth studying. The larger the K_a value, the stronger the acid.

   EXAMPLE: List the acids in order of increasing strength: HCN, HCl, $HClO_2$, HNO_2.

   ANSWER: HCN, HNO_2, $HClO_2$, and HCl. The first three weak acids are listed in order of increasing K_a values. HCl is stronger than all of the weak acids given. HCl is a strong acid.

   EXAMPLE: Arrange the following species in order of increasing base strength: NO_2^-, ClO_2^-, CN^-, Cl^-

ANSWER: Cl^-, ClO_2^-, NO_2^-, CN^-; The bases are listed in reverse order of their conjugate acids in the previous example because the stronger an acid, the weaker its conjugate base.

Strong bases include group I and II hydroxides such as NaOH. Weak bases such as NH_3, ammonia, are not group I or II hydroxides. The base dissociation constant, K_b, measures the extent to which a base reacts with water. The reaction of a weak base, B, with water and its corresponding equilibrium expression is

$$B\ (aq) + HOH(l) \rightleftharpoons BH^+(aq) + OH^-(aq);\ \ K_b = \frac{[BH^+]\,[OH^-]}{[B]}.$$

AP tip

Writing the reaction for base dissociation can be tricky. Always remember to react the base with water. Use the Arrhenius and Brønsted-Lowry theories to help write the reaction. For example, the base must react with water to produce hydroxide ions according to the Arrhenius theory. The base accepts an H^+ ion according to the Brønsted-Lowry theory. Remember to check the charges of the reactants and products and be sure that the sums of the charges on both sides of the reaction are equal.

THE EFFECT OF STRUCTURE ON ACID-BASE PROPERTIES

(*Chemistry* 7th ed. pages 661–662 / 8th ed. pages 667–678)

For binary acids, HX, the strength of the H–X bond and the polarity of the bond will determine the behavior of the acid. The polarity of the bonds in hydrogen halides become less polar going down a group. The strength of the H-F bond is what makes it a weak acid, whereas the rest of the hydrogen halides are strong acids: HI>HBr>HCl.

For a given series of oxoacids such as $HClO_4$, $HClO_3$, $HClO_2$, and $HClO$, the acid strength increases with increasing number of oxygens attached to the central atom. $HClO_4$ is a strong acid. The remaining oxoacids are listed in order of decreasing strength (decreasing number of oxygen atoms). The O–H bond becomes more polarized and weakened due to the electron density drawn toward the highly electronegative oxygen atoms.

CALCULATING THE pH OF STRONG ACIDS AND BASES

(*Chemistry* 7th ed. pages 634–635, 645–646 / 8th ed. pages 650–651, 662–663)

The pH of a strong acid can be calculated directly from the hydrogen ion concentration,

$$pH = -\log [H^+]$$

$[H^+]$, the molar concentration of the hydrogen ion, is obtained from the molarity of the acid.

EXAMPLE: Calculate the pH of 0.010 M HCl.

ANSWER: The pH equals 2.00; –log (0.010) = 2.00.

The pH of a strong base can be calculated from its hydroxide ion concentration.

$$pOH = -\log [OH^-]$$

$$pH + pOH = 14.00$$

The concentration of a strong acid or strong base can be determined from the solution's pH.

EXAMPLE: The pH of a $Sr(OH)_2$ solution is 13.50. Calculate the concentration of $Sr(OH)_2$.

ANSWER: pOH = 14 – pH = 14 – 13.50 = 0.50

$[OH^-]$ = inv log (–pOH); inv log (–0.50) = 0.32 M OH^-

$$\frac{0.32 \text{ mol OH}}{1 \text{ L}} \times \frac{1 \text{ mol Sr(OH)}_2}{2 \text{ mol OH}^-} = 0.16 \text{ M Sr(OH)}_2$$

AP tip

The number of significant figures in a pH measurement is equal to the number of decimal places in the pH. For example, 1.70 has 2 significant figures.

CALCULATING THE pH OF WEAK ACIDS

(*Chemistry* 7th ed. pages 635–640 / 8th ed. pages 651–656)

The pH of a weak acid cannot be calculated directly from the concentration of the acid since all of the acid does not dissociate to form H^+. The equilibrium reaction of the acid must be considered.

EXAMPLE: Calculate the pH of 0.25 M HCN.

ANSWER: First, write the reaction of the acid with water. Use the Arrhenius and Brønsted-Lowry theories to help you write the products. Check that you have the correct charges on the products.

$$HCN + H_2O \rightleftharpoons H_3O^+ + CN^-$$

Second, set up an "ICE" chart as you did in the previous chapter for equilibrium problems.

	HCN	+ H₂O ⇌	H₃O⁺	+ CN⁻
I	0.25		0	0
C	– x		+x	+x
E	0.25 – x		x	x

Third, write the equilibrium expression for K_a in the same manner as you did in the last chapter. Plug in the values from the equilibrium line of the ICE chart.

If x is very small compared to the concentration of the acid you are subtracting it from, then you can assume [HA – x] is approximately equal to [HA]. One way to determine if this approximation is valid is to compare the magnitude to K to [HA]; if [HA] is greater than K by a factor of 10^3 or more then x can be safely ignored. You can always assume that x is small and then check the value of x you calculate to see if [HA – x] is within 5% of [HA].

$$K_a = 6.2 \times 10^{-10} = \frac{[H_3O^+][CN^-]}{[HCN]} ; \frac{x^2}{(0.25 - x)} \cong \frac{x^2}{0.25}.$$

Fourth, solve for x which equals H_3O^+.

$$(6.2 \times 10^{-10})(0.25) = x^2 ; x = 1.2 \times 10^{-5}\,M.$$

Finally calculate the pH from the value of x, the H_3O^+ concentration.

$$pH = -\log(1.2 \times 10^{-5}) = 4.90$$

AP tips

Always follow these steps when performing calculations involving the disassociation of weak acids or weak bases.
1. Write the reaction of the acid or base with water. (Use the acid-base theories and check charges.)
2. Set up ICE chart.
3. Write the equilibrium expression in terms of reactants and products (without numbers).
4. Solve for x, using the method of approximation. Test approximation.
5. Solve for pH. (Be careful for base equilibria, $x = OH^-$. You need to find pOH and then pH.)
6. Always remember that for weak acids and bases, at equilibrium, $pH = -\log[H_3O^+]$.

CALCULATING THE PERCENT DISSOCIATION

(*Chemistry* 7th ed. pages 641–644 / 8th ed. pages 657–660)

The percent dissociation of an acid (or a base) is the amount of the acid, HA, which has dissociated, x, divided by the acid's initial concentration, HA_o, multiplied by 100.

$$\frac{x}{[HA_o]} \times 100\%$$

When making assumptions in an equilibrium calculation, it is best to test the assumption by making sure that the percent dissociation is less than or equal to 5%. The test for the assumption is the same as the calculation for the percent dissociation.

EXAMPLE: The percent dissociation of an acid, HA, which is 0.100 M is 2.5 %. Calculate the K_a of the acid.

ANSWER: x/0.100 M × 100% = 2.5% ; x = 2.5 × 10^{-3} M.

$$K_a = \frac{[H_3O^+][A^-]}{[HA]} ; K_a = \frac{(2.5 \times 10^{-3})^2}{(0.100 - 2.5 \times 10^{-3})} = 6.4 \times 10^{-5}.$$

CALCULATING THE pH OF WEAK BASES

(*Chemistry* 7th ed. pages 646–650 / 8th ed. pages 662–666)

The calculations involving weak base equilibria are similar to the weak acid equilibria problems except that the equation is written for a base reacting with water and the calculation initially involves finding $[OH^-]$. You will need to find the pOH and then the pH.

EXAMPLE: The pH of a 0.20 M solution of H_2NNH_2 is 11.38. Calculate K_b for H_2NNH_2.

Write the reaction with water. Bases accept H^+. Watch charges! One hint to help in writing the reaction is that the pH is 11.38. The basic pH indicates that OH^- must be one of the products.

Fill out the ICE Chart under the reaction.

	H_2NNH_2 + HOH ⇌	OH^- +	$H_2NNH_3^+$
I	0.20	0	0
C	$-x$	$+x$	$+x$
E	$0.20 - x$	x	x

You are given the pH, but x equals $[OH^-]$.

Find pOH ; pH + pOH = 14.00 ; pOH = 14.00 − 11.38 = 2.62

pOH = −log $[OH^-]$; Find $[OH^-]$ = inv log (−pOH) ;

inv log (−2.62) = 2.4 × 10^{-3} M. (You can also type this into your calculator as 10^{-pOH}.)

Plug this value of x into the K_b expression:

$K_b = [OH^-] [H_2NNH_3^+] / [H_2NNH_2] = x^2 / (0.20 - x)$.

$K_b = (2.4 \times 10^{-3})^2 / 0.20 = 2.9 \times 10^{-5}$

POLYPROTIC ACIDS

(*Chemistry* 7th ed. pages 650–655 / 8th ed. pages 666–671)

Polyprotic acids can donate more than one proton, H^+, and dissociate by losing 1 H^+ at a time.

EXAMPLE: Calculate the [H⁺] of a 0.20 M solution.

Also determine the concentrations of H_3AsO_4, $H_2AsO_4^-$, $HAsO_4^{2-}$, and AsO_4^{3-}.

For H_3AsO_4 $K_{a_1} = 5 \times 10^{-3}$, $K_{a_2} = 8 \times 10^{-8}$, $K_{a_3} = 6 \times 10^{-10}$.

	H_3AsO_4	$+ H_2O$	$\rightleftharpoons$	H_3O^+	$H_2AsO_4^-$
I	0.20			0	0
C	$-x$			$+x$	$+x$
E	$0.20 - x$			x	x

$$H_3AsO_4 + H_3O^+ \rightleftharpoons H_3O^+ \quad H_2AsO_4^-$$

$$K_{a_1} = \frac{[H_3O^+][H_2AsO_4^-]}{[H_3AsO_4]}$$

$$5 \times 10^{-3} = \frac{x^2}{0.20 - x}$$

Note: You cannot assume x is small since the K and 0.20 only differ by a factor of 10^2.

$x = 3 \times 10^{-2}$ M (as a result of solving quadratics)

$[H_3O^+] = [H_2AsO_4^-] = 3 \times 10^{-2}$ M

$[H_3AsO_4] = 0.20 - 0.03 = 0.17$ M

Since $K_{a_3} <<< K_{a_2} <<< K_{a_1}$, very little of $H_2AsO_4^-$ and $HAsO_4^{2-}$ dissociates compared to H_3AsO_4, so $[H_3O^+]$ and $[H_2AsO_4^-]$ will not change very much by the K_{a_2} dissociation, and we can use their concentrations to find the concentration of $HAsO_4^{2-}$.

	$H_2AsO_4^-$	$+ H_2O$	$\rightleftharpoons$	H_3O^+	$+$	$HAsO_4^{2-}$
I	3×10^{-2}			3×10^{-2}		0
C	$-x$			$+x$		$+x$
E	$3 \times 10^{-2} - x$			$3 \times 10^{-2} + x$		x

$$K_{a_2} = 8 \times 10^{-8} = \frac{(3 \times 10^{-2})[HAsO_4^{2-}]}{(3 \times 10^{-2})}$$

$[HAsO_4^{2-}] = 8 \times 10^{-8}$ M; the assumption that K_{a_2} does not contribute significantly to $[H_3O^+]$ and $[H_2AsO_4^-]$ is good.

Repeat the process to find $[AsO_4^{3-}]$.

	$HAsO_4^{2-}$	$+$	H_2O	$\rightleftharpoons$	H_3O^+	$+$	AsO_4^{3-}

I	8×10^{-8}	3×10^{-2}	0
C	$-x$	$+x$	$+x$
E	$8 \times 10^{-8} - x$	$3 \times 10^{-2} + x$	x

$$K_{a_3} = 6 \times 10^{-10} = \frac{(3 \times 10^{-2})\,[AsO_4^{3-}]}{(8 \times 10^{-8})}$$

$[AsO_4^{3-}] = 2 \times 10^{-15}$ M. Assumption that x is small is good.

ACID-BASE PROPERTIES OF SALTS

(*Chemistry* 7th ed. pages 655–660 / 8th ed. pages 671–677)

PREDICTING THE pH OF SALTS

You are to determine if a salt is acidic, basic, or neutral by looking at its chemical formula. This process involves two steps as outlined in the flowcharts which follow. In the first flowchart, you will determine the acidity or basicity of the individual ions. Then, using the second chart and the results from the first flowchart, you can determine if the salt is acidic, basic, or neutral.

Determining the Approximate pH of Ions in a Salt

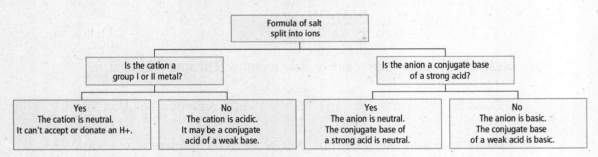

Determining the Approximate pH of a Salt

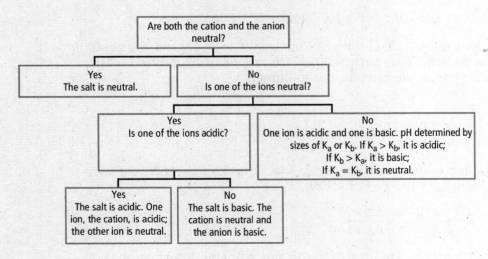

EXAMPLE: Determine whether an aqueous solution of $KC_2H_3O_2$ is acidic, basic, or neutral.

ANSWER: KCH_3O_2 is basic. Use the method of asking questions outlined above.

Split the salt into its cation, K^+, and its anion, $C_2H_3O_2^-$.

Is the cation a Group I metal? Yes; the cation does not affect the pH.

Is the anion a conjugate base of a strong acid? No; $C_2H_3O_2^-$ is the conjugate base of a weak acid, $HC_2H_3O_2$, which makes the solution basic.

Because we have a salt with a cation which doesn't affect the pH and a basic anion, an aqueous solution of the salt is basic.

CALCULATING THE pH OF SALTS

To calculate the pH of a salt, first you must decide whether the salt is acidic, basic, or neutral. If the salt is basic, then its anion is the conjugate base of a weak acid. The anion will undergo hydrolysis. We will need to write an equation for the reaction of that ion with water to form the acid and OH^- ions: $A^- + H_2O = HA + OH^-$. Then we will write the equilibrium expression

$$K_b = \frac{K_w}{K_a} = \frac{[HA][OH^-]}{[A^-]}.$$

If the salt is basic, then the hydrolysis reaction will produce an acid.

EXAMPLE: Determine the pH of a 0.100 M aqueous solution of NaCN. The K_a for HCN is 5.8×10^{-10}.

ANSWER: In aqueous solutions, NaCN ionizes completely into Na^+ and CN^-. However, the cyanide ions ionize in water according to the following equation:

$$CN^- (aq) + H_2O(l) \rightleftharpoons HCN (aq) + OH^- (aq).$$

Na^+ ions do not affect the pH. Ions from Group IA and IIA never undergo hydrolysis because they are cations of strong bases. The equilibrium expression for the solution is

$$K_b = \frac{[OH^-][HCN]}{[CN^-]}$$

We need a value for K_b. Since we have K_a for HCN, we can calculate the value of K_b for CN^-.

$$K_b = \frac{1.0 \times 10^{-14}}{K_a} = \frac{1.0 \times 10^{-14}}{5.8 \times 10^{-10}}$$

Constructing a table for filling in the available information, we get

$$\text{CN}^- \quad + \text{H}_2\text{O} \quad \rightleftharpoons \quad \text{HCN} + \quad \text{OH}^-$$

I	0.100		0	0
C	$-x$		$+x$	$+x$
E	$0.100 - x$		x	x

Since K is much smaller than 0.100, assume that $0.100 - x \approx 0.100$.

$$K_b = \frac{[\text{OH}^-][\text{HCN}]}{[\text{CN}^-]}; \quad 1.7 \times 10^{-5} = \frac{x^2}{0.100}$$

Solving for x, we have

$x = [\text{OH}^-] = 1.3 \times 10^{-3}$

$\text{pOH} = 2.89$; $\text{pH} = 14.00 - \text{pOH}$

$\text{pH} = 11.11$.

BUFFERS

(*Chemistry* 7th ed. pages 684–696 / 8th ed. pages 701–6713)

Buffers resist changes in pH when acids or bases are added. The components of a buffer are summarized in the next table.

Buffer	
Components	**Examples**
Weak acid + salt containing the conjugate base	HCN and NaCN
Weak base + salt containing the conjugate acid	CH_3NH_2 and CH_3NH_3Cl
Weak acid + excess strong base or Weak base + excess strong acid	2 mol of HCN + 1 mol NaOH react to yield 1 mol HCN and 1 mol NaCN 2 mol NH_3 + 1 mol HCl react to yield 1 mol NH_3 and 1 mol NH_4Cl

CALCULATING THE PH OF A BUFFER

The calculation to find the pH of a buffer is similar to all equilibrium calculations EXCEPT there are now two initial concentrations, one for each part of the buffer pair.

Example: Calculate the pH of a solution that is 0.60 M HF and 1.00 M KF. K_a for HF is 7.2×10^{-4}.

Answer: First, write the reaction of the acid with water. (You are given K_a, and the buffer is a made of a weak acid and its conjugate base.)

$$HF + H_2O \rightleftharpoons H_3O^+ + F^-$$

Second, set up an ICE chart.

	HF	$+ H_2O \rightleftharpoons$	H_3O^+	$+ F^-$
I	0.60		0	1.00
C	$-x$		$+x$	$+x$
E	$0.60 - x$		x	$1.00 + x$

Third, write the equilibrium expression for K_a.

Plug in the values from the equilibrium line of the ICE chart. Check to see if x is small.

$$K_a = 7.2 \times 10^{-4} = \frac{[H_3O^+][F^-]}{[HF]} = \frac{x(1.00 - x)}{(0.60 - x)} = \frac{x(1.00)}{0.60}$$

Fourth, solve for x which equals H_3O^+.

$$K_a = 7.2 \times 10^{-4} = \frac{x(1.00)}{0.60}$$

$$x = [H_3O^+] = 4.3 \times 10^{-4}\ M$$

$$pH = -\log [H_3O^+]$$

$$pH = -\log (4.3 \times 10^{-4}) = 3.37.$$

Preparation of a Buffer

A buffer can be made from a weak acid and a salt containing its conjugate base or from a weak base and a salt containing its conjugate acid.

Example: Calculate the mass of $NaC_2H_3O_2$ required to prepare a buffer of pH 4.55 when added to 0.500 L of 0.67 M acetic acid. (Assume no change in volume.) $K_a = 1.8 \times 10^{-5}$ for $HC_2H_3O_2$.

Answer: 1.8 g $NaC_2H_3O_2$

This problem can be solved using an ICE chart or by using the Henderson-Hasselbalch equation.

$$\boxed{\begin{array}{c} \text{Henderson-Hasselbalch} \\ pH = pK_a + \log (\,[\text{conjugate base}] / [\text{acid}]\,) \end{array}}$$

$$pK_a = -\log K_a$$

Brackets [ ] represent the molar concentration of the substance.

$$pH = pKa + \log\left(\frac{[C_2H_3O_2^-]}{[HC_2H_3O_2]}\right)$$

$$4.55 = 4.74 + \log\left(\frac{[C_2H_3O_2^-]}{0.67\ M}\right)$$

$$-0.19 = \log\left(\frac{[C_2H_3O_2^-]}{0.67\ M}\right)$$

$$0.65 = \frac{[C_2H_3O_2^-]}{0.67\ M}$$

$$\text{mass } NaC_2H_3O_3 = \frac{0.44\ mol\ [C_2H_3O_2^-]}{L} \times \frac{1\ mol\ NaC_2H_3O_2}{1\ mol\ [C_2H_3O_2^-]} \times \frac{0.500\ L \times 82.0\ g}{1\ mol\ NaC_2H_3O_2}$$

ADDITION OF STRONG ACID AND STRONG BASE TO A BUFFER

The next example summarizes the steps for this type of problem. The problem is solved in the same manner as the buffer problem on the previous page with HF and NaF, except for the first three steps which are new.

EXAMPLE: Calculate the pH when 100.0 mL of 0.50 M HCl are added to the buffer consisting of 20.0 g $HC_2H_3O_2$ and 18.0 g $NaC_2H_3O_2$ dissolved in 5.00 x 10^2 mL of water.

ANSWER: Calculate the initial moles of weak acid and conjugate acid present in the buffer. Write these amounts under the reaction, above the ICE table.

	$HC_2H_3O_2$ + H_2O $\rightleftharpoons$	H_3O^+	$C_2H_3O_2^-$
Initial moles	0.33	0	0.21
Add mol H^+	+0.05	0	–0.05
New Initial Mol	0.38	0	0.16
I	0.63		0.27
C	$-x$	$+x$	$+x$
E	$0.63 - x$	x	$0.27 + x$

Next, calculate the moles of strong acid, H^+, added. The added strong acid will react with the basic part of the buffer, $C_2H_3O_2^-$., to produce the other part of the buffer $HC_2H_3O_2$.

$$H^+ + C_2H_3O_2^- \longrightarrow HC_2H_3O_2$$

Make a new line under the initial moles called "Add mol H^+." Note: The minus sign shows the amount of conjugate base decreasing upon added H^+ and the plus sign shows the amount of acid increasing.

Calculate the new initial moles of weak acid and its conjugate base, after the conjugate base has reacted with the strong acid, by adding or subtracting the number of moles H$^+$ added.

Fill out the ICE chart, computing the initial concentration of weak acid and conjugate base by dividing the moles of weak acid and conjugate base by the total volume (volume of buffer + volume of added strong acid, in liters).

$HC_2H_3O_2 = 0.38$ mol / 0.600 L $= 0.63$ M

$C_2H_3O_2^- = 0.16$ mol / 0.600 L $= 0.27$ M

Proceed to fill out the ICE chart and plug the values into the K_a expression (or solve Henderson-Hasselbalch) as in a normal buffer problem.

$$K_a = \frac{[H^+]\,[C_2H_3O_2]}{[HC_2H_3O_2]}$$

$$1.8 \times 10^{-5} = \frac{[H^+]\,(0.27 + x)}{0.63 + x} \approx \frac{[H^+]\,0.27}{0.63}$$

4.2×10^{-5} M $= [H^+]$ pH $= -\log\,[H^+] = 4.38$

Alternatively, you can use the Henderson-Hasselbalch equation. Volume cancels.

pH $= pKa + \log\,[0.16$ mol$/0.38$ mol$] = 4.74 + (-0.38) = 4.36$

ADDITION OF STRONG BASE TO A BUFFER

The calculation for addition of a strong base to a buffer is similar to the previous problem.

EXAMPLE: Calculate the pH when 100.0 mL 0.50 M NaOH are added to the buffer consisting of 20.0 g of $HC_2H_3O_2$ and 18.0 g of $NaC_2H_3O_2$ dissolved in 5.00×10^2 mL of water.

ANSWER: Calculate the initial moles of weak acid and conjugate acid that are present in the buffer. Write these amounts under the reaction, above the ICE table.

	$HC_2H_3O_2 + H_2O$	$\rightleftharpoons$	H_3O^+	$C_2H_3O_2^-$
Initial moles	0.33		0	0.21
Add mol OH$^-$	−0.05		0	+0.05
New initial mol	0.28		0	0.26
I	0.47			0.43
C	−x		+x	+x
E	0.47 − x		x	0.43 + x

Next, calculate the moles of strong base, OH$^-$, added. The added strong base will react with the acidic part of the buffer, $HC_2H_3O_2$, to produce the other part of the buffer ($C_2H_3O_2^-$).

$$OH^- + HC_2H_3O_2 \longrightarrow HOH + C_2H_3O_2^-$$

Make a new line under the initial moles called "Add mol OH⁻." Note, the minus sign shows the amount of acid decreasing upon added OH⁻ and the plus sign shows the amount of conjugate base increasing.

Calculate the new initial concentration of the acid and its conjugate base and fill out the ICE chart as in the previous problem.

The answer is: $[H^+] = 2.0 \times 10^{-5} M$; pH = 4.71.

TITRATION

(*Chemistry* 7th ed. pages 696–711 / 8th ed. pages 713–728)

A titration can be used to determine the concentration of an unknown solution. The calculations involve stochiometric principles.

The table below summarizes the other types of problems involving acid-base titrations. Examples of each type of titration problem follow the table.

Characteristics of Titrations			
Titrations	Before equivalence point	At equivalence point	After equivalence point
Strong acid titrated with a strong base	Use excess $[H^+]$ ion to calculate the pH pH < 7	pH = 7	Use excess $[OH^-]$ ion to calculate the pH pH > 7
Species in solution which affect the pH (in addition to H_2O)	H^+	H_2O	OH^-
Weak acid titrated with a strong base	Weak acid + conjugate base (a buffer problem) pH < 7	Salt hydrolysis problem pH > 7	Use excess OH⁻ ion to calculate the pH pH > 7
Species in solution which affect the pH (in addition to H_2O)	HA, A⁻	A⁻	OH⁻
Weak base titrated with a strong acid	Weak base + conjugate acid (a buffer problem) pH > 7	Salt hydrolysis pH < 7	Excess H^+ ion pH < 7
Species in solution which affect the pH (in addition to H_2O)	B, HB⁺	HB⁺	H^+

TITRATION OF A STRONG ACID WITH A STRONG BASE

(*Chemistry* 7th ed. pages 696–699 / 8th ed. pages 713–717)

At any point during the titration of a strong acid with a strong base (or a strong base with a strong acid), the pH can be calculated from the molarity of the ion present in excess after the complete reaction.

At the equivalence point of any type of titration,

$$\text{moles } H^+ = \text{moles } OH^-.$$

For a strong acid-strong base titration, the pH equals 7.

EXAMPLE: Consider the titration of 40.0 mL of 0.200 M HBr by 0.100 M KOH. Calculate the pH of the resulting solution when the following volumes of KOH have been added.

a. 10.0 mL   b. 80.0 mL   c. 100.0 mL

ANSWER: First, using the volume and molarity of the HBr, calculate the moles of H^+ to be titrated. The resulting moles of H^+ appear in the first column in the following table.

Next, for the titration in question, calculate the moles of OH^- added from the volume and concentration of KOH. The resulting moles of OH^- used in each titration appear in the second column of the table.

Calculate the moles of ion in excess, H^+ or OH^-. The results for each titration appear in column four in the table below.

Using the total volume, in column five of the table, and the moles of ion in excess in column four, determine the molarity of the ion in excess. The molarity of the ion in excess for each of the three titrations is in column six.

It is the molarity of the ion in excess which determines the pH. In part b of the example, neither ion is in excess. The pH equals 7 because the titration is at the equivalence point.

Be careful, in part c of the example, the excess ion is the OH^- ion. To calculate the pH, you must determine pOH and then the pH.

Problem	Mol H^+	Mol OH^-	Mol Ion in Excess	Total Volume (When acid and base react)	Molarity Ion In Excess	pH
a	8.00×10^{-3}	1.00×10^{-3}	7.00×10^{-3} (H^+)	0.0500 L	0.140 M (H^+)	0.854
b	8.00×10^{-3}	8.00×10^{-3}	None Equivalence point	Not needed	None	7.00
c	8.00×10^{-3}	10.0×10^{-3}	2.0×10^{-3} (OH^-)	0.140 L	0.014 M (OH^-)	12.15

AP tip

When you are calculating the pH during a titration, be careful to take the negative log of the concentration of H^+, not just the moles of H^+ which is a common mistake. And if the excess ion or the value of x is OH^-, be sure that you calculate the pOH and then the pH.

TITRATION OF A WEAK ACID WITH A STRONG BASE

There are three characteristic types of calculations for the titration of a weak acid with a strong base: before the equivalence point, at the equivalence point, and after the equivalence point.

> EXAMPLE: A 25.0 mL sample of 0.100 M $HC_3H_5O_2$ is titrated with the 0.100 M NaOH. Calculate the pH when the following volumes of 0.100 M NaOH are added: $K_a = 1.3 \times 10^{-5}$ for $HC_3H_5O_2$.
>
> a) 8.0 mL   b) 12.5 mL   c) 25.0 mL   d) 30.0 mL

CALCULATION OF THE pH BEFORE THE EQUIVALENCE POINT

Part a) of the example above involves the titration before the equivalence point. Perform the stoichiometry and record the results in a table under the reaction. The first three lines in the table on the next page are explained below.

Calculate the initial moles of the weak acid present in the sample.

$0.0250 \, L \times 0.100 \, mol/L = 2.50 \times 10^{-3} \, mol \; HC_3H_5O_2$

Calculate the moles of strong base, OH^-, added.

$0.0080 \, L \times 0.100 \, mo/L = 8.0 \times 10^{-4} \, mol$

Calculate the new initial moles of weak acid and its conjugate base present after the OH^- has reacted.
   This problem is now identical to a "buffer" problem.
   Make a new line under the initial moles called "mol OH^- added." Note the minus sign to show the amount of acid decreasing upon added OH^- and the plus sign to show the amount of conjugate base increasing.
   Calculate the new initial moles of weak acid and its conjugate base, after the weak acid has reacted with the strong base, by subtracting or adding the number of moles OH^- added.

	HC₃H₅O₂ + H₂O	⇌	H₃O⁺ +	C₃H₅O₂⁻

	HC₃H₅O₂ + H₂O	⇌ H₃O⁺ +	C₃H₅O₂⁻
Initial moles	2.50×10^{-3}	0	0
Add mol OH⁻	-8.0×10^{-4}	0	$+8.0 \times 10^{-4}$
New Initial Mol	1.70×10^{-3}	0	8.0×10^{-4} mol
I	1.70×10^{-3} / 0.0330 L $= 0.0515$ M		8.0×10^{-4} / 0.0330 L $= 0.0242$ M
C	$-x$	$+x$	$+x$
E	$0.0515 - x$	x	$0.0242 + x$

The stoichiometric calculations are described on the previous page. You can now perform the equilibrium calculation.

To begin filling out the ICE chart, compute the initial concentration of weak acid and conjugate base by dividing the moles of weak acid and conjugate base by the total volume (volume of acid titrated + volume of added strong base, in liters).

$$K_a = \frac{[H^+][C_3H_5O_2^-]}{[HC_3H_5O_2]}$$

$$1.3 \times 10^{-5} = \frac{[H^+](0.0242 + x)}{0.0515 + x} \approx \frac{[H^+]0.0242}{0.0515}$$

2.77×10^{-5} M = [H⁺]; pH = −log [H⁺] = 4.56

Alternatively, you can use the Henderson-Hasselbalch equation.

HALFWAY TO THE EQUIVALENCE POINT

In part b of the example on the previous page, a 25.00 mL sample of 0.100 M HC₃H₅O₂ is titrated with 12.5 of 0.100 M NaOH. K_a is 1.3×10^{-5} for HC₃H₅O₂.

Half of the acid being titrated is neutralized. This point is halfway to the equivalence point.

Since half the acid is neutralized and half remains, [HA] = [A⁻].

$K_a = \frac{[H^+][A^-]}{[HA]}$

$pK_a = pH$

$pK_a = -\log K_a = -\log (1.3 \times 10^{-5}) = 4.89$

AT THE EQUIVALENCE POINT

In part c of the example on the previous page, a 25.00 mL sample of 0.100 M HC₃H₅O₂ is titrated with 25.0 mL of 0.100 M NaOH.

K_a is 1.3×10^{-5} for HC₃H₅O₂.

Calculate the initial moles of the weak acid in the sample.

Calculate the moles of strong base, OH⁻, added.

0.0250 L × 0.100 mol/ L = 2.50 × 10^{-3} mol OH^- = mol $C_3H_5O_2^-$

There is no weak acid present once the strong base added completely reacts. All that is present is A^- ($C_3H_5O_2^-$), the conjugate base of weak acid, HA ($HC_3H_5O_2$).

~~HA~~ + ~~NaOH~~ ⟶ NaA + HOH

The problem from this point on is a salt hydrolysis problem.

$C_3H_5O_2^-$ + H_2O ⇌ OH^- + $HC_3H_5O_2$

I   2.50 × 10^{-3}/ 0.0500 L

0.0500 M   0   0

C $-x$ $+x$ $+x$

E   0.0500 – x x x

$$K_b = \frac{K_w}{K_a} = \frac{1.0 \times 10^{-14}}{1.3 \times 10^{-5}} = 7.7 \times 10^{-10}$$

$$K_b = \frac{[OH^-][\,HC_3H_5O_2]}{[C_3H_5O_2^-]} = \frac{x^2}{0.0500 - x} \approx \frac{x^2}{0.0500}$$

$x = [\,(7.7 \times 10^{-10})(0.0500)\,]^{\frac{1}{2}} = 6.2 \times 10^{-6}$ M OH^-

pOH = 5.21 ; pH = 14.00 – 5.21 = 8.79

AFTER THE EQUIVALENCE POINT

The calculations involved here are identical to those in a strong acid-strong base titration.

In part d of the example on the previous page, a 25.00 mL sample of 0.100 M $HC_3H_5O_2$ is titrated with 30.0 mL of 0.100 M NaOH. K_a is 1.3 × 10^{-5} for $HC_3H_5O_2$.

0.0300 L × 0.100 mol/L = 0.00300 mol NaOH

0.02500 L × 0.100 mol/L = 0.002500 mol $HC_3H_5O_2$

Excess OH^- = 0.00300 mol – 0.002500 mol/0.0500 L = 0.010 M OH^-

pOH = 2.00 ; pH = 14 – pOH = 12.00

TITRATION OF A WEAK BASE WITH A STRONG ACID

The calculations for a weak base-strong acid titration are very similar to the weak acid-strong base titration.

The differences are these:
- ◼ The pH before the equivalence point is greater than 7.
- ◼ At the equivalence point, the pH is less than 7 since the salt formed is acidic.
- ◼ After the equivalence point, the pH is determined by the molarity of the excess H^+.

CHARACTERISTICS OF TITRATION CURVES

It is important for you to be able to sketch and identify the key points on titration curves for the AP Exam. The next example will help you to review titration curves for the different types of titrations discussed in this section.

EXAMPLE: The questions that follow refer to the following three titration curves. All solutions are equimolar.

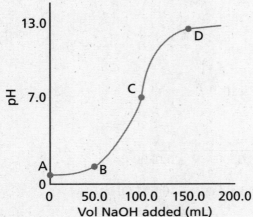

Figure 1:

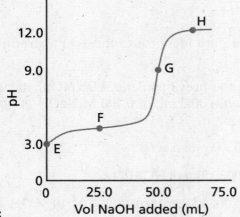

Figure 2:

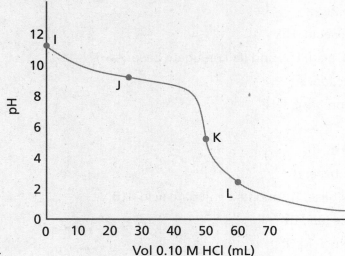

Figure 3:

(a) Which titration curve represents a strong acid titrated by a strong base? Explain your answer.
(b) Which titration curve represents a weak acid titrated by a strong base? Explain your answer.
(c) Which titration curve represents a weak base titrated by a strong acid? Explain your answer.
(d) For each lettered part of each curve, A through L, identify the major species which affects the pH.

ANSWER: (a) Figure 1 represents a strong acid titrated by a strong base. The pH is very low when no base has been added. The pH at the equivalence point is 7.00.

ANSWER: (b) Figure 2 represents a weak acid titrated by a strong base. The pH is acidic, but slightly higher than the titration curve in figure 1. The equivalence point occurs at pH 9, signifying the presence of the conjugate base of the weak acid being titrated.

ANSWER: (c) Figure 3 represents a weak base titrated by a strong acid. When no acid has been added, the pH is basic.

ANSWER: (d) The points are labeled as follows:

Figure 1:

A strong acid, H^+

B strong acid, H^+

C water

D strong base, OH^-

Figure 2:

E weak acid, HA

F weak acid, HA and its conjugate base, A⁻

G basic ion, A⁻

H strong base, OH⁻

Figure 3:

I weak base, B

J weak base, B , and its conjugate acid, HB⁺

K acidic ion HB⁺

L strong acid H⁺

MULTIPLE-CHOICE QUESTIONS

Calculators may not be used on this part of the exam.

1. The calculation of concentration and pH for weak acids is more complex than for strong acids due to
   (A) the incomplete ionization of weak acids.
   (B) the low K_a value for strong acids.
   (C) the more complex atomic structures of strong acids.
   (D) the low percent ionization of strong acids.
   (E) the inconsistent K_b value for strong acids.

2. The general reaction of an acid dissolving in water may be shown as

   $$HA(aq) + HOH(l) \rightleftharpoons H_3O^+(aq) + A^-(aq).$$

   A conjugate acid base pair for this reaction is
   (A) HA and HOH.
   (B) HA and A⁻.
   (C) HOH and A⁻.
   (D) H_3O^+ and A⁻.
   (E) HA and H_3O^+.

3. Strong acids are those which
   (A) have an equilibrium lying far to the left.
   (B) yield a weak conjugate base when reacting with water.
   (C) have a conjugate base which is a stronger base than water.
   (D) readily remove the H⁺ ions from water.
   (E) are only slightly dissociated (ionized) at equilibrium.

4. When calculating the pOH of a hydrofluoric acid solution ($K_a = 7.2 \times 10^{-4}$) from its concentration, the contribution of water ionizing ($K_w = 1.0 \times 10^{-14}$) is usually ignored because
   (A) hydrofluoric acid is such a weak acid.
   (B) hydrofluoric acid can dissolve glass.
   (C) the ionization of water provides relatively few H^+ ions.
   (D) the [OH⁻] for pure water is unknown.
   (E) the conjugate base of HF is such a strong base.

5. The percent dissociation (percent ionization) for weak acids
   (A) is always the same for a given acid, no matter what the concentration.
   (B) usually increases as the acid becomes more concentrated.
   (C) compares the amount of acid that has dissociated at equilibrium with the initial concentration of the acid.
   (D) may only be used to express the dissociation of weak acids.
   (E) has no meaning for polyprotic acids.

6. The [OH⁻] of a certain aqueous solution is 1.0×10^{-5}M. The pH of this same solution must be
   (A) 1.0×10^{-14}.
   (B) 5.00.
   (C) 7.00.
   (D) 9.00.
   (E) 12.00.

7. In many calculations for the pH of a weak acid from the concentration of the acid, an assumption is made that often takes the form $[HA]_o - x = [HA]_o$.
   This
   (A) is valid because x is very small compared to the initial concentration of the weak acid.
   (B) is valid because the concentration of the acid changes by such large amounts.
   (C) is valid because actual value of x cannot be known.
   (D) is valid because pH is not dependent upon the concentration of the weak acid.
   (E) approximation is always shown to be valid and so need not be checked.

8. HA is a weak acid which is 4.0% dissociated at 0.100M. Determine the K_a for this acid.
   (A) 0.0040
   (B) 0.00016
   (C) 0.040
   (D) 1.6
   (E) 16 5

9. Sulfur trioxide is an acidic oxide due to the
   (A) high electronegativity of sulfur in the O–S bond, forming strong covalent bonds.
   (B) low electronegativity of sulfur in the O–S bond, forming strong covalent bonds.
   (C) high electronegativity of sulfur in the O–S bond, forming strong ionic bonds.
   (D) low electronegativity of sulfur in the O–S bond, forming strong ionic bonds.
   (E) attraction for the H–O bonds by the H^+ ion in water.

10. Ionic substances known as salts can form acidic, basic, and neutral solutions when dissolved in water. When dissolved in water
    (A) KNO_3 forms a basic solution.
    (B) NaCl forms an acidic solution.
    (C) $NaNO_3$ forms an acidic solution.
    (D) NaF forms a basic solution.
    (E) $KClO_4$ forms an acidic solution.

11. The net ionic equation representing the equilibrium established when potassium sulfide dissolves in water is shown. Label the species present as either acid or base according to Brønsted-Lowry theory.

S^{2-}	+	H_2O	⇌	HS^-	+	OH^-
(A) acid		base		acid		base
(B) base		acid		acid		base
(C) acid		base		base		acid
(D) base		acid		base		acid
(E) acid		acid		base		base

12. Which one of the following is a Lewis acid, but not a Brønsted-Lowry acid?
    (A) HCl
    (B) BBr_3
    (C) NH_3
    (D) KOH
    (E) CH_4

13. Which of the following metal ions would you expect to show the greatest acidic properties in water?
    (A) K^+
    (B) Cs^+
    (C) Ca^{2+}
    (D) Cu^{2+}
    (E) Al^{3+}

14. Which of the following binary hydrides is the most basic?
    (A) HCl
    (B) H_2S
    (C) PH_3
    (D) SiH_4
    (E) NaH

15. Which statements describing a buffer solution are correct?
    1. A solution containing a weak base and its conjugate acid is a buffer
    2. A strong acid combined with its conjugate base constitutes a buffer.
    3. Buffers resist changes in pH upon addition of strong acids or bases.
    4. Solutions having equal concentrations of a weak acid and its conjugate base have pH = pK_a.
    (A) only 1
    (B) 1, 3, and 4
    (C) only 1 and 4
    (D) 2, 3 and 4
    (E) all are correct

FREE-RESPONSE QUESTIONS

Calculators may be used for this part of the exam.

1. The K_w for water at 25°C is 1.0×10^{-14}, but is 1.0×10^{-13} at 60°C.
   (a) Give the chemical equation for the autoionization of water.
   (b) Determine the [OH⁻] for water at 60°C.
   (c) Determine the pH of water at 60°C.
   (d) Is the reaction of the autoionization of water endothermic or exothermic? Support your answer with data and explanation.

2. Phosphoric acid, H_3PO_4, is a triprotic acid.
   (a) Show the three equations involved in the dissociation of this substance.
   (b) Illustrate how these three equations might be combined to show the complete dissociation of phosphoric acid.
   (c) If a 7.0 M H_3PO_4 solution dissociated, calculate the pH of the solution.
      $K_{a_1} = 7.5 \times 10^{-3}$ $K_{a_2} = 6.2 \times 10^{-8}$ $K_{a_3} = 4.8 \times 10^{-13}$.
   (d) Determine the concentration for the ions $H_2PO_4^-$, HPO_4^{2-}, and $PO_4^{3-}(aq)$ in the dissociated 7.0 M H_3PO_4 solution.
   (e) Determine the pOH for this same 7.0 M H_3PO_4 solution.

Answers

MULTIPLE CHOICE

1. **A** Strong acids ionize 100%, therefore the [H⁺] is equal to the initial concentration of the strong acid. This is not the case for weak acids. Since not all of the acid ionizes, it is necessary to determine both how much of the acid forms H⁺ and how much is left in molecular form (*Chemistry* 7th ed. pages 634–639 / 7th ed. pages 650–655).

2. **B** The conjugate acid differs from its conjugate base by a proton, (H⁺). The acid form has the proton (HA in this example) and the base form has lost the proton (A⁻ in this case) (*Chemistry* 7th ed. page 624 / 8th ed. page 640).

3. **B** Strong acids give up their protons (H⁺ ions) easily. If their conjugate base form were strong, then the acid formed would hold strongly to the H⁺, which would not be a characteristic of a strong acid (*Chemistry* 7th ed. pages 626–628 / 8th ed. pages 642–644).

4. **C** By comparing the K_a for HF with K_w (for water), you can see that even though hydrofluoric acid is a very weak acid it is a much stronger acid than is water, so much so that the H⁺ ion contribution of water may be ignored (*Chemistry* 7th ed. pages 635–638 / 8th ed. pages 651–654).

5. **ANSWER: C** The percent dissociation does change with the concentration of the acid (see sample exercise 14.10, page 672–673, 6th edition / page 641–642, 7th edition). For example, 1.0 M acetic acid does not ionize nearly as much as 0.10 M acetic acid. The term applies to the amount of acid dissociated/the initial concentration, expressed as a percent (*Chemistry* 7th ed. pages 641–642 / 8th ed. pages 657–658).

6. **D** Referring to pH + pOH = 14.00 at 25°C, if [OH⁻] = 1 × 10⁻⁵M, pOH = 5.0, pH = 14.00 − 5.0 = 9.0 (*Chemistry* 7th ed. pages 630–633 / 8th ed. pages 646–649).

7. **A** We can expect x to be very small (as indicated by the low value for K_a for weak acids), so that the concentration of the weak acid does not significantly change from its initial value. As the final step in such problems, you must determine that the change in the initial [HA] is less than 5% for the approximation to be considered valid (*Chemistry* 7th ed. pages 638–641 / 8th ed. pages 654–657).

8. **B**

[HA]₀ =0.100M

[H⁺] = [A⁻]= 0.004 M

K_a = [H⁺][A⁻] / [HA] = (0.004 × 0.004) / 0.100 =

(seen more simply for our purposes here)

$$(4 \times 10^{-3})(4 \times 10^{-3}) / (10^{-1}) = 1.6 \times 10^{-4}$$

Note that although the actual value for [HA] is 0.100 − 0.004 = 0.096M, this approximation is within the 5% rule, and students do not have calculators for this part of the test (*Chemistry* 7th ed. pages 641–643 / 8th ed. pages 657–659).

9. **A** As oxides dissolve and form hydrates, one must compare the electronegativity difference of the X–O bond to the H–O bond. If the electronegativity difference between atoms X and O is small, this bond is primarily covalent, which means that the O–H bond is

easily broken to form acidic solutions (*Chemistry* 7th ed. pages 661–663 / 8th ed. pages 677–679).

10. **D** When NaF is dissolved in water the Na^+ has little affinity for H^+ but the F^-, the anion of a weak acid, does have an attraction for the H^+ in water, leaving an excess of OH^- in solution. This kind of reaction with water is known as a hydrolysis reaction and, in this case, forms a basic solution (*Chemistry* 7th ed. pages 654–657 / 8th ed. pages 670–673).

11. **B** According to the Brønsted-Lowry theory, an acid is a proton donor and a base is a proton, or hydrogen ion, acceptor (*Chemistry* 7th ed. pages 623–625 / 8th ed. pages 639–642).

12. **B** The B atom can accept a pair of electrons, which by definition, makes it a Lewis acid; yet BBr_3 has no ionizable protons (*Chemistry* 7th ed. pages 663–665 / 8th ed. pages 679–681).

13. **E** Al^{3+} has the largest charge/ionic radius ratio; therefore it should react to the greatest extent with polar water to form $H^+(aq)$ (*Chemistry* 7th ed. pages 658–659,661 / 8th ed. pages 675, 677–678).

14. **E** Alkali and alkaline-earth binary hydrides contain the H^- ion, which is a basic ion (*Chemistry* 7th ed. page 665–660, 880–882 / 8th ed. pages 671–678).

15. **B** (*Chemistry* 7th ed. pages 684–698 / 8th ed. pages 701–713)

FREE RESPONSE

1. (a) $H_2O(l) + H_2O(l) \rightleftharpoons H_3O^+(aq) + OH^-(aq)$

   (*Chemistry* 7th ed. pages 629–630 / 8th ed. pages 645–647)

   (b) Since $[H_3O^+] = [OH^-]$

   and $K_w = [H_3O^+][OH^-] = 1.0 \times 10^{-13}$

   $[OH^-] = \sqrt{1.0 \times 10^{-13}}\ 1.0 \times 10^{-13} = 3.2 \times 10^{-7}\ M$

   (*Chemistry* 7th ed. pages 629–631 / 8th ed. pages 645–647)

   (c) $pH = -\log[H_3O^+] = -\log 3.2 \times 10^{-7} = \underline{6.49}$. (*Chemistry* 7th ed. pages 630–633 / 8th ed. pages 646–649)

   (d) Since raising the temperature from 25°C to 60°C causes the K_w to increase, more product must be forming. If raising the temperature causes the reaction to shift to the right (toward products), then the reaction must be endothermic. (You could also approach this as a bond breaking reaction; bond breaking is always endothermic) (*Chemistry* 7th ed. pages 631–633 / 8th ed. pages 647–649).

2. (a) $H_3PO_4(aq) \rightleftharpoons H^+(aq) + H_2PO_4^-(aq)$

   $H_2PO_4^-(aq) \rightleftharpoons H^+(aq) + HPO_4^{2-}(aq)$

$$HPO_4^{2-}(aq) \rightleftharpoons H^+(aq) + PO_4^{3-}(aq)$$

(b) $H_3PO_4(aq) \rightleftharpoons H^+(aq) + \cancel{H_2PO_4^-}(aq)$

$\cancel{H_2PO_4^-}(aq) \rightleftharpoons H^+(aq) + \cancel{HPO_4^{2-}}(aq)$

$\cancel{HPO_4^{2-}}(aq) \rightleftharpoons H^+(aq) + PO_4^{3-}(aq)$

$H_3PO_4(aq) \rightleftharpoons 3H^+(aq) + PO_4^{3-}(aq)$

(c) Note that only the first dissociation forms significant amounts of H^+, so use:

$K_{a_1} = [H^+][H_2PO_4^-] / [H_3PO_4] = 7.5 \times 10^{-3}$

If $x = [H^+] = [H_2PO_4^-]$

Assume $[H_3PO_4]_o = 7.0 - x \sim 7.0$ M.

$(x)(x) / 7.0 = 7.35 \times 10^{-3}$

$x = \sqrt{(7.35 \times 10^{-3} \times 7.0)} = 0.23$ M $= [H^+]$

pH $= -\log[H^+] = -\log 0.23 = \underline{0.64}$

Note: Check to see that $(7.0 - x) = (7.0 - 0.23)$ is about equal to 7.0, (this is within the allowable 5%, as the rule suggests, since 5% of 7.0 is 0.35), therefore the assumption is acceptable.

(d) From part (c), $[H^+] = \underline{0.23M} = [H_2PO_4^-]$.

In $H_2PO_4^-(aq) \rightleftharpoons H^+(aq) + HPO_4^{2-}(aq)$ $K_a = 6.2 \times 10^{-8}$

$[H^+][HPO_4^{2-}] / [H_2PO_4] = (0.23)[HPO_4^{2-}] / 0.23 = 6.2 \times 10^{-8}$

$[HPO_4^{2-}] = \underline{6.2 \times 10^{-8}}$ M

For $[PO_4^{3-}]$:

$HPO_4^{2-}(aq) \rightleftharpoons H^+(aq) + PO_4^{3-}(aq)$ $K_a = 4.8 \times 10^{-13}$

$[H^+] = 0.23$ M (from above)

$[HPO_4^{2-}] = 6.2 \times 10^{-8}$ M (also, from above)

$$\frac{[H^+][PO_4^{3-}]}{[HPO_4^{2-}]} = \frac{(0.23)[PO_4^{3-}]}{6.2 \times 10^{-8}} = 4.8 \times 10^{-13}$$

$$\left[PO_4^{3-}\right] = \frac{(4.8 \times 10^{-13})(6.2 \times 10^{-8})}{(0.23)} = \underline{1.3 \times 10^{-19}} \text{ M}$$

From part c, pH = 0.64.

pH + pOH = 14.00

pH = 14.00 − 0.64 = <u>13.36</u>

(*Chemistry* 7th ed. pages 650–655, especially sample exercises 14.15, pages 651–652 and 14.16, page 653 / 8th ed. pages 666–671, especially sample exercises 14.15, pages 667–668 and 14.16, page 669)

14

SOLUBILITY

In this chapter, you will apply basic equilibrium concepts involving solids dissolving to form aqueous solutions.

You should be able to

- Write balanced equations for the dissolution of a salt and its corresponding solubility product expression.
- Predict the relative solubilities of salts which dissolve to give the same number of ions from their K_{sp} values.
- Calculate the K_{sp} value from the solubility of a salt and also calculate the solubility of the salt in units of mol/L or g/L from the given K_{sp} value.
- Predict the effect of a common ion on the solubility of a salt and perform calculations.
- Perform calculations to predict if a precipitate will form when two solutions are mixed.
- Do problems involving selective precipitation.
- Perform calculations involving complex ions and solubility.
- Use qualitative analysis to separate a mixture of ions.

AP tip

Solubility problems would appear in the first free-response question, since they would be considered an equilibrium problem. Awareness of the types of problems outlined above and the methods used to solve them will lead you to succeed in this topic on the AP exam.

SOLUBILITY PRODUCT

(*Chemistry* 7th ed. pages 717–718, 721 / 8th ed. pages 744–745, 748)

You already have memorized the solubility rules and know which salts are soluble. For slightly soluble or insoluble salts, equilibrium exists between the solid and its aqueous ions. For example, lead(II) chloride dissolves in water as follows:

$$PbCl_2(s) \rightleftharpoons Pb^{2+}(aq) + 2\ Cl^-(aq)$$

At first, when the salt is added to the water, there are no ions present. As the solid dissolves, the concentration of the ions increases. A simultaneous competing process is the reverse of the dissolution, that is, the reforming of the solid called crystallization. At some point, the maximum amount of dissolution is achieved, which is called the saturation point. However, remember that on a molecular level, a dynamic equilibrium exists between dissolved solute and undissolved solid. (The rate of dissolution equals the rate of crystallization.) The solution is saturated when no more solid dissolves and equilibrium is reached.

The equilibrium expression for the dissolution of lead(II) chloride is

$$K_{sp} = [Pb^{2+}]\ [Cl^-]^2.$$

The constant, K_{sp}, is the solubility product constant. For salts producing the same number of ions, the K_{sp} value can be used to measure the extent to which the solid dissolves. The larger the K_{sp} value, the more soluble the salt.

EXAMPLE: Given the following salts and their K_{sp} values, which salt is the most soluble? Which salt is the least soluble?

Formula	K_{sp}
$NiCO_3$	1.4×10^{-7}
MnS	2.3×10^{-13}
$CaSO_4$	6.1×10^{-5}

The most soluble salt is the salt with the largest K_{sp} value, $CaSO_4$.

The least soluble salt is the salt with the lowest value of K_{sp}, MnS. You are able to compare the K_{sp} values to determine the relative solubilities of the salts because they all produce the same number of ions.

CALCULATIONS INVOLVING SOLUBILITY

(*Chemistry* 7th ed. pages 718–721 / 8th ed. pages 743–749)

The solubility of a salt is the amount of salt that will dissolve in 1 liter of water. The solubility of a salt can be given in units of mol/L or g/L. The solubility of a salt can be used to determine the K_{sp} value for the salt.

CALCULATING K_{SP} FROM SOLUBILITY

EXAMPLE: The solubility of $Pb_3(PO_4)_2$ is 6.2×10^{-12} M. Calculate the K_{sp} value for the solid.

K_{sp} equals 9.9×10^{-55}.

Step 1: Write the reaction for the dissolution of the solid.

$Pb_3(PO_4)_2 (s) \rightleftharpoons 3\ Pb^{2+} (aq) + 2\ PO_4^{3-} (aq)$.

Step 2: Underneath the reaction, make an ICE chart.

	$Pb_3(PO_4)_2(s)$	$\rightleftharpoons$	$3\ Pb^{2+} (aq) +$	$2\ PO_4^{3-} (aq)$
Initial			0	0
Change			$+3x$	$+2x$
Equilibrium			$3x$	$2x$

In the ICE chart above, x represents x mol/L of $Pb_3(PO_4)_2(s)$ dissolving to reach equilibrium which equals 6.2×10^{-12} M.

For every 1 mol per liter of $Pb_3(PO_4)_2$ which dissolves, 3 moles per liter of Pb^{2+} and 2 moles per liter of PO_4^{3-} form.

$3x$ is the mol/L of $Pb^{2+} (aq)$ produced when the solid, $Pb_3(PO_4)_2$, dissolves.

$2x$ is the mol/L of $PO_4^{3-} (aq)$ produced when the solid , $Pb_3(PO_4)_2$, dissolves.

Step 3: Write the equilibrium expression for the reaction and plug in the values from the equilibrium line of the ICE chart.

$K_{sp} = [Pb^{2+}]^3 [PO_4^{3-}]^2 = (3x)^3 (2x)^2 = 108\ x^5$.

The value of x is the solubility of $Pb_3(PO_4)_2$, which equals 6.2×10^{-12} M.

$K_{sp} = 108\ (6.2 \times 10^{-12})^5 = 9.9 \times 10^{-55}$.

Alternate solution:

$$6.2 \times 10^{-12}\ \frac{mol\ Pb_3(PO_4)_2}{1\ L} \times \frac{3\ mol\ Pb^{2+}}{1\ mol\ Pb_3(PO_4)_2} = 1.9 \times 10^{-11} M\ Pb^{2+}.$$

$$6.2 \times 10^{-12} \frac{\text{mol Pb}_3(\text{PO}_4)_2}{1 \text{ L}} \times \frac{2 \text{ mol Pb}_4^{3-}}{1 \text{ mol Pb}_3(\text{PO}_4)_2} = 1.2 \times 10^{-11} \text{M PO}_4^{3-}$$

Plug these values into the K_{sp} expression and solve for K_{sp}.

$K_{sp} = [\text{Pb}^{2+}]^3 [\text{PO}_4^{3-}]^2 = (1.9 \times 10^{-11})^3 (1.2 \times 10^{-11})^2 = 9.9 \times 10^{-55}$.

CALCULATING SOLUBILITY FROM K_{SP}

If you are given a group of salts which do not all have the same cation to anion ratio and asked which is more soluble, you must perform a calculation to determine the solubility of each salt.

EXAMPLE: Given the two salts in the table below, which is more soluble? Show calculations to support your answer.

	K_{sp}
FeC_2O_4	2.1×10^{-7}
$Cu(IO_4)_2$	1.4×10^{-7}

You cannot directly compare the K_{sp} values to predict which is more soluble because the salts dissolve to produce a different number of ions. FeC_2O_4 dissolves to produce two ions and $Cu(IO_4)_2$ produces three ions. You must calculate the solubility for each salt from its K_{sp} value.

Step 1: Write the reaction for the dissolution of the solid.

$FeC_2O_4 (s) \rightleftharpoons Fe^{2+} (aq) + C_2O_4^{2-} (aq)$.

Step 2: Underneath the reaction, make an ICE chart.

	$FeC_2O_4 (s)$ $\rightleftharpoons$	$Fe^{2+} (aq) +$	$C_2O_4^{2-} (aq)$
Initial		0	0
Change		$+x$	$+x$
Equilibrium		x	x

Step 3: Write the equilibrium expression for K_{sp}, plug in the equilibrium line, and solve for x.

$K_{sp} = [\text{Fe}^{2+}][\text{C}_2\text{O}_4^{2-}]$; $2.1 \times 10^{-7} = x^2$; $x = 4.6 \times 10^{-4}$ mol Fe^{2+}/L.

$$4.6 \times 10^{-4} \text{mol Fe}^{2+} \times \frac{1 \text{ mol FeC}_2\text{O}_4}{1 \text{ L}} = 4.6 \times 10^{-4} \text{mol FeC}_2\text{O}_4/\text{L}.$$

Repeat the same calculations for the next salt.

Step 1: Write the reaction for the dissolution of the solid.

$Cu(IO_4)_2 (s) \rightleftharpoons Cu^{2+} (aq) + 2 IO_4^- (aq)$.

Step 2: Underneath the reaction, make an ICE chart.

$$Cu(IO_4)_2\ (s) \rightleftharpoons Cu^{2+}\ (aq)\ +\ 2\ IO_4^-\ (aq)$$

	Cu^{2+}	2 IO$_4^-$
Initial	0	0
Change	+x	+2x
Equilibrium	x	2x

Step 3: Write the equilibrium expression for K_{sp}, plug in the equilibrium line, and solve for x.

$K_{sp} = [Cu^{2+}][\ IO_4^-]^2$; $1.4 \times 10^{-7} = x(2x)^2$; $4x^3 = 1.4 \times 10^{-7}$;

$x = 3.3 \times 10^{-3}$ mol/L.

$$3.3 \times 10^{-3}\ \text{mol Cu}^{2+} \times \frac{1\ \text{mol Cu(IO}_4)_2}{1\ \text{L}} = \frac{3.3 \times 10^{-3}\ \text{mol Cu(IO}_4)_2}{\text{L}}$$

3.3×10^{-3} mol Cu(IO$_4$)$_2$ /L is greater than 4.6×10^{-4} mol FeC$_2$O$_4$ /L.

Therefore Cu(IO$_4$)$_2$ is more soluble even though K_{sp} for FeC$_2$O$_4$ is slightly larger than K_{sp} for Cu(IO$_4$)$_2$.

COMMON ION EFFECT

(Chemistry 7th ed. pages 722–724 / 8th ed. pages 750–752)

When a salt is dissolved in water containing a common ion, its solubility is decreased.

For example, consider the solubility equilibrium of silver sulfate.

$$Ag_2SO_4(s) \rightleftharpoons 2\ Ag^+(aq)\ +\ SO_4^{2-}\ (aq).$$

When silver sulfate is dissolved in 0.100 M AgNO$_3$, the Ag$^+$ ion from silver nitrate causes the equilibrium to shift to the left, decreasing the solubility of silver sulfate.

EXAMPLE: Calculate the molar solubility of Ag$_2$SO$_4$ in 0.10 M AgNO$_3$.

K_{sp} for Ag$_2$SO$_4(s) = 1.2 \times 10^{-5}$.

Fill out the ICE chart. Be sure to include the initial concentration of the ions from the soluble salt, AgNO$_3$.

$$Ag_2SO_4(s) \rightleftharpoons 2\ Ag^+\ (aq)\ +\ SO_4^{2-}\ (aq)$$

	2 Ag$^+$	SO$_4^{2-}$
Initial	0.10	0
Change	+2x	+x
Equilibrium	0.10 + 2x	x

Plug in the equilibrium line into the expression:

$K_{sp} = [Ag^+]^2\ [SO_4^{2-}] = (0.10 + 2x)^2(x)$.

Assume that $0.10 + 2x$ is about 0.10, since K_{sp} is small you can assume that the change ($2x$) from the initial concentration (0.10) is negligible.

$K_{sp} = 1.2 \times 10^{-5} = (0.10)^2(x)$; $x = 1.2 \times 10^{-3}$ mol SO_4^{2-}/L.

$$1.2 \times 10^{-3} \frac{mol}{L} SO_4^{2-} \times \frac{1 \text{ mol } Ag_2SO_4}{1 \text{ mol } SO_4^{2-}} = 1.2 \times 10^{-3} \frac{mol}{L} Ag_2SO_4.$$

The solubility of silver sulfate in 0.100 M silver nitrate, 1.2×10^{-3} mol/L, is less than the solubility of silver sulfate in pure water, 1.4×10^{-2} mol/L.

pH AND SOLUBILITY

Chromium(III) hydroxide dissolves according to the equilibrium

$$Cr(OH)_3 (s) \rightleftharpoons Cr^{3+}(aq) + 3\,OH^-(aq).$$

An increase in pH, caused by the addition of OH^- ions, will shift the equilibrium to the left, decreasing the solubility of $Cr(OH)_3$.

A decrease in pH, caused by the addition of H^+ ions, will shift the equilibrium to the right, increasing the solubility of $Cr(OH)_3$. The H^+ ions remove the OH^- ions from the solution.

A salt with the general formula, MX, will show increased solubility in acidic solution if the anion, X^-, is an effective base (if HX is a weak acid). Common anions that make effective bases include S^{2-}, OH^- and CO_3^{2-}.

EXAMPLE: Calculate the solubility of $Fe(OH)_3$ in a solution with a pH equal to 5.0. $K_{sp} = 4.0 \times 10^{-38}$.

Fill out an ICE chart. Calculate the initial hydroxide concentration from the pH.

pOH = 14 –pH = 14.0 – 5.0 = 9.0 ; $[OH^-] = 1.0 \times 10^{-9}$ M.

	$Fe(OH)_3(s)$	$\rightleftharpoons$	$Fe^{3+}(aq)$ +	$3OH^-(aq)$
Initial			0	1.0×10^{-9}
Change			x	$+3x$
Equilibrium			x	1.0×10^{-9}

$K_{sp} = 4.0 \times 10^{-38} = [Fe^{3+}][\,OH^-]^3 = x(1.0 \times 10^{-9})^3$

Because the pH is 5.0, the equilibrium [OH] must be equal to 1.0×10^{-9}.

$x = 4.0 \times 10^{-11}$ M.

PRECIPITATE FORMATION

(*Chemistry* 7th ed. pages 724–727 / 8th ed. pages 752–755)

A precipitate may or may not form when two solutions are mixed, depending on the concentrations of the ions involved in the formation of the solid.

ION PRODUCT

The ion product, Q, is written in the same way as the K_{sp} expression. For lead(II) chloride, $Q = [Pb^{2+}][Cl^-]^2$

Calculation of the value, Q, involves the use of the initial concentrations of the solutions mixed, $[Pb^{2+}]_0$ and $[Cl^-]_0$, instead of the equilibrium concentrations.

A comparison of the value of Q to K_{sp} determines if a precipitate is formed.

$Q > K_{sp}$: precipitation occurs

$Q < K_{sp}$: no precipitation occurs

$Q = K_{sp}$: the solution is saturated

EXAMPLE: Will a precipitate form when 100.0 mL of 4.0×10^{-4} M $Mg(NO_3)_2$ is added to 100.0 mL of 2.0×10^{-4} M NaOH?

Step 1: Determine the identity of the precipitate formed.

$Mg(OH)_2$ is the possible precipitate. $NaNO_3$ is always soluble.

Step 2: Determine the concentration of the ions after they are mixed and before any reaction occurs.

Determine the moles of concentration of each solute present. Be sure to divide by the total volume of the two solutions mixed.

$[Mg^{2+}]_0 = (0.1000 \text{ L} \times 4.0 \times 10^{-4} \text{ mol/L}) / 0.2000 \text{ L} = 2.0 \times 10^{-4} M$.

$[OH^-]_0 = (0.1000 \text{ L} \times 2.0 \times 10^{-4} \text{ mol/L}) / 0.2000 \text{ L} = 1.0 \times 10^{-4} M$.

Step 3: Write the ion product expression, calculate its value, and compare it to K_{sp}, which equals 8.8×10^{-12}.

$Q = [Mg^{2+}][OH^-]^2 = (2.0 \times 10^{-4} M)(1.0 \times 10^{-4} M)^2 = 2.0 \times 10^{-12}$

Since $Q < K_{sp}$, no precipitate will form.

SELECTIVE PRECIPITATION

(*Chemistry* 7th ed. pages 727–729 / 8th ed. pages 755–757)

A reagent is added to a mixture of metal ions, thus forming a precipitate. One metal ion will precipitate before the other, allowing the mixture to be separated.

EXAMPLE: A solution contains 0.25 M $Ni(NO_3)_2$ and 0.25 M $Cu(NO_3)_2$. A solution of Na_2CO_3 is slowly added to this solution.

a) Will $NiCO_3$ (K_{sp} = 1.4 × 10^{-7}) or $CuCO_3$ (K_{sp} = 2.5 × 10^{-10}) precipitate first?

b) Calculate the concentration of CO_3^{2-} necessary to begin the precipitation of each salt.

c) Determine the concentration of Cu^{2+} when $NiCO_3$ begins to precipitate.

ANSWERS:

a) $CuCO_3$ will precipitate first because its K_{sp} is smaller.

b) For $CuCO_3$, precipitation begins when

$[CO_3^{2-}] = K_{sp}/[Cu^{2+}] = 2.5 × 10^{-10}/0.25\ M = 1.0 × 10^{-9}\ M.$

For $NiCO_3$, precipitation begins when

$[CO_3^{2-}] = K_{sp}/[Ni^{2+}] = 1.4 × 10^{-7}/0.25\ M = 5.6 × 10^{-7}\ M.$

c) The concentration of $[Cu^{2+}]$ when $NiCO_3$ begins to precipitate

$[Cu^{2+}] = K_{sp}/[CO_3^{2-}] = 2.5 × 10^{-10}/5.6 × 10^{-7} M = 4.5 × 10^{-4}\ M.$

COMPLEX ION EQUILIBRIA

(*Chemistry* 7th ed. pages 731–736 / 8th ed. pages 759–764)

A complex ion consists of a metal ion surrounded by ligands which are Lewis bases such as H_2O, OH^-, NH_3, Cl^- and CN^-.

The coordination number is the number of ligands which attach to the transition metal ion. Common coordination numbers include 4 for Cu^{2+} and Co^{2+} and 2 for Ag^+.

The ligands attach one at a time to the metal ion; each step has a formation constant, K_f:

$Ag^+ + S_2O_3^{2-}$ ⇌ $Ag(S_2O_3)^-$ $\qquad K_{f_1} = 7.4 × 10^8$

$Ag(S_2O_3)^- + S_2O_3^{2-}$ ⇌ $Ag(S_2O_3)_2^{3-}$ $\qquad K_{f_2} = 3.9 × 10^4$

Formation of a complex ion causes a precipitate to dissolve. The equilibrium, $AgBr(s) ⇌ Ag^+ + Br^-$ is not affected by the addition of H^+. But the concentration of Ag^+ can be lowered by the addition of excess $S_2O_3^{2-}$ forming the complex ion, $Ag(S_2O_3)_2^{3-}$.

EXAMPLE: Calculate the mass of AgBr that can dissolve in 1.00 L of 0.500 M $Na_2S_2O_3$. K_{sp} for AgBr = 5.0 × 10^{-13}.

Step1: Determine the overall reaction by adding up the reactions for the solubility equilibrium and the stepwise formation of the complex ion. Remember, to get the equilibrium constant for the overall reaction, K', multiply together the K for each step.

	AgBr(s)	⇌	$Ag^+(aq)$ +	$Br^-(aq)$		$K_{sp} = 5.0 \times 10^{-13}$
Ag^+ +	$S_2O_3^{2-}$	⇌	$Ag(S_2O_3)^-$			$K_{f_1} = 7.4 \times 10^8$
$Ag(S_2O_3)^-$ +	$S_2O_3^{2-}$	⇌	$Ag(S_2O_3)_2^{3-}$			$K_{f_2} = 3.9 \times 10^4$

AgBr(s)	$2\,S_2O_3^{2-}$	⇌	$Ag(S_2O_3)_2^{3-}$ +	Br^-	$K' = 14.4$

Step 2: Complete an ICE chart with the overall reaction:

	AgBr(s)	$2\,S_2O_3^{2-}$	⇌	$Ag(S_2O_3)_2^{3-}$ +	Br^-
Initial		0.500		0	0
Change		$-2x$		$+x$	$+x$
Equilibrium		$0.500 - 2x$		x	x

$K' = 14.4 = [Ag(S_2O_3)_2^{3-}][Br^-]/[S_2O_3^{2-}]^2$

$= x^2 / (0.500 - 2x)^2$; take the square root of both sides

$3.79 = x / (0.500 - 2x)$

$x = 1.90 - 7.58x$; $x = 0.221\ M\ Br^- = 0.221\ M\ AgBr(s)$

$1.00\ L \times \dfrac{0.221\ mol\ AgBr}{1\ L} \times \dfrac{187.8\ g}{1\ mol} = 41.5\ g\ AgBr = 42\ g\ AgBr$

QUALITATIVE ANALYSIS

(*Chemistry* 7th ed. pages 729–731 / 8th ed. pages 757–759)

Qualitative analysis involves separating a mixture of cations or anions based on their solubilities. Cations can be separated into five major groups based on their solubilities: Group I: insoluble chlorides; Group II: sulfides soluble in acidic solution; Group II: sulfides insoluble in basic solution; Group IV: insoluble carbonates; and Group V: alkali metal and ammonium ions. Each of these groups can be treated further to separate and identify the individual ions.

EXAMPLE: Separate a mixture containing the Group I cations: Ag^+, Hg_2^{2+} and Pb^{2+}.

The flowchart below shows the separation of Group I ions using qualitative analysis.

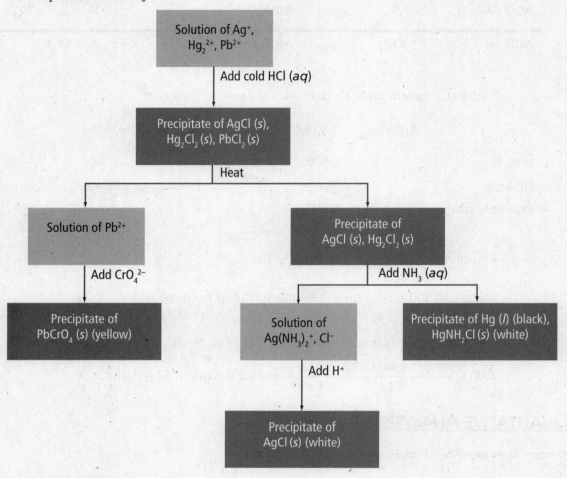

MULTIPLE-CHOICE QUESTIONS

No calculators are to be used in this section.

1. Temperature is often given with K_{sp} values because
   (A) the solubility of solids always increases with increasing temperature.
   (B) the solubility of solids varies with temperature changes.
   (C) solubility changes with temperature but K_{sp} values do not.
   (D) K_{sp} varies with temperature even though concentrations do not.
   (E) the number of ions varies with the kind of salt that is dissolving.

2. The K_{sp} expression for silver phosphate is
   (A) $K_{sp} = [Ag^+][PO_4^{3-}]$.
   (B) $K_{sp} = [Ag^+]^2[PO_4^{3-}]$.
   (C) $K_{sp} = [Ag^+]^3[PO_4^{3-}]$.
   (D) $K_{sp} = [Ag^+][PO_4^{3-}]^3$.
   (E) $K_{sp} = [Ag^+][PO_4^{3-}] / [Ag_3PO_4]$.

3. The solubility of nickel(II) hydroxide, $K_{sp} = 1.6 \times 10^{-16}$, is about
   (A) $\sqrt[3]{4.0 \times 10^{-17}}$.
   (B) $\sqrt[3]{\left(\dfrac{1.6 \times 10^{-16}}{2}\right)}$.
   (C) $\sqrt{1.6 \times 10^{-16}}$.
   (D) $\sqrt{(1.6 \times 10^{-16} \times 3)}$.
   (E) $\sqrt{(1.6 \times 10^{-16} \times 4)}$.

4. Determine the solubility of lead(II) fluoride, $K_{sp} = 4.0 \times 10^{-8}$, in a 0.0040 M lead(II) nitrate solution.
   (A) $2.0 \times 10^{-3} M$
   (B) $2.0 \times 10^{-2} M$
   (C) $\sqrt{4.0 \times 10^{-5}}\ M$
   (D) $\dfrac{\sqrt{1.0 \times 10^{-5}}}{2}\ M$
   (E) $(4.0 \times 10^{-8} / 4.0 \times 10^{-3})\ M$

5. The solubility of magnesium hydroxide ($K_{sp} = 8.0 \times 10^{-12}$) in a buffered solution of pH = 11.00 would be
   (A) $1.0 \times 10^{-11} M$.
   (B) $8.0 \times 10^{10} M$.
   (C) $8.0 \times 10^{-12} M$.
   (D) $8.0 \times 10^{-11} M$.
   (E) $8.0 \times 10^{-6} M$.

6. The solubility of silver sulfide is $8.0 \times 10^{-17} M$. Determine the K_{sp} of this salt.
   (A) 6.4×10^{-50}
   (B) $(8.0 \times 10^{-7})(16.0 \times 10^{-17})$
   (C) $(8.0 \times 10^{-7})(16.0 \times 10^{-17})^2$
   (D) $(8.0 \times 10^{-7})(4.0 \times 10^{-17})$
   (E) $(8.0 \times 10^{-7})(4.0 \times 10^{-17})^2$

7. Equal volumes of $1.6 \times 10^{-5} M$ KCl and $1.6 \times 10^{-5} M$ AgNO$_3$ are mixed. The K_{sp} for silver chloride is 1.6×10^{-10}. As these two solutions intermingle,
   (A) a precipitate of AgCl forms.
   (B) there is no precipitate formed.
   (C) NaCl will precipitate.
   (D) AgNO$_3$ will precipitate.
   (E) the [Na$^+$] will become 0.020 M.

8. A saturated solution of the strong base, MOH, has a pH of 11.00. Determine the K_{sp} for this base.
   (A) 1.0×10^{-6}
   (B) 1.0×10^{-11}
   (C) 5.0×10^{-11}
   (D) 2.5×10^{-12}
   (E) 1.0×10^{-22}

9. Reactions like $Ag^+(aq) + 2NH_3(aq) \rightleftharpoons Ag(NH_3)_2^+(aq)$ show the formation of a complex ion. If the formation constant, K_f, is given,
   (A) the relative stability of the complex ion is known.
   (B) the stability of the Lewis acid is predictable.
   (C) the stability of the metal cation is predictable.
   (D) the pH effects are also indicated.
   (E) the stability of many complex ions may be calculated from these data.

10. The K_{sp} of the metal hydroxide, MOH, is 1.0×10^{-8}. What is the pH of a saturated solution of MOH?
    (A) pH = 4.00.
    (B) pH = 8.00.
    (C) pH = 10.00.
    (D) pH = 12.00.
    (E) pH = 14.00.

11. The pH of a standard solution of H_2S is too low to precipitate ZnS. Which of the following could be added to the solution to cause the ZnS to precipitate?
    (A) HCl
    (B) HNO_3
    (C) H_2S
    (D) NaCl
    (E) NaOH

12. Which of the following describes what happens to the solubility of a slightly soluble ionic compound when a common ion is added to the solution?
    (A) The solubility of the ionic compound is reduced.
    (B) The solubility of the ionic compound is increased.
    (C) There is no effect on the solubility of the ionic compound.
    (D) More of the ionic compound dissolves.
    (E) The ionic compound dissolves more rapidly.

13. The metal center in a complex ion is
    (A) a Lewis base.
    (B) a Lewis acid.
    (C) a ligand.
    (D) insoluble in the solvent.
    (E) a Brønsted-Lowry base.

14. A solution is to be examined for the presence of metal ions. In the first step, aqueous HCl is added to a portion of the solution in order to
    (A) eliminate the carbonates as carbon dioxide and water.
    (B) separate mercury, silver and lead from the rest of the metals.
    (C) precipitate the metals into acidic and basic groups.
    (D) separate cadmium, sodium and copper from the rest of the metals.
    (E) precipitate the salts of sodium of sodium and potassium.

15. Two salts, **AX** and **BX$_2$**, have identical K_{sp} values at a given temperature. We can say
    (A) the salts are more soluble in 0.1M NaX than in water.
    (B) the molar solubility of **AX** is identical to **BX$_2$**.
    (C) addition of NaX will not affect the solubilities of the salts.
    (D) the molar solubility of **AX** is greater than that of **BX$_2$**.
    (E) the molar solubility of **BX$_2$** is greater than that of **AX**.

FREE-RESPONSE QUESTIONS

1. Lead(II) chromate has a K_{sp} of 2.0×10^{-16}. Exactly 4.0 mL of 0.0040 M lead(II) nitrate is mixed with 2.0 mL of 0.00020 M sodium chromate.
    (a) Give the solubility equilibrium reaction.
    (b) Show the K_{sp} expression for this reaction.
    (c) Will a precipitate form? Support your answer.
    (d) What would be the effect on the solubility equilibrium system if a concentrated potassium chromate solution is added dropwise?
    (e) Which is more soluble, AgCl, K_{sp} 2×10^{-10}, or Ag$_2$CO$_3$, $K_{sp} = 8 \times 10^{-12}$?

    Support your answer with calculations.

2. (a) Compare the solubility of solid $Mg(OH)_2$ in water, with a solution of high pH, and with a solution of low pH.
   (b) Apply similar reasoning and explain why copper(II) sulfide is more soluble in acidic solution than in water.
   (c) In the qualitative analysis method of selective precipitation, Group 3 Cations (Co^{2+}, Fe^{2+}, Mn^{2+}, Ni^{2+}, and Zn^{2+}) are removed from an acidified solution by the addition of sodium hydroxide solution, following the removal of Group 2 Cations with H_2S. Explain how solid sulfides of these ions (not hydroxides) are formed.
   (d) Given

   $$Ag^+ + S_2O_3^{2-} \rightleftharpoons Ag(S_2O_3)^- \qquad K = 7.4 \times 10^8$$

   $$Ag(S_2O_3)^- + S_2O_3^{2-} \rightleftharpoons Ag(S_2O_3)_2^{3-} \quad K = 4.0 \times 10^4$$

   Calculate $[Ag(S_2O_3)_2^{3-}]$ when 0.500 L of 2.0 M $Na_2S_2O_3$ is mixed with 0.500 L of 0.040 M $AgNO_3$.

Answers

MULTIPLE CHOICE

1. **B** The solubility of ionic salts is related to temperature. Though the solubility most often increases with increasing temperature, this is not always the case. Since K_{sp} is calculated from solubility, this value is temperature dependent, as are all equilibrium constants (*Chemistry* 7th ed. pages 718–719 / 8th ed. pages 745–747).

2. **C** Begin by writing the equation for dissolving the ionic salt, often called the solubility equilibrium reaction. Recall that K_{sp} is just a special case of the equilibrium constant, which is [products]/[reactants], each raised to a power expressed by its coefficients, and that solids are not shown in the expression. In this case:

   $$Ag_3PO_4(s) \rightleftharpoons 3Ag^+ (aq) + PO_4^{3-} (aq) \quad K_{sp} = [Ag^+]^3[PO_4^{3-}].$$

   (*Chemistry* 7th ed. pages 717–718 / 8th ed. pages 745–747).

3. **A** First show the solubility equilibrium expression:

   $$Ni(OH)_2(s) \rightleftharpoons Ni^{2+}(aq) + 2OH^-(aq).$$

   If x = mol/L of solid dissolving, the $[Ni^{2+}] = x$ and $[OH^-] = 2x$.

   From $K_{sp} = [Ni^{2+}][OH^-]^2 = (x)(2x)^2 = 4x^3$,

   $$\text{solubility} = x = \sqrt[3]{\left(\frac{K_{sp}}{4}\right)}$$

   (*Chemistry* 7th ed. pages 720–721 / 8th ed. pages 748–749)

4. **D** The solubility equilibrium expression is $PbF_2(s) \rightleftharpoons Pb^{2+}(aq) + 2F^-(aq)$.

$K_{sp} = [Pb^{2+}][F^-]^2 = 4.0 \times 10^{-8}$.

$[Pb^{2+}] = 4.0 \times 10^{-3} M$.

Since each mole of $PbF_2(s)$ that dissolves produces two moles of F^- (aq), the solubility of lead(II) fluoride = $[F^-] / 2$.

$$[F^-] = \sqrt{\frac{K_{sp}}{[Pb^{2+}]}} = \sqrt{\left(\frac{4.0 \times 10^{-8}}{4.0 \times 10^{-3}}\right)} = \sqrt{1.0 \times 10^{-5}} \, M.$$

So the solubility of the PbF_2 will equal one-half that amount (*Chemistry* 7th ed. pages 722–724 / 8th ed. pages 750–752).

5. **E** Begin by writing the solubility equilibrium expression:

$Mg(OH)_2(s) \rightleftharpoons Mg^{2+}(aq) + 2OH^-(aq)$.

With a pH=11.00, pOH = 14.00–11.00 = 3.00, and $[OH^-] = 1.0 \times 10^{-3} M$.

$K_{sp} = [Mg^{2+}][OH^-]^2 = 8.0 \times 10^{-12}$

Solubility = $[Mg^{2+}] = K_{sp} / [OH^-]^2 = 8.0 \times 10^{-12} / (1.0 \times 10^{-3})^2 = 8.0 \times 10^{-6} \, M$.

(*Chemistry* 7th ed. page 724 / 8th ed. page 752)

6. **C** The equation for the solubility reaction is

$Ag_2S(s) \rightleftharpoons 2Ag^+(aq) + S^{2-}(aq)$.

$K_{sp} = [Ag^+]^2[S^{-2}]$.

Keeping in mind that there are two silver ions formed for each sulfide ion formed, $K_{sp} = (16. \times 10^{-7})^2 (8.0 \times 10^{-7})$. (*Chemistry* 7th ed. pages 634–637 / 8th ed. pages 650–653)

7. **B** When the equal volumes are mixed but before there is any reaction, $[Ag^+] = 0.80 \times 10^{-5} M$ and $[Cl^-] = 0.80 \times 10^{-5} M$.

So the ion product, Q, = $(0.80 \times 10^{-5})(0.80 \times 10^{-5}) = 6.4 \times 10^{-11}$.

Since the K_{sp} value of 1.6×10^{-10} is not exceeded by Q, a precipitate of AgCl will not form (*Chemistry* 7th ed. pages 638–639 / 8th ed. pages 654–655).

8. **A** The reaction: $MOH(s) \rightleftharpoons M^+(aq) + OH^-(aq)$. If pH = 11.00, then $[OH^-] = 1.0 \times 10^{-3} M$, which is also the value for $[M^+]$. $K_{sp} = [M^+][OH^-] = (1.0 \times 10^{-3})(1.0 \times 10^{-3}) = 1.0 \times 10^{-6}$ (*Chemistry* 7th ed. pages 642–644 / 8th ed. pages 658–661).

9. **A** If the K_f value is very large, then the complex ion is quite stable in solution, so this is a good predictor of stability for this complex ion. Since K_f values (formation constants) are given for specific complex ions, this only allows questions about the one equilibrium system. (Note: the Lewis acid in this reaction is the metal cation.) (*Chemistry* 7th ed. pages 647–648 / 8th ed. pages 663–664)

10. **C** First determine the solubility of MOH, which will lead to $[OH^-]$.

   The reaction $MOH(s) \rightleftharpoons M^+(aq) + OH^-(aq)$.

   $K_{sp} = [M^+][OH^-] = 1.0 \times 10^{-8}$, and $[M^+] = [OH^-] =$ the solubility of MOH.

   The solubility then is $\sqrt{1.0 \times 10^{-8}} = 1.0 \times 10^{-4}\,M = [OH^-]$.

   Therefore pOH $= -\log[OH^-] = -\log(1.0 \times 10^{-4}) = 4.00$ from pH + pOH = 14.00, pH = 14.00 − 4.00 = 10.00 (*Chemistry* 7th ed. pages 642–644 / 8th ed. pages 658–660).

11. **E** The anion of a weak acid will show increased solubility in an acidic solution and thus salts containing the anions of weak acids are more soluble in acidic solutions than in pure water. Thus the only way to decrease solubility is to raise the pH and in this case, add a base, NaOH (*Chemistry* 7th ed. page 724 / 8th ed. page 752).

12. **A** The solubility of a slightly soluble ionic compound is always lowered whenever the solution already contains ions common in the solid or ions common to the solid that are added after a solution is prepared. The common-ion effect is an example of LeChâtelier's principle (*Chemistry* 7th ed. pages 682, 722–723 / 8th ed. pages 750–752).

13. **B** A Lewis acid is an electron-pair acceptor and all metallic centers accept the electron pairs from the ligands (*Chemistry* 7th ed. pages 663, 731 / 8th ed. pages 679, 759).

14. **B** One should recognize that these three common metals are the ones which form the insoluble chlorides (*Chemistry* 7th ed. pages 729–730 / 8th ed. pages 757–758).

15. **D** For each salt, the respective solubility equilibrium expressions are:

   $$AX_{(s)} \rightleftharpoons A^+_{(aq)} + X^-_{(aq)} \quad \text{and} \quad BX_{2(s)} \rightleftharpoons B^+_{(aq)} + 2X^-_{(aq)}$$

   Since they have the same K_{sp} values, let us call that value N.

   Let "a" represent the solubility of **AX**. Thus, $(a)(a) = N$ or $a = \sqrt{N}$

   Let "b" represent the solubility of **BX₂**. Thus $(b)(2b)^2 = N$ or $b = \sqrt[3]{N/4}$

   Since a>b the molar solubility of **AX** is greater than that of **BX₂** (*Chemistry* 7th ed. pages 717–721 / 8th ed. pages 748–750).

FREE RESPONSE

1   (a) $PbCrO_4(s) \rightleftharpoons Pb^{2+}(aq) + CrO_4^{2-}(aq)$

   (b) $K_{sp} = [Pb^{2+}][CrO_4^{2-}]$

   (c) $Q = \left[\left(\dfrac{0.0040\,M}{0.0060\,L} \right) 0.0040\,L \right] \left[\left(\dfrac{0.00020\,M}{0.0060\,L} \right) 0.0020\,L \right] = 1.8 \times 10^{-7}.$

   Since the K_{sp} (2.0×10^{-16}) has been exceeded, a precipitate does form.

   (d) This would cause an equilibrium shift to the left, and more solid $PbCrO_4$ will form.

   (e) With a 1:1 salt, like AgCl, solubility = $\sqrt{K_{sp}}$, while with a 2:1 salt, solubility = $\sqrt[3]{\left(\dfrac{K_{sp}}{4} \right)}$.

   Using the above gives solubility of 1×10^{-5} M for AgCl and $2 \times 10^{-4} M$ for $PbCrO_4$, so it can be seen that lead(II) chromate is more soluble even though it has the lower K_{sp} value. (K_{sp} can only be used for a direct comparison of solubility if a 1:1 salt is compared to a 1:1 salt, or if a 2:1 salt is compared to a 2:1 salt, etc.) (*Chemistry* 7th ed. pages 638–639 / 8th ed. pages 654–655)

2. (a) First consider the solubility equilibrium reaction:

   $$Mg(OH)_2(s) \rightleftharpoons Mg^{2+}(aq) + 2OH^-(aq).$$

   If pH < 7, H^+ reacts with OH^- to form water; this causes a shift in the solubility equilibrium toward the right, increasing the solubility of $Mg(OH)_2$.

   If pH > 7, the OH^- ions cause a shift in the solubility equilibrium to the left, forming more solid, decreasing the solubility of magnesium hydroxide.

   Summary: $Mg(OH)_2$ will be most soluble in acidic solution, less in water, and even less in basic solution.

   (b) Again, first consider the solubility equilibrium reaction,

   $$CuS(s) \rightleftharpoons Cu^{2+}(aq) + S^{2-}(aq)$$

   and the reactions in acidic solution,

   $$S^{2-}(aq) + H^+(aq) \rightleftharpoons HS^-(aq)$$

   $$HS^-(aq) + H^+(aq) \rightleftharpoons H_2S(aq).$$

The reaction of the S^{2-} with H^+ (and the HS^- with H^+) cause the solution equilibrium to shift to the right allowing more of the CuS to dissolve.

(c) When the system is acidified, the reaction with metal ions, M^{2+}, is $H_2S\ (aq) + M^{2+}(aq) \rightleftharpoons MS(s) + 2H^+(aq)$. The addition of the acid has kept this equilibrium shifted to the left. Adding a base forms water, (from $H^+ + OH^- \longrightarrow H_2O$) shifting the equilibrium toward the right, forming insoluble sulfides.

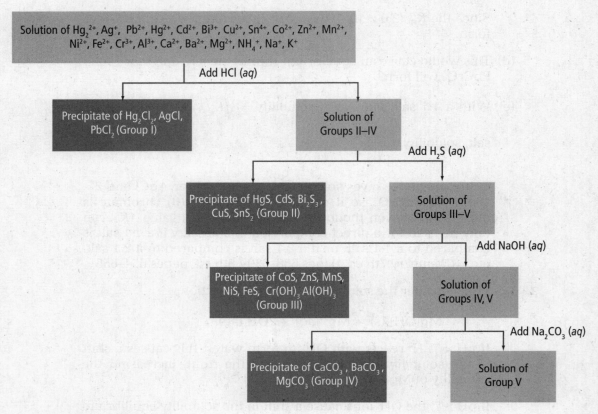

(d) 0.020 M

With the large excess of $S_2O_3^{2-}$ ions these two reactions go to completion (i.e. equilibrium is far toward the products). This means that essentially all of the Ag^+ goes to form the complex ion, $Ag(S_2O_3)_2^{3-}$; hence

$$[Ag+] = \frac{(0.040\ M \times 0.500\ L)}{(0.500\ L + 0.500\ L)} =$$

$[Ag(S_2O_3)_2^{3-}]$ after reaction = 0.0200 M.

(Chemistry 7th ed. pages 724–734 / 8th ed. pages 752–762)

15

ELECTROCHEMISTRY

Electrochemistry is the study of the interchange of electrical and chemical energy. There are two types of electrochemical cells, galvanic and electrolytic. In galvanic cells, spontaneous oxidation-reduction (redox) reactions generate electric current. In electrolytic cells, a nonspontaneous chemical reaction occurs with the application of an electric current.

You should be able to

- Identify and compare the two types of electrochemical cells: galvanic and electrolytic.
- Draw and label a galvanic cell, including labeling the electrodes, the flow of electrons, and the flow of ions.
- Write half-reactions and determine which reaction occurs at the anode and which reaction occurs at the cathode.
- Give the line notation for a galvanic cell or write a balanced redox reaction from the given line notation.
- Calculate the cell potential for a galvanic cell and an electrolytic cell.
- Determine if a reaction is spontaneous from its cell potential.
- Calculate the cell potential under nonstandard conditions when the solutions are not 1M. This involves the use of the Nernst equation.
- Determine the strengths of oxidizing agents and reducing agents.
- Draw and label an electrolytic cell.
- Determine the reactions which occur at the anode and the cathode during electrolysis.
- Perform stoichiometric calculations involving electrolysis.

GALVANIC CELLS

COMPONENTS OF THE GALVANIC CELL

(*Chemistry* 7th ed. pages 791–793 / 8th ed. pages 823–825)

In a galvanic cell, a spontaneous chemical reaction is used to produce electrical energy. The current produced by the cell is measured in volts by a voltmeter. One compartment of the galvanic cell contains the anode, where oxidation, a loss of electrons, occurs. The other compartment contains the cathode where reduction, a gain of electrons, occurs. The electrodes, the anode, and the cathode are immersed in respective solutions containing metal cations. The two compartments of a galvanic cell are connected by a salt bridge or a porous disk. This is shown in the figure to the right.

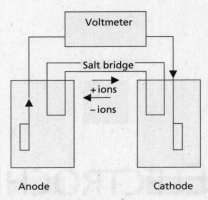

The oxidizing agent is the substance being reduced at the cathode.
The reducing agent is the substance being oxidized at the anode.

LINE NOTATION

(*Chemistry* 7th ed. pages 798–799 / 8th ed. pages 831–832)

A galvanic cell can be abbreviated with line notation.

reactant /product ‖ reactant/product
anode reaction   cathode reaction

The salt bridge is indicated by the symbol ‖.

EXAMPLE: Give the correct line notation for the galvanic cell pictured below.

ANSWER: $Zn/Zn^{2+}\|H^+/H_2$

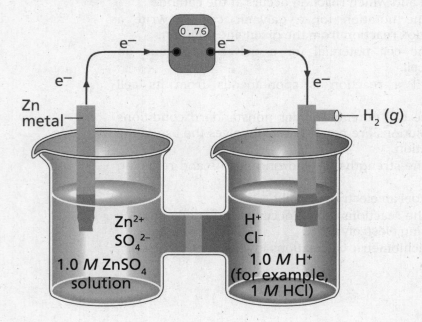

AP tips

Here are some mnemonic devices to help you remember some facts about electrochemistry and redox reactions.

■ "LEO" goes "GER" means Loss of Electrons is Oxidation and Gain of Electrons is Reduction.

■ To recall what happens at the anode and the cathode: RedCat and AnOx means reduction occurs at the cathode and oxidation occurs at the anode.

■ To know the migration of ions toward the electrodes for both types of cells, "CAT" ions move to the "CAT" ode and "AN" ions move to the "AN" ode.

You will be provided with a table of standard reduction potentials on the AP test.

You should be able to sketch a galvanic cell and label the electrodes, the flow of electrons, and the flow of ions.

STANDARD REDUCTION POTENTIALS

(*Chemistry* 7th ed. pages 794–797 / 8th ed. pages 826–830)

The cell potential, E_{cell}, is the potential of the cell to do work on its surroundings by driving an electric current through a wire. By definition, a potential of 1 volt is produced when 1 joule of energy moves 1 coulomb of electric charge across a potential. The magnitude of the cell potential is a measure of the driving force behind an electrochemical reaction. Sometimes it is referred to as the electromotive force or emf. Tables of reduction potentials give standard voltages for reduction half-reactions measured at standard conditions of 1 atmosphere, 1 molar solution, and 25°C.

CALCULATING THE CELL POTENTIAL OF A GALVANIC CELL

(*Chemistry* 7th ed. pages 797–798 / 8th ed. pages 830–831)

The reaction occurring in a galvanic cell can be broken down into an oxidation half-reaction and a reduction half-reaction. Using the Table of Standard Reduction Potentials on page 30 of this book, you can calculate the cell potential of the overall reaction.

EXAMPLE: Consider a galvanic cell based on the reaction:

$$Al + Ni^{2+} \longrightarrow Al^{3+} + Ni$$

Give the balanced cell reaction and calculate the cell potential, $E°_{cell}$, for the reaction.

Step 1: Write the oxidation and reduction half-reactions.

Oxidation: $Al \longrightarrow Al^{3+} + 3 e^-$ Reduction: $Ni^{2+} + 2 e^- \longrightarrow Ni$

Step 2: For the reduction half-reaction, look up the potential in the table.

$Ni^{2+} + 2e^- \longrightarrow Ni$ $E^\circ_{red} = -0.23$ V.

Step 3: For the oxidation half-reaction, $E^\circ_{ox} = -E^\circ_{red}$.

The oxidation half-reaction must be the reverse of the reduction reaction that you find in the table for the value of E°_{red}.

Oxidation: $Al \longrightarrow Al^{3+} + 3e^-$ $E^\circ_{ox} = -E^\circ_{red} = -(-1.66V) = +1.66$ V.

Step 4: The cell potental for the overall reaction is equal to the sum of the reduction potential, E°_{red}, and the oxidation potential, E°_{ox}.

$E^\circ_{cell} = E^\circ_{ox} + E^\circ_{red}$; $E^\circ_{cell} = -0.23$ V $+ 1.66$ V $= 1.43$ V.

To obtain the balanced cell reaction, you must make sure that the electrons lost equal the electrons gained. When multiplying the half-reactions through by a coefficient, do not change the value of E°.

$3 (Ni^{2+} + 2e^- \longrightarrow Ni)$ $\qquad E^\circ_{red} = -0.23$ V

$2 (Al \longrightarrow Al^{3+} + 3 e^-)$ $\qquad E^\circ_{ox} = +1.66$ V

$3 Ni^{2+} + 2 Al \longrightarrow 3 Ni + 2 Al^{3+}$ $E^\circ_{cell} = 1.43$ V

SPONTANEOUS REACTIONS

(*Chemistry* 7th ed. pages 800–803 / 8th ed. pages 833–836)

Gibbs free energy, ΔG°, can be calculated from the cell potential, E°_{cell}.
$$\Delta G^\circ = -nFE^\circ_{cell}$$
Faraday's constant, F, has a value of 96,485 C/mol e⁻.
The number of moles of electrons transferred in a redox reaction is represented by n.
A spontaneous reaction is one that has a negative value for ΔG° or a positive value for E°_{cell}.

EXAMPLE: Will 1 M HCl dissolve silver metal and form Ag^+ solution?

Write the half reactions and calculate E°_{cell}.

$2H^+ + 2 e^- \longrightarrow 2H_2$ $\qquad E^\circ_{red} = 0.00$ V

$2Ag \longrightarrow 2Ag^+ + 2e^-$ $\qquad E^\circ_{ox} = -0.80$ V

$2H^+ + 2 Ag \longrightarrow H_2 + 2 Ag^+$ $\qquad E^\circ_{cell} = -0.80$ V

The negative value for E°_{cell} indicates that the reaction will not occur.

You may be asked if an element or ionic species is capable of reducing another element or ion. To determine if the reaction will occur, write the half-reactions and calculate the cell potential as in the previous example.

EXAMPLE: Bromine, Br_2, can oxidize iodide, I^-, to iodine, I_2. However, Br_2 cannot oxidize chloride, Cl^-, to chlorine, Cl_2. Explain why the first reaction occurs yet, the second one does not.

Begin by writing the appropriate half-reactions. Then calculate the cell potential for the overall reaction.

First the reaction in which Br_2 oxidizes I^-:

$$Br_2 + 2e^- \longrightarrow 2Br^- \qquad E^\circ_{red} = 1.09 \text{ V}$$

$$\underline{2I^- \longrightarrow I_2 + 2e^- \qquad E^\circ_{ox} = -0.54 \text{ V}}$$

$$Br_2 + 2I^- \longrightarrow 2Br^- + I_2 \qquad E^\circ_{cell} = 0.55 \text{ V}$$

This reaction occurs; E°_{cell} is positive.

$$Br_2 + 2e^- \longrightarrow 2Br^- \qquad E^\circ_{red} = 1.09 \text{ V}$$

$$\underline{2Cl^- \longrightarrow Cl_2 + 2e^- \qquad E^\circ_{ox} = -1.36 \text{ V}}$$

$$Br_2 + 2Cl^- \longrightarrow 2Br^- + Cl_2 \qquad E^\circ_{cell} = -0.27 \text{ V}$$

This reaction does not occur; E°_{cell} is negative

CELL DEPENDENCE ON CONCENTRATION

(*Chemistry* 7th ed. pages 803–806 / 8th ed. pages 836–840)

The galvanic cell represented by the reaction
$$3 Ni^{2+} + 2 Al \longrightarrow 3 Ni + 2 Al^{3+}$$
has a cell potential, E°_{cell}, equal to 1.43 V under standard conditions (all solutions are 1 molar).

Increasing the concentration of Ni^{2+} will shift the reaction to the right by Le Châtelier's principle, increasing the driving force on the electrons and increasing the cell potential.

The relationship between the cell potential and concentrations at 25°C is given by the Nernst equation:

$$E_{cell} = E^\circ_{cell} - (0.0591/n) \log Q$$

The cell potential, E_{cell}, is for nonstandard conditions.
The moles of electrons transferred are represented by n.
The mass action quotient is represented by Q.

At equilibrium, E_{cell} equals 0.

EXAMPLE: Calculate the cell potential for the reaction
$$3 Ni^{2+} + 2 Al \rightarrow 3 Ni + 2 Al^{3+}$$
in which $[Al^{3+}] = 2.00 \, M$ and $[Ni^{2+}] = 0.750 \, M$.

You already know that $E^\circ_{cell} = 1.43$ V and the number of moles of electrons transferred, n, equals 6.

$$Q = [Al^{3+}]^2 / [Ni^{2+}]^3 = (2.00)^3/(0.750)^2 = 14.2$$

$$E_{cell} = 1.43 \text{ V} - (0.0591/6) \log 14.2 = 1.42 \text{ V}$$

DETERMINING THE STRENGTH OF OXIDIZING AND REDUCING AGENTS

(*Chemistry* 7th ed. pages 819–820 / 8th ed. pages 851–852)

You may be asked to list atoms or ions in order of increasing strength as reducing agents or oxidizing agents.

For a substance to be oxidized, it must lose electrons and another substance must gain electrons because oxidation and reduction always occur together. The substance that causes another substance to be oxidized is called an oxidizing agent. An oxidizing agent is reduced; it is the reactant in the reduction half-reaction. The larger (more positive) $E°_{red}$ indicates the stronger oxidizing agent.

A reducing agent is oxidized; it is the reactant in the oxidation half-reaction. The larger (more positive) the $E°_{ox}$ indicates the stronger reducing agent.

> **EXAMPLE:** Classify each of the following as an `oxidizing agent, reducing agent, or both. Within each list, arrange in order of increasing strength as oxidizing agents and reducing agents.
>
> Br_2, Mg, Fe^{2+}, I_2, Cl^-, Cu^{2+}
>
> Oxidizing agents : $Fe^{2+} < Cu^{2+} < I_2 < Br_2$
>
> Reducing agents: $Cl^- < Fe^{2+} < Mg$

To be an oxidizing agent, a substance must be capable of gaining electrons or being reduced. Of the species listed, Mg and Cl^- are the only ones listed that cannot have a lower oxidation state. For the oxidizing agents listed above, the respective reduction potentials are –0.44 V, 0.16 V or 0.34 V for Cu^{2+} (which can be reduced to Cu^0 or Cu^+), 0.54 V, and 1.09 V. The more positive the cell potential indicates the stronger oxidizing agent.

Reducing agents must be capable of being oxidized to a higher oxidation state. Cl^- and Mg can go to Cl^0 and Mg^{2+}. Fe^{2+} can exist as Fe^{3+} or Fe^0 so it can act as an oxidizing agent or reducing agent. For the reducing agents listed above, the corresponding oxidation potentials are –1.36 V, –0.77 V, +2.37 V. The more positive the cell potential indicates the stronger reducing agent.

ELECTROLYTIC CELLS

(*Chemistry* 7th ed. pages 816–818 / 8 h ed. pages 847–850)

In an electrolytic cell, a nonspontaneous reaction is made to occur by forcing an electric current through the cell. In an earlier example, it was shown that the following reaction is spontaneous:

$$3 \text{ Ni}^{2+} + 2\text{Al} \longrightarrow 3 \text{ Ni} + 2 \text{ Al}^{3+}$$

The reverse of this reaction: $3 \text{ Ni} + 2\text{Al}^{3+} \longrightarrow 3 \text{ Ni}^{2+} + 2\text{Al}$ is nonspontaneous and can be made to occur by the addition of an external power source. This electrolytic cell can be set up with two compartments just like the galvanic cell, with the replacement of a power supply for the voltmeter.

In the process of electroplating, the electrolytic cell can also be set up using only one compartment, as shown in the figure below.

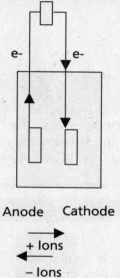

For example, if an object is to be plated with copper, make it the cathode and immerse it into a copper(II) sulfate solution. At the cathode, the reaction that will occur and deposit copper onto the object is $Cu^{2+} + 2e^- \longrightarrow Cu$. The anode can also be made of copper. The oxidation of copper occurs at this anode.

REACTIONS THAT OCCUR IN AN ELECTROLYTIC CELL

To determine which reaction occurs at the anode and the cathode during electrolysis, you must consider all possible reactions and their reduction and oxidation potentials. If the reaction takes place in an aqueous solution, the oxidation and reduction of water must be considered.

EXAMPLE: A solution of copper (II) sulfate is electrolyzed. Calculate the cell potential of the reaction, E^o_{cell}.

	Possible reactions[1]	Cell potential, E^o(V)
Cathode	$Cu^{2+} + 2e^- \longrightarrow Cu$	0.34
	$SO_4^{2-} + 4H^+ + 2e^- \longrightarrow$	0.20
	$H_2SO_3 + H_2O$	
	$2H_2O + 2e^- \longrightarrow H_2 + 2OH^-$	−0.83
Anode	$Cu \longrightarrow Cu^{2+} + 2e^-$	−0.34
	$2H_2O \longrightarrow O_2 + 4H^+ + 4e^-$	−1.23

---

[1]Note: There is no half-reaction for the oxidation of SO_4^{2-}. S in SO_4^{2-} is in its highest oxidation state, 6+, and cannot be oxidized further.

For each electrode, the reaction with the more positive potential will occur. At the cathode, Cu^{2+} will be reduced. At the anode, Cu will be oxidized.

$$Cu^{2+} + 2e^- \longrightarrow Cu$$

$$Cu \longrightarrow Cu^{2+} + 2e^-$$

AP tip

Frequently, the electrodes are inert for electrolysis. For example, during the electrolysis of KI (aq) K^+, I^-, and H_2O are the only species present. Only I^- and H_2O are present to be oxidized at anode. Note: In aqueous KI, there is no K(s) to be oxidized.

STOICHIOMETRY OF ELECTROLYTIC PROCESSES

(*Chemistry* 7th ed. page 818 / 8th ed. pages 849–850)

In this section, you will review how much chemical change occurs with the flow of a given current for a specified time. You might be asked how much metal was plated (formed) or how long an electroplating process will take or how much current is required to produce a specified amount of metal over a period of time.

Some units to be familiar with include A, amperes; 1 A = 1 C/s; coulombs, C; Faraday's constant is 96,485 C = 1 mol e^-.

EXAMPLE: A current of 10.0 A is passed through a solution containing M^{2+} for 30.0 min. It produces 5.94 g of metal, M. Determine the identity of metal, M.

10.0 A = 10.0 C/s × 30.0 min × 60 s /min = 1.80×10^4 C

1.80×10^4 C × 1 mol e^- / 96, 485 C × 1 mol M / 2 mol e^-

= 9.33×10^{-2} mol M

Molar mass of M = g M / mol M = 5.94 g M / 9.33×10^{-2} mol M = 63.7 g/mol is Cu.

This can also be shown in one step:

$$5.94 \text{ g M} \times \frac{2 \text{ mole } e^-}{1 \text{ mol } M} \times \frac{96485 \text{ C}}{10.0 \text{ C}} \times \frac{1 \text{ s}}{\text{mole } e^-} \times \frac{1 \text{ min}}{60 \text{ s}} \times \frac{1}{30.0 \text{ min}} = 63.7 \text{ g/mol}$$

COMPARISON OF GALVANIC AND ELECTROLYTIC CELLS

Galvanic and electrolytic cells have a few features in common. For both types of cells, reduction always occurs at the cathode and oxidation at the anode. In an electrolytic cell, electrons travel from the battery to the cathode. In both cases, electrons travel in the wire, but you wouldn't say the electrons travel from the anode to the cathode in an electrolytic cell. Positive ions are always attracted to the cathode whether the cell is electrolytic or galvanic. The table below compares the galvanic cell to the electrolytic cell.

	Galvanic	Electrolytic
Sign of the cathode	+	–
Sign of the anode	–	+
Ions attracted to the cathode	Cations (+)	Cations (+)
Ions attracted to the anode	Anions (–)	Anions(–)
Sign of $E^o{}_{cell}$	+	–
Spontaneity	Spontaneous	Nonspontaneous

In both types of cells, the + ions or cations move toward the cathode because there is an excess of negative ions at the cathode caused by the reduction of + ions in solution. Likewise, oxidation at the anode produces + ions, so negative ions or anions in the salt bridge must move to the anode to maintain electrical neutrality.

MULTIPLE-CHOICE QUESTIONS

No calculators are to be used on this part of the exam.

1. In the reaction $MnO_4^- + 5Fe^{2+} + 8H^+ \longrightarrow Mn^{2+} + 5Fe^{3+} + 4H_2O$,
    (A) Fe^{3+} is oxidized and Mn^{2+} is reduced.
    (B) H^+ is oxidized and MnO_4^- is reduced.
    (C) Fe^{2+} is oxidized and MnO_4^- is reduced.
    (D) electrons are transferred from Mn^{2+} to Fe^{2+}.
    (E) electrons are transferred from H^+ to H_2O.

AP tip

Remember that you may be asked to write a balanced equation such as the one used in question 1, and then answer a related question about the chemical species being oxidized or reduced or about the change in oxidation of one of the elements.

2. The $E°$ for Mg $\longrightarrow$ Mg^{2+} + $2e^-$ is +2.37 volts.

   The $E°$ for $2Mg^{2+} + 4e^- \longrightarrow 2$ Mg is
   (A) 2.37 volts.
   (B) –2.37 volts.
   (C) 4.74 volts.
   (D) –4.74 volts.
   (E) zero.

3. The function of a salt bridge (or porous barrier) in an electrochemical cell (galvanic cell) is to allow ions to flow
   (A) though the wire from reducing agent to oxidizing agent.
   (B) to encourage charge building up on both sides of the cell.
   (C) to keep the net charge on each side at zero.
   (D) from the oxidizing side of the cell to the reducing side.
   (E) from the reducing agent side of the cell to the oxidizing agent side.

4. What is the value of $\Delta G°$ for the reaction written below:

   $2MnO_4^-(aq)+16H_3O^+(aq)+5Zn(s)\rightarrow 2Mn^{2+}(aq)+24H_2O(l)+ 5\ Zn^{2+}(aq)$

   You are given the following standard reduction potentials for the two half reactions:

   $E°(MnO_4^- \mid Mn^{2+}) = +1.49$ V

   $E°(Zn^{2+} \mid Zn) = -0.76$V
   (A) -704 kJ
   (B) -1.34 MJ
   (C) -1.92 MJ
   (D) -2.17 MJ
   (E) -4.71 MJ

5. Given: $Mg^{2+} +2e^- \longrightarrow$ Mg $\qquad$ –2.37 V

   $\qquad Fe^{3+} + 1e^- \longrightarrow Fe^{2+} \qquad + 0.77$V.

   When the reaction $2Mg + Fe^{3+} \longrightarrow 2Mg^{2+} + Fe^{2+}$ comes to

   equilibrium, the $E°_{cell}$ value (cell potential) becomes
   (A) 3.91 V.
   (B) 3.14 V.
   (C) 1.60 V.
   (D) 0.83 V.
   (E) 0.00 V.

6. For the reaction $Cu^{2+}(aq) + Zn(s) \longrightarrow Zn^{2+}(aq) + Cu(s)$ $E° = +1.10$ V. How many faradays are transferred as 1.00 mole of zinc is consumed?
   (A) $96,500 \times 2.00 \times 6.023 \times 10^{23}$
   (B) 96,500
   (C) $96,500 \times 2$
   (D) 2.00
   (E) zero

7. The galvanic cell based on $Ag^+ + 1e^- \longrightarrow Ag$ ($E° = +0.80$ V) and $Fe^{3+} + 1e^- \longrightarrow Fe^{2+}$ ($E° = +0.77$ V)
   (A) has electrons lost by Ag^+ and gained by Fe^{2+}.
   (B) has the mass of the silver electrode decreasing as this reaction proceeds.
   (C) has an overall cell potential difference (voltage) of +1.57 V.
   (D) has electrons flowing from Fe^{2+} to Ag^+.
   (E) has the compartment containing the Ag^+ as the anode.

8. Cathodic protection is a method often used to protect iron. Selecting from the list of four metals, Mg, Zn, Sn, and Cu, a chemist could protect iron with
   (A) Zn or Mg.
   (B) Sn or Cu.
   (C) Zn or Ag.
   (D) Cu or Mg.
   (E) Zn or Cu.

9. The Nernst equation is helpful in calculating $E°_{cell}$ values for electrochemical cells in which
   (A) the temperature is not 0°C.
   (B) the temperature is not 25°C.
   (C) the concentrations of the solutions are not all 1.00 M.
   (D) equilibrium has been reached.
   (E) the value of K_C is very large.

10. Copper may be used for electroplating, with a half-reaction of $Cu^{2+} + 2e^- \longrightarrow Cu$. If a current of 10.0 A is applied to a Cu^{2+} solution for 60.0 minutes, the mass of copper plated out can be calculated as
    (A) $10.0 \times 3600. \times 1 / 96,500 \times 2.00 / 1 \times 63.5$.
    (B) $10.00 \times 3600. \times 6.02 \times 10^{23} \times 1 / 2.00 \times 63.5$.
    (C) $96,500 \times 1 / 10.0 \times 1 / 2.00 \times 1 / 3600. \times 63.5$.
    (D) $96,500 \times 1 / 10.0 \times 1 / 2.00 \times 1 / 3600. \times 1 / 63.5$.
    (E) $96,500 \times 10.0 \times 3600. \times 2.00 \times 1 / 63.5$.

11. Given the following data:

$$Ca^{2+}(aq) + 2e^- \rightarrow Ca\ (s)\quad E^\circ = -2.87\ V$$

$$Zn^{2+}(aq) + 2e^- \rightarrow Zn(s)\quad E^\circ = -0.76\ V$$

$$Co^{2+}(aq) + 2e^- \rightarrow Co(s)\quad E^\circ = -0.28\ V$$

$$Sn^{2+}(aq) + 2e^- \rightarrow Sn(s)\quad E^\circ = -0.14\ V$$

$$Pb^{2+}(aq) + 2e^- \rightarrow Pb(s)\quad E^\circ = -0.13\ V$$

Which of the following correctly describes the ease of oxidation of the substances listed under standard state conditions?
(A) $Ca^{2+} > Zn^{2+} > Co^{2+} > Sn^{2+} > Pb^{2+}$
(B) $Pb^{2+} > Sn^{2+} > Co^{2+} > Zn^{2+} > Ca^{2+}$
(C) $Ca > Zn > Co > Sn > Pb$
(D) $Pb > Sn > Co > Zn > Ca$
(E) $Pb > Ca^{2+} > Zn > Sn^{2+} > Co^{2+}$

Use the following data to answer questions 12, 13 and 14.

$$Ag^+ + e^- \rightarrow Ag\quad E^\circ = 0.80\ V$$

$$Al^{3+} + 3e^- \rightarrow Al\quad E^\circ = -1.66V$$

12. What is the E°_{cell} for a voltaic cell using the two half-reactions at 25°C?
(A) $-2.46\ V$
(B) $-0.74\ V$
(C) $+0.74\ V$
(D) $+0.86\ V$
(E) $+2.46\ V$

13. What would happen to the cell emf if NH_3 were added to the silver cell and $Ag(NH_3)_2^+$ forms?
(A) no change
(B) increased
(C) reduced

14. What is the value of K for the reaction at 25°C?
(A) 10^{125}
(B) 10^{42}
(C) 10^{-125}
(D) 10^{-42}
(E) 1

15. All of the following statements concerning galvanic cells are true
except
(A) the two half-cells are connected by a salt bridge.
(B) electrons flow from the anode to the cathode.
(C) oxidation occurs at the cathode.
(D) voltaic cells can be used as a source of energy.
(E) a voltaic cell consists of two half-cells.

FREE-RESPONSE QUESTIONS

Calculators may be used on this part of the exam, as well as Tables and Equations.

1. (a) Which occurs more often in nature, oxidation or reduction? Discuss your answer.
   (b) The half-reaction for the hydrogen electrode, $2H^+ + 2e^- \longrightarrow$ H_2, has been given the $E°$ value of zero. Explain why such a standard is necessary.
   (c) The system $Mg(s) + Cu^{2+}(aq) \longrightarrow Mg^{2+}(aq) + Cu(s)$ has a value of $E°_{cell}$ of +2.71 V. What would be the $E°_{cell}$ value if $[Mg^{2+}]$ = 1.00 M and $[Cu^{2+}]$ = 0.050 M with no other changes in conditions?

2. Consider the following galvanic cell: $Mg(s) \,|\, Mg^{2+}(aq) \,\|\, Al^{3+}(aq) \,|\, Al(s)$

   Standard Reduction Potentials : $Mg^{2+} + 2e^- \longrightarrow Mg$ –2.37 V

   $Al^{3+} + 3e^- \longrightarrow Al$ –1.66 V

   (a) Give the equation for the spontaneous cell reaction which produces charge flow.
   (b) Indicate the oxidizing agent and the reducing agent in this reaction. Label the anode and the cathode electrodes.
   (c) Calculate the $E°_{cell}$.
   (d) How will the potential difference (voltage) of the galvanic cell $Cu(s) + 2Ag^+(aq) \longrightarrow Cu^{2+}(aq) + 2Ag(s)$ be affected by the addition of NaCl(aq)? Explain.
   (e) How does increasing the size of the anode affect the voltage? Justify your response.
   (f) The electrolysis of aqueous NaCl does not produce metallic sodium. Explain.

Answers

MULTIPLE CHOICE

1. **C** In changing from Fe^{2+} to Fe^{3+}, iron(II) ions lose an electron, which is oxidation; MnO_4^- changing to Mn^{2+} is a gain of electrons, which is reduction (*Chemistry* 7th ed. page 791 / 8th ed. page 823).

2. **B** When the half-reaction is reversed, the sign of the potential is also reversed. However, the $E°$ does not change. The potential is an intensive property, which means that it is not dependent upon how many times the reaction takes place (*Chemistry* 7th ed. pages 796–797 / 8th ed. pages 829–830).

3. **C** While electrons flow through the wire from reducing agent to oxidizing agent, it is ions which flow from one side of the cell through the salt bridge to the other side to keep the net charge at zero. If this were not done, a large amount of energy would be required to 'force' electrons to move from a positive environment into a negative environment; with such charge differences due to the charge buildup on the two sides the cell would not function; the current would be zero (*Chemistry* 7th ed. pages 791–793 / 8th ed. pages 823–825).

4. **D** In examination of the balanced oxidation-reduction reaction equation written above, you can quickly determine that the number of electrons transferred is 10 ($5 Zn \rightarrow 5 Zn^{2+}$) and that the cell potential is determined by $E°_{cell} = E°(oxid) + E°(red) = +0.76 +1.49 = 2.25V$. Using $\Delta G° = -nFE°$ where $n = 10$ and F is 96,485 coulombs, $= -(10)(96485)(2.52) = -2.17 \times 10^6 J$ or -2.17 MJ (*Chemistry* 7th ed. pages 800-803 / 8th ed. pages 835-836)

5. **E** When the cell reaction reaches equilibrium, no electrons flow and $E°_{cell}$ = zero (*Chemistry* 7th ed. pages 798–802, 805 / 8th ed. pages 831-835, 838-839).

6. **D** The charge on one mole of electrons is called the faraday, (F, named for the Englishman Michael Faraday). In this case, two moles of electrons are transferred, hence 2.00 faradays of charge are transferred. (One faraday is 96,500 coulombs of charge, the charge on one mole of electrons.) (*Chemistry* 7th ed. pages 800–802 / 8th ed. pages 833–835)

7. **D** The overall equation for this cell reaction is $Fe^{2+}(aq) + Ag^+(aq) \longrightarrow Fe^{3+}(aq) + Ag(s)$ in which Fe^{2+} is oxidized (loses electrons) and is therefore the anode. Because Fe^{2+} loses electrons, they flow from Fe^{2+} to Ag^+, which is reduced to $Ag(s)$, and the mass of the silver electrode increases (*Chemistry* 7th ed. pages 798–800 / 8th ed. pages 831–833).

8. **A** Look for metals which are better reducing agents than iron so that the metal furnishes the electrons. Such metals have an $E°$ more negative than iron (*Chemistry* 7th ed. pages 814–815 / 8th ed. pages 846–847).

9. **C** The Nernst equation gives the relationship between cell potential and the concentration of the cell components: $E_{cell} = E°_{cell} - (0.0591/n) \log Q$, at 25°C, where n is the number of moles of electrons transferred (*Chemistry* 7th ed. pages 814–815 / 8th ed. pages 846–847).

10. **A** Adding units to the style of responses where a setup is given is often very helpful. In this case it might better look like: (10.0C/s)(3600/s)(1 mole e⁻/96,500 C)(2.00 mole e⁻/1 mole Cu)(63.5g/mol Cu). (*Chemistry* 7th ed. pages 816–818 / 8th ed. pages 847–850)

11. **C** Oxidation is the loss of electrons, and an oxidation half-reactions would be the reverse of each of the above equations. This means the potential for each of the oxidation half-reactions would have a positive value; thus calcium would be most easily oxidized and lead is the least easily oxidized (*Chemistry* 7th ed. pages 794–797 / 8th ed. pages 828-830).

12. **E** Reverse the $Al^{3+} \longrightarrow Al$ to give = +1.66 V, and add to the potential for $Ag^+ \longrightarrow Ag$ of +0.80 V to give +2.46 V (*Chemistry* 7th ed. pages 794–798 / 8th ed. pages 830–831).

13. **C** If you reduce the [Ag⁺] by formation of $Ag(NH_3)_2^+$ then the reaction $3Ag^+ + Al \rightarrow 3Ag + Al^{3+}$ will be driven to the left and the cell emf will be reduced (*Chemistry* 7th ed. page 803 / 8th ed. page 836).

14. **A** Using the Nernst equation as modified for equilibrium conditions, $\log K = n\, E°/0.0591$. $\log K = (3)(2.46V)/0.0591 = 125$

    So $K = $ antilog$(125) = 10^{125}$

    (*Chemistry* 7th ed. pages 807–808 / 8th ed. pages 841–842)

15. **C** Reduction occurs at the cathode and oxidation at the anode (*Chemistry* 7th ed. pages 791–792 / 8th ed. pages 831–833).

FREE RESPONSE

1 (a) They occur equally; for every electron lost by one substance an electron must be gained by another substance (*Chemistry* 7th ed. page 796 / 8th ed. page 829).

   (b) Because the potential of any half-reaction cannot be measured directly, some standard is required to rank the potentials of all half-reactions. Using the hydrogen half-reaction as an arbitrary standard allows such calculations. When we say, for example, that $Zn^{2+} + 2e^- \rightarrow Zn$ has a potential of –0.76 V, we really mean with respect to the potential for the hydrogen half-reaction. A good comparison might be assigning sea level as the zero measure for altitude and then comparing mountain heights and ocean depths to this standard marker (*Chemistry* 7th ed. page 794 / 8th ed. page 826).

   (c) From the Nernst equation, $E_{cell} = E°_{cell} - (0.0591/n) \log Q$, the value of Q is $[Mg^{2+}] / [Cu^{2+}] = 1.00 / 0.050 = 20$. Since $\log Q$ increases ($\log 20. = 1.3$), from $E_{cell} = E°_{cell} - (0.0591/n) \log 20.$, E_{cell} decreases by 0.038, to 2.67 volts (*Chemistry* 7th ed. pages 814–815 / 8th ed. pages 846–847).

2. (a) $3Mg(s) + 2Al^{3+}(aq) \longrightarrow 3Mg^{2+}(aq) + 2Al(s)$. (*Chemistry* 7th ed. pages 796–806 / 8th ed. pages 829–838)

    (b) Mg loses e⁻, therefore is oxidized; it is the reducing agent and is the site of oxidation called the anode. In like manner, Al^{3+}, which is gaining these same electrons, is the oxidizing agent, and the Al electrode is the site of reduction, called the cathode (*Chemistry* 7th ed. pages 796–806 / 8th ed. pages 829–838).

    (c)   +2.37V

    <u>   −1.66V</u>
        +0.71 V

    (*Chemistry* 7th ed. pages 796–806 / 8th ed. pages 829–838)

    (d) The added Cl⁻(aq) will react with the Ag⁺(aq), precipitating as AgCl(s). This lowers the silver ion concentration, which shifts the equilibrium to the left, making the silver ions less available to react with the solid copper, and <u>lowers</u> the voltage. This could also be supported via the Nernst equation, where it would have been seen that Q ([Cu²⁺] / [Ag⁺]²) is increasing, and the overall cell voltage is thus decreasing (*Chemistry* 7th ed. pages 796–806 / 8th ed. pages 829–838).

    (e) If the size of the anode (site of oxidation) is changed there will be no effect on the voltage. The oxidation and reduction potentials are determined by the nature of the metal and not the amount of the metal. You might be able to "run" the cell longer by changing the size of certain electrodes, but would not produce more voltage (*Chemistry* 7th ed. pages 796–806 / 8th ed. pages 829–838).

    (f) When the standard reduction potentials for the two possible reactions are compared,

    $Na^+ + 1e^- \longrightarrow Na$ $E° = -2.71$ V

    $2H_2O + 2e^- \longrightarrow H_2 + 2OH^-$ $E° = -0.83$ V,

    it is evident that water is the better competitor for electrons of these two (E° for water is the more positive value (*Chemistry* 7th ed. pages 796–806 / 8th ed. pages 829–838).

16

NUCLEAR CHEMISTRY

This chapter deals with nuclear transformations and the kinetics of radioactive decay.

In this chapter, you should be able to
- Identify the products of nuclear transformation processes.
- Interconvert between the half-life of a nuclide and the amount of that nuclide remaining after a specific amount of time.
- Be familiar with critical mass, fusion/fission, mass defect, binding energy, nucleon, strong nuclear force, differences in penetrating power, and the ionizing ability of alpha, beta, and gamma radiation.
- Use a graph to predict the type of decay that will occur.

RADIOACTIVE DECAY

Radioactive decay occurs when a nucleus undergoes decomposition to form a different nucleus and additional particles. Alpha-particle and gamma-ray production involve a change in the mass number of the decaying nucleus. The mass number (the sum of the protons and neutrons) of the decaying nucleus does not change during β-particle production, positron production or during electron capture.

ALPHA-PARTICLE PRODUCTION

(*Chemistry* 7th ed. pages 843–844 / 8th ed. pages 875–876)

An alpha particle, α particle, is a helium nucleus: $_2^4He$

Heavy radioactive nuclides, isotopes, undergo decay to form alpha particles.

EXAMPLE: Write a balanced equation for the decay of uranium-235 by α– particle production.

$$^{235}_{92}U \longrightarrow \, ^{4}_{2}He + \, ^{231}_{90}Th$$

The product can be identified by knowing that the total of the mass number and the atomic number must be the same on both sides of the equation. Thus the mass number for the product must be 235 – 4 = 231. The atomic number for the product must be 92 – 2 = 90 which is the atomic number for thorium.

GAMMA-RAY PRODUCTION

(*Chemistry* 7th ed. pages 843–844 / 8th ed. pages 875–876)

A gamma ray, γ, is a high-energy photon often produced during nuclear decay. A nucleus with excess energy can relax to its ground state by emitting a gamma ray: $^{235}_{92}U \longrightarrow \, ^{4}_{2}He + \, ^{231}_{90}Th + \, ^{0}_{0}\gamma$

BETA-PARTICLE PRODUCTION

(*Chemistry* 7th ed. pages 843–844 / 8th ed. pages 875–876)

A beta particle, β particle, is an electron represented in a nuclear reaction by the symbol: $^{0}_{-1}e$

The β particle has a mass number 0. Its mass is very small compared to a proton or neutron. The mass number and atomic number must be conserved as seen in alpha-particle production.

EXAMPLE: Write a balanced equation for the decay of carbon-14 by β-particle production.

$$^{14}_{6}C \longrightarrow \, ^{0}_{-1}e + \, ^{14}_{7}N$$

Thus the mass number for the product must be 14 – 0 = 14. The atomic number for the product must be 6 – (–1)= 7 which is the atomic number for nitrogen.

POSITRON PRODUCTION

(*Chemistry* 7th ed. page 845 / 8th ed. page 877)

A positron is a particle which has the same mass as an electron but opposite charge.

EXAMPLE: Write a balanced equation for the decay of bismuth-83 by positron production.

$$^{205}_{83}Bi \longrightarrow \, ^{0}_{+1}e + \, ^{205}_{82}Pb$$

Thus the mass number for the product must be 205 – 0 = 205. The atomic number for the product must be 83 – 1 = 82, which is the atomic number for lead.

ELECTRON CAPTURE

(*Chemistry* 7th ed. page 845 / 8th ed. page 877)

Electron capture occurs when a nucleus captures an inner orbital electron.

$$^{241}_{96}\text{Cm} + \, ^{0}_{-1}\text{e} \longrightarrow \, ^{241}_{95}\text{Am}$$

THE KINETICS OF RADIOACTIVE DECAY

(*Chemistry* 7th ed. pages 846–849 / 8th ed. pages 878–888)

The decay of nuclides is represented by $\ln\left(\dfrac{N}{N_0}\right) = -kt$

where N_o = the original mass of nuclides
N = the mass remaining at time t
k = the first-order rate constant
t = time.

HALF-LIFE

(*Chemistry* 7th ed. page 847 / 8th ed. page 879)

The half-life, $t_{1/2}$, is the time required for the number of nuclides to reach half of their original value. It can be determined by the equation:

$$t_{1/2} = 0.693/k.$$

EXAMPLE: Iodine-131, used in the diagnosis and treatment of thyroid disease, has a half-life of 8.1 days. If a patient with thyroid disease consumes a sample containing 10 µg of iodine-131, how long will it take for the amount of iodine-131 to decrease to 1/100 of the original amount?

Using the equation: $\ln (N/N_o) = -kt$

Substitute 0.010 N_o for N, because $N = 0.010\ N_o$.

Also for k, substitute 0.693 / 8.1 days because $t_{1/2} = 0.693/k$ and $t_{1/2} = 8.1$ days.

Solving for t, you get

$\ln(0.010) = -0.693t/8.1$ days

$t = 54$ days.

Multiple-choice Questions

1. Isotopes of the same element are nuclides with
   (A) the same number of protons and the same atomic number (Z).
   (B) the same number of protons and the same number of neutrons.
   (C) the same mass number (A) and the same number of electrons.
   (D) the same mass number (A) and the same number of protons.
   (E) the same sum of protons and neutrons as well as the same mass number (A).

2. When carbon-14 undergoes radioactive decay, it forms one nitrogen-14 and one
   (A) carbon-13.
   (B) carbon-12.
   (C) nitrogen-13.
   (D) nitrogen-12.
   (E) beta particle.

3. When potassium-40 undergoes decay by electron capture, the product is
   (A) ^{39}K
   (B) ^{40}K
   (C) ^{41}K
   (D) ^{39}Ar
   (E) ^{40}Ar

4. When aluminum undergoes bombardment by an alpha particle, the products are phosphorus-30 and
   (A) a proton.
   (B) an electron.
   (C) a neutron.
   (D) a helium nucleus.
   (E) a 2+ nucleus.

5. The atomic number is decreased by one by
   (A) electron capture only.
   (B) positron emission only.
   (C) both electron capture and positron emission.
   (D) both gamma and beta emission.
   (E) proton bombardment.

6. When $^{214}_{82}Pb$ undergoes decay producing two beta particles, the other product is
   (A) $^{214}_{82}Pb$.
   (B) $^{214}_{84}Po$.
   (C) $^{214}_{80}Pb$.
   (D) $^{212}_{84}Po$.
   (E) $^{214}_{82}Po$.

7. Referring to the graph which considers neutron to proton ratios, you can conclude that when the neutron to proton ratio is too low
   (A) positron emission is likely.
   (B) beta emission is likely.
   (C) positron will be absorbed by the nucleus.
   (D) alpha particle production is expected.
   (E) electron capture is expected.

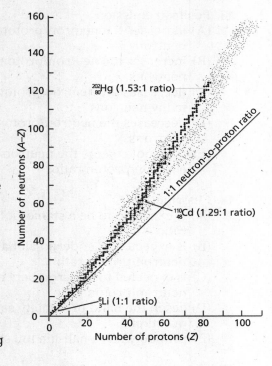

8. The number of particles of a given nuclide decays from 120 units to 15 units in 60. minutes, indicating a half-life of
   (A) 10. minutes.
   (B) 15. minutes.
   (C) 20. minutes.
   (D) 30. minutes.
   (E) 60. minutes.

9. The number of neutrons emitted when U-235 is bombarded with a neutron, forming Sm-160 and Zn-72, is
   (A) 1.
   (B) 2.
   (C) 3.
   (D) 4.
   (E) 5.

10. U-238 emits an alpha particle to produce a nucleus which then decays by emitting a beta particle. The mass number and atomic number of the two successive products are
   (A) 90-234 and 91-235.
   (B) 90-234 and 91-234.
   (C) 91-233 and 91-234.
   (D) 90-234 and 92-238.
   (E) 92-234 and 89-234.

11. Which of the following particles cannot be accelerated in a cyclotron?
   (A) alpha particle
   (B) beta particle
   (C) proton
   (D) neutron
   (E) positron

12. How is the presence of radioactivity detected in a scintillation counter?
   (A) by fluorescence of ZnS upon interaction with radiation.
   (B) by interaction of the radiation with a photographic plate.
   (C) by ionization of a gas.
   (D) by precipitation of a radioactive substance.
   (E) by gas chromatography techniques.

13. Positron emission
   (A) increases the neutron/proton ratio in nuclei that have too many neutrons.
   (B) increases the neutron/proton ratio in nuclei that have too few neutrons.
   (C) decreases the neutron/proton ratio in nuclei that have too many neutrons.
   (D) decreases the neutron/proton ratio in nuclei that have too few neutrons.
   (E) does not change the neutron/proton ratio, only the electron/proton ratio.

14. Fluorine-20
   (A) is expected to be a stable isotope because its neutron/proton ratio is 11/9.
   (B) is expected to undergo alpha decay to increase the neutron/proton ratio.
   (C) is expected to undergo beta decay to decrease the neutron/proton ratio.
   (D) lies in the band of stability as shown in the graph used in question 7.
   (E) probably has a half-life too short to enable it to be detected.

15. How is the mass defect for a nucleus calculated?
   (A) Atomic weight of the element minus mass number.
   (B) Atomic weight of the element minus mass of nucleons.
   (C) Nuclear mass minus mass number.
   (D) Nuclear mass minus mass of nucleons.
   (E) none of the above

FREE-RESPONSE QUESTIONS

1. Refer to the graph in multiple-choice question #7: the graph of the number of neutrons plotted vs. the number of protons shows the "zone of stability."
   (a) What is meant by the "zone of stability?
   (b) What are the characteristics of those nuclides found in the "zone of stability"?
   (c) Discuss ways in which such a graph is helpful to the nuclear chemist.

2. Contrast nuclear reactions with ordinary chemical reactions. Include in your discussion or outline
   (i)   the effects of temperature, pressure, concentration, and the addition of a catalyst;
   (ii) the relative amounts of energy involved; and
   (iii) what is changed or converted in each type of reaction.

Answers

MULTIPLE CHOICE

1. **A** "Isotope" is the term given to a group of nuclides that have the same atomic number, therefore the same number of protons (individual atoms are known as nuclides). For example, chlorine-35 and chlorine-37 both have an atomic number of 17, with 17 protons in the nucleus, but differ in mass due to a different number of neutrons (18 vs. 20) (*Chemistry* 7th ed. pages 840–841 / 8th ed. pages 872–873).

2. **E** Perhaps this is most easily seen as $^{14}_{6}C \longrightarrow {}^{14}_{7}N + {}^{0}_{-1}e$.

   Note that both Z and A values (Z represents atomic number and A stands for atomic mass) are conserved in nuclear reactions; that is, they give the same sum on both sides of the equation (*Chemistry* 7th ed. pages 843–845 / 8th ed. pages 875–877).

3. **E** The nuclear reaction is $^{40}_{19}K + {}^{0}_{-1}e \longrightarrow {}^{40}_{18}Ar$

   This will only make sense to you if you understand what is meant by electron capture (*Chemistry* 7th ed. page 845 / 8th ed. page 877).

4. **C** The nuclear reaction is $^{27}_{13}Al + {}^{4}_{2}He \longrightarrow {}^{30}_{15}P + {}^{1}_{0}n$.

   Note that $^{1}_{0}n$ is the symbol for a neutron, with a mass = 1 and a charge = 0 (*Chemistry* 7th ed. pages 843–844 / 7th ed. pages 875–876).

5. **C** The process of electron capture involves an inner orbital electron being captured by the nucleus, causing the atomic number to decrease by one. Positron production involves emission of a particle which has the same mass as an electron but the opposite charge (*Chemistry* 7th ed. pages 843–844 / 8th ed. pages 875–876).

6. **B** This time be sure that you understand that the beta particle is $^{0}_{-1}e$, and that you form two of them: $^{214}_{82}Pb \longrightarrow {}^{0}_{-1}e + {}^{0}_{-1}e + {}^{214}_{84}Po$. (*Chemistry* 7th ed. pages 843–844 / 8th ed. pages 875–876)

7. **A** If the neutron/proton is too low, this suggests too many protons, which is an unstable condition. This condition can be relieved by positron production. You will find more on this topic in free-response question #1 (*Chemistry* 7th ed. pages 841–843 / 8th ed. pages 873–875).

8. **C** The change in the number of nuclide particles from 120 to 60 to 30 to 15 units suggests a change through three half-lives. Since three half-lives took 60. minutes, one half-life would be 60./3 = 20. minutes (*Chemistry* 7th ed. pages 846–847 / 8th ed. pages 878–879).

9. **D**

$$^{235}_{92}U + ^{1}_{0}n \longrightarrow ^{160}_{62}Sm + ^{72}_{30}Zn + ? \, ^{1}_{0}n.$$ Since the total mass on the right equals the total mass on the left = 236 and Sm – 160 + Zn – 72 contribute a mass of (160 + 72) = 232, four neutrons are necessary to balance this expression. Note that a neutron, usually shown as $^{1}_{0}n$, has a mass of one and a charge of zero (*Chemistry* 7th ed. pages 843–845 / 8th ed. pages 875–877).

10. **B**

$$^{238}_{92}U \longrightarrow ^{4}_{2}He + ^{234}_{90}X$$

$$^{234}_{90}X \longrightarrow ^{0}_{-1}e + ^{234}_{91}Y$$

(*Chemistry* 7th ed. pages 843–845 / 8th ed. pages 875–877)

11. **D** In order for a particle to be accelerated in a cyclotron it must have a charge. Since a neutron has no charge, it cannot be accelerated (*Chemistry* 7th ed. pages 851–852 / 8th ed. pages 881–882).

12. **A** (*Chemistry* 7th ed. pages 852–853 / 8th ed. pages 883–884)

13. **B** Positron emission involves the conversion of a proton to a neutron, thereby increasing the neutron/proton ratio in the new isotope. (*Chemistry* 7th ed. page 844 / 8th ed. page 876).

14. **C** The ideal neutron/proton ratio for the lighter elements is 1:1. Thus, fluorine-20 can achieve this ratio by beta decay, which converts a neutron to a proton and an electron (*Chemistry* 7th ed. pages 841–844 / 8th ed. pages 873–875).

15. **D** The mass defect is the change in mass associated with the difference in the mass of the nucleus of the particular isotope of an element and the sum of the masses of the protons and neutrons that make up the nucleus (*Chemistry* 7th ed. pages 856–857 / 8th ed. pages 887–888).

FREE RESPONSE

1. (a) This is the area in which stable nuclides reside.

   (b) As light nuclides, the number of neutrons and protons are about equal. With more massive nuclides the number of neutrons must exceed the number of protons. Nuclides with an even number of both neutrons and protons seem to have a higher incidence of stability (you could also note the so called "magic numbers" of 2, 8, 20, 28, 50, etc. for the number of protons in very stable nuclides).

   (c) The comparison of the stability of a given nuclide to the neutron to proton ratio helps to predict the nature of decay that unstable substances will most likely experience. Either beta emission or positron emission usually leads to the formation of a more stable substance (*Chemistry* 7th ed. pages 841–843 / 8th ed. pages 873–875).

2. i Chemical Reactions: The rate of reaction is influenced by changes in temperature, pressure, concentration, and by the addition of a suitable catalyst.

Nuclear Reactions: These are not influenced by changes in pressure, temperature, concentration, or the addition of a catalyst. (Note: Concentration may affect the amount of available critical mass in certain cases).

Both types of reactions can be very fast, very slow, or somewhere between these extremes.

ii Chemical Reactions: These usually involve a process that gains or loses around 400 kJ or less.

Nuclear Reactions: These can involve very large amounts of energy, perhaps as much as 10^{10} kJ.

iii Chemical Reactions: These reactions involve gaining, losing, or sharing electrons. The outermost electron configurations are most important.

Nuclear Reactions: As the name suggests, these involve rearrangement of the nucleus of the atom. Often a different nucleus is formed through a process called radioactive decay. The relative number of protons and neutrons is especially important (*Chemistry* 7th ed. pages 840–844 / 8th ed. pages 872–876).

Part III

Practice Tests

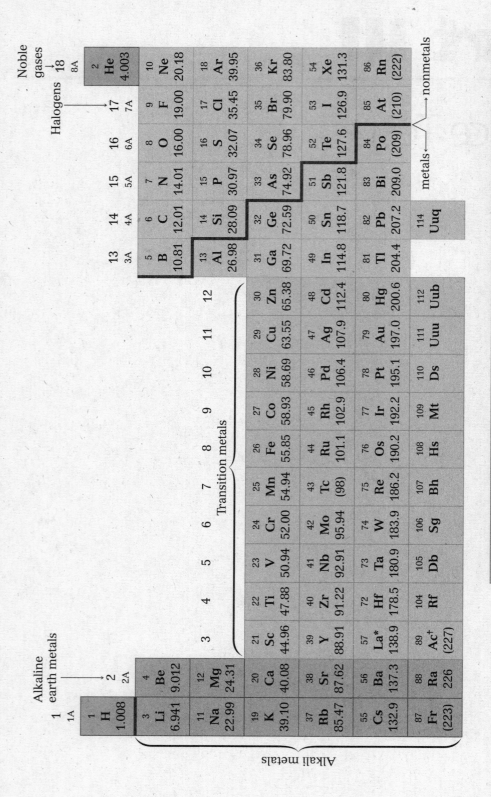

Group numbers 1–18 represent the system recommended by the International Union of Pure and Applied Chemistry.

Practice Test 1

These questions are representative of the AP Chemistry examination, but keep in mind that it is impossible to predict exactly how well you will do on the actual exam. The first section of this test is 50% of your total test grade. Time yourself so that you finish this part in 90 minutes. Remember that on the actual examination, there is a guess factor; scoring is the number of questions answered correctly minus one-fourth the number wrong.

AP CHEMISTRY EXAMINATION
Section I: Multiple-Choice Questions
Time: 90 minutes
Number of Questions: 75

No calculators are to be used in this section; no tables (except periodic table) permitted.

Part A

Directions: The following questions consist of five lettered headings followed by a listing of numbered phrases. For each phrase you are to select the one heading that is most closely related to it. Headings may be used once, more than once, or not at all.

Questions 1–4:
(A)   0.10 M $Ca(NO_3)_2$
(B)   0.10 M $NaC_2H_3O_2$
(C)   0.10 M $CuSO_4$
(D)   0.10 M $HC_2H_3O_2$
(E)   0.10 M C_2H_5OH

1. Which of the above would be expected to have the lowest pH?

2. Which of the above would be expected to be colored?

3. Which of the above would be expected to have the lowest freezing temperature?

4. Which of the above would be expected to have the highest pH?

Questions 5–7
(A) XeF_4
(B) PCl_4^-
(C)   Zn
(D)   Fe
(E) F_2O

5. Which of the above would be expected to have a tetrahedral shape?

6. Which of the above would be expected to form a protective coating in the galvanizing of steel?

7. Which of the above would be expected to have a bent (angular) shape?

GO ON TO NEXT PAGE

Part B

<u>Directions:</u> Each of the questions or incomplete statements below is followed by five suggested answers or completions. Select the one that is best in each case.

8. Ideal gases differ from real gases in which of the following ways?
   I.   Ideal gases have zero molecular volume.
   II. Ideal gases have no intermolecular forces.
   III. $PV = nRT$ describes both ideal and real gases at all temperatures and pressures.
   (A) I only
   (B) II only
   (C) III only
   (D) I and II only
   (E) I, II, and III

9. The VSEPR model predicts that the structure of XeF_4 is square planar. This is due to
   (A) the tendency of four electron pairs always to form a square.
   (B) the tendency of four electron pairs always to form a tetrahedron.
   (C) bond angles of 90°.
   (D) the high degree of repulsion between the two lone pairs.
   (E) a high nuclear charge.

10. Which of the following salts will form a basic solution in water solution?
    (A) KCl
    (B) Na_2SO_4
    (C) $CuCl_2$
    (D) Na_2CO_3
    (E) NH_4NO_3

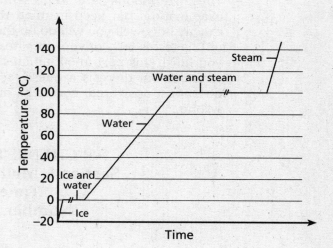

11. A heating curve for water (ice → water → steam) shows that the slope in the section for water heating is different than the slope for steam heating. This difference in slope occurs because
    (A) it takes more energy to vaporize water than to melt ice.
    (B) little energy is needed to cool steam at 100°C.
    (C) little energy is needed to cool ice at 0°C.
    (D) water and steam have different molar heat capacities.
    (E) raising the temperature of ice is not involved.

12. The phase diagram for water is not typical of most substances. Carbon dioxide has a phase diagram more typical of most substances. The main nontypical feature of the phase diagram for water is
(A) the critical point is at a higher temperature than the triple point.
(B) the triple point is below a pressure of one atm.
(C) the solid/liquid line extends indefinitely.
(D) water can exist in all three physical states.
(E) the solid/liquid line has a negative slope.

13. A sample of helium gas has a volume of 5.6 L at STP. The number of moles of He present are
(A) 0.25 mole.
(B) 0.50 mole.
(C) $0.25 \times 6.02 \times 10^{23}$.
(D) $0.50 \times 6.02 \times 10^{23}$.
(E) $4.0 \times 6.02 \times 10^{23}$.

14. The K_{sp} expression for lead (II) bromide is
(A) $[Pb^{2+}][Br^-]$.
(B) $[Pb^{2+}][2Br^-]$.
(C) $[Pb^{2+}]^2[Br^-]$.
(D) $[Pb^{2+}][Br^-]^2$.
(E) $[Pb^{2+}]^2[Br^-]^2$.

15. Grinding an ionic salt into a fine powder accomplishes which of the following:
I.   increases its solubility.
II. increases the value of the K_{sp}.
III. increases the rate of dissolving.
(A) I only
(B) II only
(C) III only
(D) I and II only
(E) I, II, and III

16. It is possible to predict if two ionic solutions will form a precipitate by noting
(A) that the ion product (Q) is a high value.
(B) that the ion product (Q) is a low value.
(C) that the ion product (Q) exceeds the K_{sp} value.
(D) that the solubility product exceeds the ion product.
(E) temperature changes as a solid forms.

17. A buffered solution is one which can be described by which of the following:
I.   a solution which resists pH change.
II. may contain a weak acid and its salt.
III. contains species which can react with both H^+ and OH^-.
(A) I only
(B) II only
(C) III only
(D) I and II only
(E) I, II, and III

18. Entropy is greater in which of the following:
I.   solid salt than salt solution.
II. ice formed on a cold window than water vapor.
III. graphite rather than diamond.
(A) I only
(B) II only
(C) III only
(D) I and II only
(E) I, II, and III

19. The vapor pressure of a heavier substance (MM) will be lower than that of a lighter substance, all other things being equal, because
(A) of weaker intermolecular forces.
(B) the molecules are more polarizable.
(C) hydrogen "bonding" isn't considered.
(D) vapor pressure is a constant for a given substance.
(E) the enthalpy of vaporization is lower.

GO ON TO NEXT PAGE

20. For the reaction $O_2(g) \rightarrow 2O(g)$
    (A) ΔH would be +, ΔS would be +.
    (B) ΔH would be –, ΔS would be –.
    (C) ΔH would be +, ΔS would be –.
    (D) ΔH would be –, ΔS would be +.
    (E) would be spontaneous at low temperatures.

21. The purpose of the salt bridge in an electrochemical cell is
    (A) to make possible electron flow through the external circuit.
    (B) to provide for ion flow through the cell.
    (C) to provide for oxidation at the cathode.
    (D) to provide for reduction at the anode.
    (E) to reach equilibrium sooner.

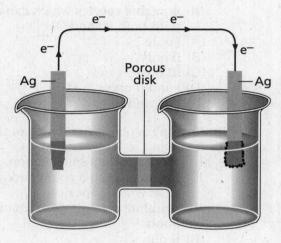

22. A concentration cell of Ag / Ag⁺ has Ag⁺ of 1.0 M on the right side of the cell and 0.100 M Ag⁺ on the left side.
    (A) Electrons would flow (external circuit) from left to right.
    (B) The Ag on the left side is the cathode.
    (C) The Ag on the right side will decrease in mass.
    (D) The solution will gradually turn light blue.
    (E) Electrons would flow internally in the cell from left to right.

23. In the half-reaction, $AgBrO_3 \rightarrow Ag + BrO_3^-$, the number of electrons
    (A) lost is 1.
    (B) lost is 2.
    (C) gained is 1.
    (D) gained is 2.
    (E) gained is 3.

24. Of the following salts, which is insoluble?
    (A) Na_2S
    (B) $Pb(NO_3)_2$
    (C) KCl
    (D) $CaCl_2$
    (E) all of the above are soluble

25. To neutralize a solution of NaOH and use the least volume of acid, you should select
    I.   1.0M HCl.
    II. 1.0 M $HC_2H_3O_2$.
    III. 1.0 M H_2SO_4.
    (A) I only.
    (B) II only.
    (C) III only.
    (D) I and II use equal and lesser volumes.
    (E) I, II, and III use the same volumes of acids.

26. The concentration of ions in 1.2 M aluminum nitrate solution is
    (A) 0.40 M.
    (B) 1.2 M.
    (C) 2.4 M.
    (D) 3.6 M.
    (E) 4.8 M.

27. The gas with the most rapid effusion rate is
    (A) $NH_3(g)$.
    (B) $HCl(g)$.
    (C) $Cl_2(g)$.
    (D) $CH_4(g)$.
    (E) $C_3H_8(g)$.

28. For which of the following is the sign of the enthalpy change different from the other four?
    (A) $2KClO_3(s) \rightarrow 2KCl(s) + 3O_2(g)$
    (B) $H_2O(l) \rightarrow H_2O(s)$
    (C) $Br_2(g) \rightarrow 2Br(g)$
    (D) $2HgO(s) \rightarrow 2Hg(l) + O_2(g)$
    (E) $Cl(g) + e^- \rightarrow Cl^-(g)$

29. Which of the following pairs illustrates the Law of Multiple Proportions?
    (A) KCl, KBr
    (B) P_2O_5, N_2O_5
    (C) NO, NO_2
    (D) CO_2, SiO_2
    (E) NH_4F, NH_4I

30. The geometry of carbon disulfide, CS_2, can be described by the VSEPR model as
    (A) bent or angular.
    (B) linear.
    (C) trigonal planar.
    (D) trigonal pyramid.
    (E) tetrahedral.

31. A hydrocarbon, C_4H_{10}, has a critical temperature above room temperature. This means that
    (A) it can never be liquified.
    (B) the intermolecular forces are strong enough to form a liquid when pressure is applied.
    (C) the critical pressure is also high.
    (D) it will exist as a liquid at room conditions.
    (E) it is extremely flammable.

32. Instantaneous reaction rate
    I.   may not be constant over time.
    II. can be calculated from slope of a line tangent to a concentration vs. time curve.
    III. always has a positive value.
    (A) I only
    (B) II only
    (C) III only
    (D) I and II are both valid statements
    (E) I, II, and III are all valid statements

33. One function of a catalyst is to
    (A) shift the equilibrium toward products.
    (B) lower the activation energy.
    (C) increase the value of ΔH.
    (D) lower the value of the rate constant.
    (E) make the reverse reaction less spontaneous.

34. A 1.0 M solution of which of the following is orange?
    (A) KCl
    (B) $K_2Cr_2O_7$
    (C) K_2CrO_4
    (D) $CuSO_4$
    (E) $FeCrO_4$

35. Two solutions are mixed: 1.0 L of 0.20 M NaOH and 1.0 L of 0.10 M H_3PO_4. After reaction, the ions in the largest concentration are Na^+ and
    (A) OH^-.
    (B) H^+, (H_3O^+).
    (C) PO_4^{3-}.
    (D) HPO_4^{2-}.
    (E) $H_2PO_4^-$.

36. The hybridization of carbon in acetylene, C_2H_2, a triple bonded substance is
    (A) sp hybrid orbitals.
    (B) sp^2 hybrid orbitals.
    (C) sp^3 hybrid orbitals.
    (D) dsp^3 hybrid orbitals
    (E) d^2sp^3 hybrid orbitals.

37. $2Al + 6HCl \rightarrow 3H_2 + 2AlCl_3$
    Aluminum reacts with hydrochloric acid, as indicated in the equation above, to produce hydrogen gas. The H_2 produced was then collected by water displacement at 27°C (V.P. H_2O = 21 torr) and a barometric pressure of 757 torr. If 0.555 L of gas is collected, the partial pressure of the hydrogen is
    (A) $0.555 \times (273+27)$.
    (B) 0.555×757.
    (C) $757-27$.
    (D) $757-21$.
    (E) $0.555 \times 300 / 273$.

38. A saturated solution of magnesium hydroxide has added to it dropwise a solution of pH = 4. This results in
    (A) a precipitate forming.
    (B) much bubbling and frothing occurring.
    (C) magnesium metal precipitates.
    (D) the solution becoming increasingly basic.
    (E) water forming.

GO ON TO NEXT PAGE

39. The volume of base at the equivalence point on the titration curve of an acid with a base is determined by
    (A) the strength of the acid.
    (B) the relative strength of the acid and the base.
    (C) the strength of the base used.
    (D) the K_a value for the acid.
    (E) the amount of acid and base added at that point.

40. Knowing only the temperature and the K_{sp} value will not provide the relative solubility of ionic salts because
    (A) different salts sometimes produce different number of ions when dissolving.
    (B) solubility does not consider lattice energy effects.
    (C) not all ionic salts are soluble.
    (D) the two ions formed are of different sizes.
    (E) the two ions formed have different charges.

41. The K_a for acetic acid, $HC_2H_3O_2$, is 1.8×10^{-5} at 25°C. In a 0.10 M solution of potassium acetate, which of the following are true?
    I. $[HC_2H_3O_2] = [OH^-]$
    II. $[H^+] = [OH^-]$
    III. $[C_2H_3O_2^-] < 0.10$ M.
    (A) I only
    (B) II only
    (C) III only
    (D) I and II
    (E) I, II, and III

42. The kinetic molecular theory indicates that temperature is best considered
    (A) the kinetic energy of gas particles.
    (B) the random motion of gas particles.
    (C) a function of the Celsius scale.
    (D) equal to the Kelvin scale.
    (E) an indicator of the average kinetic energy of gas particles.

43. Silver chromate solid exists in equilibrium with Ag^+ and CrO_4^{2-}:

    $Ag_2CrO_4(s) \rightleftharpoons 2Ag^+(aq) + CrO_4^{2-}(aq)$.

    If silver nitrate is added to this solution
    (A) a precipitate of Ag_2CrO_4 forms.
    (B) the solution gets increasingly blue.
    (C) the concentration of CrO_4^{2-} increases.
    (D) the $[Ag_2CrO_4]$ increases.
    (E) the common ion effect does not apply.

44. 4.00 mmol of a weak acid is mixed with 2.00 mmol of NaOH and the pH is determined to be 5.00. The value of K_a for this weak acid is
    (A) $4.00 / 2.00 \times 5.00$.
    (B) $4.00 / 2.00 \times 10^5$.
    (C) $2.00 / 4.00 \times 10^{-5}$.
    (D) $\log (10 \times 10^{-5})$.
    (E) 1.0×10^{-5}.

45. For which of the following equations is the sign of entropy as shown?
    (A) $Na(s) + 1/2\ Cl(g) \rightarrow NaCl(s)$ is (+).
    (B) $H_2(g) + 3N_2(g) \rightarrow 2NH_3(g)$ is (+)
    (C) $Hg(l) \rightarrow Hg(g)$ is (+).
    (D) $CO_2(s) \rightarrow CO_2(g)$ is (–).
    (E) $C_6H_{12}O_6(aq) \rightarrow C_6H_{12}O_6(s)$ is (+).

46. At what temperature will a process with $\Delta H = -10$ kJ and $\Delta S = -100$ J/K be spontaneous?
    (A) any temperature under 100K.
    (B) any temperature over 100K.
    (C) any temperature under 10 K.
    (D) any temperature under 0.10K.
    (E) any temperature over –100K.

47. The ΔG for $S(s) + O_2(g) \rightarrow SO_2(g)$, from $S(s) + 1\ 1/2\ O_2(g) \rightarrow SO_3(g)$ $\Delta G = -370$ kJ and $O_2(g) + 2SO_2(g) \rightarrow 2SO_3(g)$ $\Delta G = -140$ kJ is
    (A) +510 kJ.
    (B) –510 kJ.
    (C) +300 kJ.
    (D) –300 kJ.
    (E) –230 kJ.

48. The process with the greatest rate of reaction would be
    (A) $\Delta G = -30$ kJ.
    (B) $\Delta G = -60$ kJ.
    (C) $\Delta G = +30$ kJ.
    (D) $\Delta G = +60$ kJ.
    (E) impossible to predict from only ΔG values.

49. The $E^o = -0.60$ V for FeS(s) → $Fe^{2+}(aq) + S^{-2}(aq)$. This indicates that
    (A) FeS has very low solubility.
    (B) K_{sp} for FeS is very large.
    (C) S^{-2} ions are unstable.
    (D) Fe^{2+} ions are unstable.
    (E) both S^{2-} and Fe^{2+} ions are unstable.

50. The net equation for an electrochemical cell made with $Al^{3+}(aq)$ and Mg(s) as reactants is
    (A) Al (s) + Mg(s) → $Al^{3+}(aq)$ + $Mg^{2-}(aq)$.
    (B) $Al^{3+}(aq)$ + Mg(s) → Al(s) + $Mg^{2-}(aq)$.
    (C) $3Al^{3+}(aq)$ + 2Mg(s) → $2Mg^{2+}(aq)$ + $3Al^{3+}(aq)$.
    (D) $2Al^{3+}(aq)$ + Mg(s) → 2Al(s) + $Mg^{2+}(aq)$.
    (E) $2Al^{3+}(aq)$ + 3Mg(s) → $3Mg^{+2}(aq)$ + 2Al(s).

51. An aqueous lead (II) chloride solution is to undergo electrolysis to deposit the lead. To determine how long this process will take, you must also be given
    (A) the mass of $PbCl_2$ and the current available.
    (B) the number of moles of electrons transferred and the voltage available.
    (C) the number of coulombs required to deposit one mol of lead.
    (D) the half reaction for the lead ion/lead pair and the voltage.
    (E) the mass of the $PbCl_2$ and the voltage measured.

Questions 52 and 53 refer to the electrochemical cell:

$$Zn(s) \mid Zn^{2+}(aq) \parallel Ag^+(aq) \mid Ag(s).$$

52. The anode for this cell is the
    (A) Zn.
    (B) Zn^{2+}.
    (C) Ag^+.
    (D) Ag.
    (E) electrode undergoing reduction.

53. As this cell functions
    (A) electrons flow from the Zn electrode to the Ag electrode.
    (B) ions flow from the Zn^{2+} side to the Ag^+ side.
    (C) the mass of the Ag electrode decreases.
    (D) the $[Ag^+]$ increases.
    (E) $E^o_{cell} = 0$.

54. To calculate the molarity of a solution, it is necessary to know
    I.   the mass of the solute
    II. molar mass of the solute
    III. volume of water added
    IV. total volume of the solution.
    (A) I only
    (B) II only
    (C) III only
    (D) I, II, and III only
    (E) I, II, and IV only

GO ON TO NEXT PAGE

55. When 1.0 L of 1.0 M lead (II) nitrate is added to 1.0 L of 1.0 M potassium nitrate the concentration of ions in solution is
    (A) 1.0 M Pb^{2+}, 1.0 M NO_3^-, 1.0M K^+.
    (B) 1.0 M Pb^{2+}, 2.0 M NO_3^-, 1.0M K^+.
    (C) 0.50 M Pb^{2+}, 2.0 M NO_3^-, 0.50M K^+.
    (D) 0.50 M Pb^{2+}, 1.5 M NO_3^-, 0.50M K^+.
    (E) 0.50 M Pb^{2+}, 0.50 M NO_3^-, 0.50M K^+.

56. Naturally occurring copper is composed of two isotopes, ^{63}Cu and ^{65}Cu. If copper is 69.1% ^{63}Cu and 30.9% ^{65}Cu, the average atomic mass could be calculated as
    (A) $(30.9 + 69.1)/2$.
    (B) $(63 + 65)/2$.
    (C) $(65 \times 30.9) + (63 \times 69.1)$.
    (D) $(0.309 \times 65) + 0.691 \times 63)$.
    (E) $[(0.309 \times 65) + 0.691 \times 63)]/2$.

57. The number of atoms in a sliver of calcium of mass 5.234 mg may be calculated as
    (A) $(5.234/1000) \times (1/40.08) \times (6.022 \times 10^{23})$.
    (B) $(5.234 \times 40.08) \times (6.022 \times 10^{23})$.
    (C) $(40.08/5.234) \times (6.022 \times 10^{23})$.
    (D) $(6.022 \times 10^{23}) \times 40.08 \times 1000$.
    (E) $5.234 \times (6.022 \times 10^{23}/40.08)$.

58. The Arrhenius definition of bases is
    (A) a substance that produces H^+ in water solution.
    (B) a substance that produces protons in water solution.
    (C) a substance that produces OH^- in water solution.
    (D) a substance that acts as a proton donor in any solution.
    (E) a substance that acts as a proton acceptor in any solution.

59. A weak electrolyte
    (A) forms little electrical energy.
    (B) is only slightly soluble.
    (C) is never a base.
    (D) forms few ions when dissolving.
    (E) all of the above apply.

60. Rate laws can be written
    (A) only from experimental data.
    (B) only for first-order reactions.
    (C) directly from the coefficients of the reactants.
    (D) as a ratio of the coefficients of reactants to coefficients of products.
    (E) only if the rate constant is known.

61. The solubility of AgCl in HCl
    (A) is low, as Cl^- ions cause an equilibrium shift toward the solid form.
    (B) is low, as Ag^+ has already reacted with Cl^-.
    (C) is high, as HCl readily ionizes.
    (D) is high, as AgCl can then form.
    (E) is high, as H_2 forms, then escapes.

Questions 62–63 refer to the reaction: A + 2B ⟶ C + 2D, and these data:

Exp. No.	Initial[A]	Initial [B]	Initial Rate $\Delta[A]/\Delta t$ in mol/L•s.
1	0.100M	0.050 M	7.00×10^{-6}
2	0.200 M	0.050 M	14.00×10^{-6}
3	0.100 M	0.100 M	7.00×10^{-6}

62. The rate law for this reaction is
(A) Rate = k[A][B].
(B) Rate = k[A][B]^2.
(C) Rate = k[A].
(D) Rate = k[B].
(E) Rate = k[B]^2.

63. For Experiment 2, the rate of appearance of product D is
(A) 3.50×10^{-6} mol/L•s.
(B) 7.00×10^{-6} mol/L•s.
(C) 14.0×10^{-6} mol/L•s.
(D) 28.0×10^{-6} mol/L•s.
(E) cannot be determined from these data.

64. The equations $\Delta G° = -RT \ln K$ and $\Delta G° = H° - TS°$ can be written as $\ln K = -\Delta H°/RT + \Delta S°/R$. This means that if ln k is plotted versus 1/T, the results will be linear and the slope will be
(A) $\Delta H°/R$.
(B) $-\Delta H°/R$.
(C) $\Delta S°/R$.
(D) $-\Delta S°/R$.
(E) $\ln K$.

65. In all cases, a reaction is spontaneous if
(A) $\Delta S < 0$ and $\Delta H > 0$.
(B) $\Delta S > 0$ and $\Delta H > 0$.
(C) $\Delta S = \Delta H$.
(D) $\Delta S > 0$ and $\Delta H < 0$.
(E) $\Delta S < 0$ and $\Delta H < 0$.

66. $PbSO_4(s) + H_2O (l) \rightarrow PbO_2(s) + Pb(s) + H^+(aq) + SO_4^{2-}(aq)$. In the above reaction the oxidizing agent is
(A) Pb.
(B) PbO_2.
(C) H_2O.
(D) SO_4^{2-}.
(E) $PbSO_4$.

67. The atomic mass for carbon given in the Periodic Table is 12.01, rather than 12.00, because
(A) carbon is the basis for all relative masses.
(B) natural occurring forms of ^{12}C, ^{13}C, and ^{14}C exist.
(C) all isotopes of carbon have six protons.
(D) carbon has many atoms of mass 12.01 a.m.u.
(E) a mole of carbon is 6.022×10^{23} atoms.

68. The iron shaft on a sailboat engine may be protected from corrosion by using a zinc collar. The zinc is called
(A) the sacrificial anode.
(B) the sacrificial cathode.
(C) surface reducer.
(D) catalytic protector.
(E) a galvanized metal.

69. The half-reaction with a standard reduction potential of zero is
(A) $H^+(aq) + 1e^- \rightarrow H(g)$.
(B) $H^+(aq) + 2e^- \rightarrow H_2(g)$.
(C) $2H^+(aq)(1.0M) + 2e^- \rightarrow H_2(g)(1.0$ atm).
(D) $Zn(s) - 2e^- \rightarrow Zn^{2+}(aq)(1M)$.
(E) $Zn^{2+}(1M) + 2e^- \rightarrow Zn(s)$.

70. For the reaction $X \rightarrow Y$, the rate for this second-order reaction has been determined to be 6.0×10^{-4} mol/L•s. when the value of [X] = 0.22 M. If the concentration of X is 0.66 M, the rate, in mol/L•s., will be
(A) 6.0×10^{-4}.
(B) $(6.0 \times 10^{-4})/9$.
(C) $(6.0 \times 10^{-4}) \times 3$.
(D) $(6.0 \times 10^{-4})^2$.
(E) $(6.0 \times 10^{-4}) \times 9$.

GO ON TO NEXT PAGE

71. The phrase "like dissolves like" refers to
    (A) molecules of like electronegativity differences.
    (B) molecules of like polarity.
    (C) molecules of like sharing of electron pairs within a bond.
    (D) compounds soluble in each other.
    (E) all of the above.

72. The oxidation state assigned to nitrogen in NO_3^- is
    (A) +1.
    (B) +2.
    (C) +5.
    (D) +6.
    (E) −1.

73. The 1.0 M solution which provides the fewest ions in solution is
    (A) CH_3COOH.
    (B) $Ca(NO_3)_2$.
    (C) $NaNO_3$.
    (D) $NaCl$.
    (E) H_2SO_4.

74. One liter quantities of 1.0 M dilute acids would usually be prepared in the laboratory, to be used as stock solutions, using
    (A) a buret.
    (B) a balance.
    (C) a test tube.
    (D) a volumetric flask.
    (E) all of the above.

75. Consider two solutions, Solution X with a pH of 7 and Solution Y with a pH of 9. The hydrogen ion ratio would be
    (A) X/Y = 7/9.
    (B) X/Y = 2/1.
    (C) X/Y = 1/3.
    (D) X/Y = 100/1.
    (E) X/Y = 1/100.

A5.5 Standard Reduction Potentials at 25°C (298 K) for Many Common Half-Reactions

Half-Reaction	e° (V)	Half-Reaction	e° (V)
$F_2 + 2e^- \rightarrow 2F^-$	2.87	$Cu^+ + e^- \rightarrow Cu$	0.52
$Ag^{2+} + e^- \rightarrow Ag^+$	1.99	$O_2 + 2H_2O + 4e^- \rightarrow 4OH^-$	0.40
$Co^{3+} + e^- \rightarrow Co^{2+}$	1.82	$Cu^{2+} + 2e^- \rightarrow Cu$	0.34
$H_2O_2 + 2H^+ + 2e^- \rightarrow 2H_2O$	1.78	$Hg_2Cl_2 + 2e^- \rightarrow 2Hg + 2Cl^-$	0.34
$Ce^{4+} + e^- \rightarrow Ce^{3+}$	1.70	$AgCl + e^- \rightarrow Ag + Cl^-$	0.22
$PbO_2 + 4H^+ + SO_4^{2-} + 2e^-$		$SO_4^{2-} + 4H^+ + 2e^- \rightarrow H_2SO_3 + H_2O$	0.20
$\quad \rightarrow PbSO_4 + 2H_2O$	1.69	$Cu^{2+} + e^- \rightarrow Cu^+$	0.16
$MnO_4^- + 4H^+ + 3e^- \rightarrow MnO_2 + 2H_2O$	1.68	$2H^+ + 2e^- \rightarrow H_2$	0.00
$2e^- + 2H^+ + IO_4^- \rightarrow IO_3^- + H_2O$	1.60	$Fe^{3+} + 3e^- \rightarrow Fe$	−0.036
$MnO_4^- + 8H^+ + 5e^- \rightarrow Mn^{2+} + 4H_2O$	1.51	$Pb^{2+} + 2e^- \rightarrow Pb$	−0.13
$Au^{3+} + 3e^- \rightarrow Au$	1.50	$Sn^{2+} + 2e^- \rightarrow Sn$	−0.14
$PbO_2 + 4H^+ + 2e^- \rightarrow Pb^{2+} + 2H_2O$	1.46	$Ni^{2+} + 2e^- \rightarrow Ni$	−0.23
$Cl_2 + 2e^- \rightarrow 2Cl^-$	1.36	$PbSO_4 + 2e^- \rightarrow Pb + SO_4^{2-}$	−0.35
$Cr_2O_7^{2-} + 14H^+ + 6e^- \rightarrow 2Cr^{3+} + 7H_2O$	1.33	$Cd^{2+} + 2e^- \rightarrow Cd$	−0.40
$O_2 + 4H^+ + 4e^- \rightarrow 2H_2O$	1.23	$Fe^{2+} + 2e^- \rightarrow Fe$	−0.44
$MnO_2 + 4H^+ + 2e^- \rightarrow Mn^{2+} + 2H_2O$	1.21	$Cr^{3+} + e^- \rightarrow Cr^{2+}$	−0.50
$IO_3^- + 6H^+ + 5e^- \rightarrow \frac{1}{2}I_2 + 3H_2O$	1.20	$Cr^{3+} + 3e^- \rightarrow Cr$	−0.73
$Br_2 + 2e^- \rightarrow 2Br^-$	1.09	$Zn^{2+} + 2e^- \rightarrow Zn$	−0.76
$VO_2^+ + 2H^+ + e^- \rightarrow VO^{2+} + H_2O$	1.00	$2H_2O + 2e^- \rightarrow H_2 + 2OH^-$	−0.83
$AuCl_4^- + 3e^- \rightarrow Au + 4Cl^-$	0.99	$Mn^{2+} + 2e^- \rightarrow Mn$	−1.18
$NO_3^- + 4H^+ + 3e^- \rightarrow NO + 2H_2O$	0.96	$Al^{3+} + 3e^- \rightarrow Al$	−1.66
$ClO_2 + e^- \rightarrow ClO_2^-$	0.954	$H_2 + 2e^- \rightarrow 2H^-$	−2.23
$2Hg^{2+} + 2e^- \rightarrow Hg_2^{2+}$	0.91	$Mg^{2+} + 2e^- \rightarrow Mg$	−2.37
$Ag^+ + e^- \rightarrow Ag$	0.80	$La^{3+} + 3e^- \rightarrow La$	−2.37
$Hg_2^{2+} + 2e^- \rightarrow 2Hg$	0.80	$Na^+ + e^- \rightarrow Na$	−2.71
$Fe^{3+} + e^- \rightarrow Fe^{2+}$	0.77	$Ca^{2+} + 2e^- \rightarrow Ca$	−2.76
$O_2 + 2H^+ + 2e^- \rightarrow H_2O_2$	0.68	$Ba^{2+} + 2e^- \rightarrow Ba$	−2.90
$MnO_4^- + e^- \rightarrow MnO_4^{2-}$	0.56	$K^+ + e^- \rightarrow K$	−2.92
$I_2 + 2e^- \rightarrow 2I^-$	0.54	$Li^+ + e^- \rightarrow Li$	−3.05

Advanced Placement Chemistry Equations and Constants

ATOMIC STRUCTURE

$$E = h\nu$$

$$\lambda = \frac{h}{m\nu}$$

$$E_n = \frac{-2.178 \times 10^{-18}}{n^2} \text{ joule}$$

$$c = \lambda\nu$$

$$p = m\nu$$

EQUILIBRIUM

$$K_a = \frac{[H^+][A^-]}{[HA]}$$

$$E^\Omega =$$

$$K_w = [OH^-][H^+] = 1.0 \times 10^{-14} \text{ at } 25°C$$

$$= K_a \times K_b$$

$$pH = -\log[H^+], \ pOH = -\log[OH^-]$$

$$14 = pH + pOH$$

$$pH = pK_a + \log\frac{[A^-]}{[HA]}$$

$$pOH = pK_b + \log\frac{[HB^+]}{[B]}$$

$$pK_a = -K_a, \ pK_b = -\log K_b$$

$$K_p = K_c(RT)^{\Delta n}$$

where $\Delta n = $ moles product gas − moles reactant gas

THERMOCHEMISTRY/KINETICS

$$\Delta S° = \sum S° \text{ products} - \sum S° \text{ reactants}$$

$$\Delta H° = \sum \Delta H_f° \text{ products} - \sum \Delta H_f° \text{ reactants}$$

$$\Delta G° = \sum \Delta G_f° \text{ products} - \sum \Delta G_f° \text{ reactants}$$

$$\Delta G° = \Delta H° - T\Delta S°$$

$$= -RT \ln K = -2.303RT \log K$$

$$= -n\Im E°$$

$$\Delta G = \Delta G° + RT \ln Q = \Delta G° + 2.303RT \log Q$$

$$q = mc\Delta T$$

$$C_p = \frac{\Delta H}{\Delta T}$$

$$\ln[A]_t - \ln[A]_0 = -kt$$

$$\frac{1}{[A]_t} - \frac{1}{[A]_0} = kt$$

$$\ln k = \frac{-E_a}{R}\left(\frac{1}{T}\right) + \ln A$$

E = energy	ν = frequency
λ = wavelength	p = momentum
ν = velocity	n = principal quantum number
m = mass	

Speed of light, $c = 3.0 \times 10^8 \text{ ms}^{-1}$

Planck's constant, $h = 6.63 \times 10^{-34} \text{ Js}$

Boltzmann's constant, $k = 1.38 \times 10^{-23} \text{ J K}^{-1}$

Avogadro's number $= 6.022 \times 10^{23} \text{ mol}^{-1}$

Electron charge, $e = -1.602 \times 10^{-19} \text{ coulomb}$

1 electron volt per atom $= 96.5 \text{ kJ mol} - 1$

EQUILIBRIUM CONSTANTS

K_a (weak acid)

K_b (weak base)

K_w (water)

K_p (gas pressure)

K_c (molar concentrations)

$S°$ = standard entropy

$H°$ = standard enthalpy

$G°$ = standard free energy

$E°$ = standard reduction potential

T = temperature

n = moles

m = mass

q = heat

c = specific heat capacity

C_p = molar heat capacity at constant pressure

E_a = activation energy

k = rate constant

A = frequency factor

Faraday's constant, $\Im = 96,500$ per mole of electrons

Gas constant, $R = 8.31 \text{ J mol}^{-1} \text{ K}^{-1}$

$= 0.0821 \text{ L atm mol}^{-1} \text{ K}^{-1}$

$= 8.31 \text{ volt coulomb mol}^{-1} \text{ K}^{-1}$

GASES, LIQUIDS, AND SOLUTIONS

$$PV = nRT$$

$$\left(P + \frac{n^2a}{V^2}\right)(V - nb) = nRT$$

$$P_A = P_{total} \times X_A, \text{ where } X_A = \frac{\text{moles A}}{\text{total moles}}$$

$$P_{total} = P_A + P_B + P_C + ...$$

$$n = \frac{m}{M}$$

$$°K = °C + 273$$

$$\frac{P_1 V_1}{T_1} = \frac{P_2 V_2}{T_2}$$

$$D = \frac{m}{V}$$

$$u_{rms} = \sqrt{\frac{3kT}{m}} = \sqrt{\frac{3RT}{M}}$$

KE per molecule $= \frac{1}{2}mv^2$

KE per mole $= \frac{3}{2}RT$

$$\frac{r_1}{r_2} = \sqrt{\frac{M_2}{M_1}}$$

molarity, M = moles solute per liter solution

molality = moles solute per kilogram solvent

$\Delta T_f = iK_f \times$ molality

$\Delta T_b = iK_b \times$ molality

$\pi = MRT$

$A = abc$

OXIDATION-REDUCTION; ELECTROCHEMISTRY

$$Q = \frac{[C]^c[D]^d}{[A]^a[B]^b}, \text{ where } a\,A + b\,B \rightarrow c\,C + d\,D$$

$$I = \frac{q}{t}$$

$$E_{cell} = E^\circ_{cell} - \frac{RT}{n\Im}\ln Q = E^\circ_{cell} - \frac{0.0592}{n}\log Q\,@25°C$$

$$\log K = \frac{nE^\circ}{0.0592}$$

P = pressure
V = volume
T = temperature
n = number of moles
D = density
m = mass
v = velocity

u_{rms} = root-mean-square speed

KE = kinetic energy

r = rate of effusion

M = molar mass

π = osmotic pressure

i = van't Hoff factor

K_f = molal freezing-point depression constant

K_b = molal boiling-point elevation constant

A = absorbance

a = molar absorptivity

b = path length

c = concentration

Q = reaction quotient

I = current (amperes)

q = charge (coulombs)

t = time (seconds)

E° = standard reduction potential

K = equilibrium constant

Gas constant, $R = 8.31$ J mol^{-1} K^{-1}

$= 0.0821$ L atm mol^{-1} K^{-1}

$= 8.31$ volt coulomb mol^{-1} K^{-1}

Boltzmann's constant, $k = 1.38 \times 10^{-23}$ J K^{-1}

K_f for H$_2$O $= 1.86$ K kg mol^{-1}

K_b for H$_2$O $= 0.512$ K kg mol^{-1}

1 atm = 760 mm Hg

= 760 torr

STP = 0.000°C and 1.000 atm

Faraday's constant, $\Im = 96{,}500$ coulombs per mole of electrons

Section II: Free-Response Questions
Time: 95 minutes
Number of Questions: 6

Allow yourself no more than 95 minutes to answer these questions. A calculator may be used only on questions 1, 2, and 3. You may refer to the Equations Sheet, the Periodic Table, and to the Standard Reduction Table throughout this section. All questions must be answered.

Part A

Each question in Part A is worth 20% of the Section II overall grade. You are allowed 55 minutes for Part A.

1. Given:

  (I) $H_2(g) + Cl_2(g) \rightleftharpoons 2HCl(g)$ $K_p = 2.5 \times 10^{33}$

  (II) $NH_3(g) + HCl(g) \rightleftharpoons NH_4Cl(s)$ $K_p = 5.1 \times 10^{15}$

  (III) $N_2(g) + 4H_2(g) + Cl_2(g) \rightleftharpoons 2NH_4Cl(s)$ $K_p = 3.9 \times 10^{70}$

  (a) Show how the equations, (I), (II), and (III) may be arranged to give the chemical equation $N_2(g) + 3H_2(g) \rightleftharpoons 2NH_3(g)$.

  (b) Determine the value for K_p for $N_2(g) + 3H_2(g) \rightleftharpoons 2NH_3(g)$.

  (c) Equation (III) has a much higher value for K_p than the K_p value for equation (I). What is the physical meaning of these differences?

  (d) Assume that reaction (II) above began with solid NH_4Cl, what would be the partial pressure of $NH_3(g)$ and of $HCl(g)$ at equilibrium?

  (e) Determine the value K_c for equation (II) at 25°C.

2.
  (a) 50.00 mL of 1.000 M HCl is titrated with 1.000 M KOH. Calculate the pH when a total of
     i.   49.99 mL of KOH has been added.
     ii. 50.00 mL of KOH has been added.
     iii. 50.01 mL of KOH has been added.
     iv. Comment on the pH values you have calculated.

  (b) A chemist wishes to make up 250. mL of a HNO_3 solution with a pH of 2.00, using a stock solution of 2.00 M HNO_3. What volume of the stock solution is needed?

  (c) A buffer is prepared by adding 20.5 g of acetic acid and 17.8 g of sodium acetate to water to make 500. mL of solution. Determine the pH of this buffer.

  The K_a of CH_3COOH is 1.8×10^{-5}.

3.
  (a) Use bond energy data to estimate the enthalpy change for this reaction:
     $H_2(g) + F_2(g) \rightarrow 2HF(g)$
     Bond energy data in kJ/mol of bonds: H–H 432   F–F 154   H–F 565

  (b) In this case the bond energy approach gives a good agreement with the measured ΔH for this reaction, but this is not always so. Why is this not always a successful approach to use for estimating reaction enthalpy?

  (c) The carbon–carbon bond length in C_2H_6 is 154 pm, whereas in C_2H_2 it is only 120 pm. Why is there a difference?

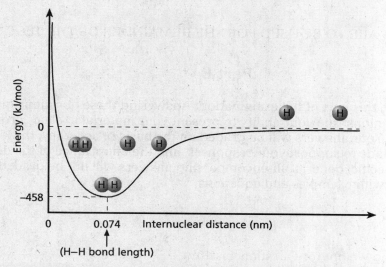

(d) Using the above diagram, explain why a H–H bond length of 0.074 nm is most reasonable.

(e) Explain what is meant by the statement "Ionic and covalent bonds are extreme bond types."

NO CALCULATORS ARE TO BE USED FOR THE REMAINDER OF THE TEST.

Part B

Spend 40 minutes on this part of the examination. Answering these questions gives you an opportunity to demonstrate your ability to present your material in clear, orderly, and convincing English. Your answers will be graded on the basis of accuracy, the kinds of information you include to support your responses, and the importance of the descriptive material used. Be specific; general, all-encompassing answers will not be graded as well as detailed answers with examples and equations.

The Section II scoring weight for Question 4 is 10%.

4. For each of the following three reactions, in part (i) write a BALANCED equation and in part (ii) answer the question about the reaction. In part (i), coefficients should be in terms of lowest whole numbers. Assume that solutions are aqueous unless otherwise indicated. Represent substances in solutions as ions if the substances are extensively ionized. Omit formulas for any ions or molecules that are unchanged by the reaction.
    a. (i) Acidified potassium dichromate solution is mixed with potassium iodide solution.
        (ii) What is the reducing agent in the above reaction?
    b. (i) 0.5 M lead (II) nitrate is added to an equal volume of 1.0 M hydrochloric acid.
        (ii) Identify the spectator ions in the reaction mixture.
    c. (i) Propanol is burned in air.
        (ii) How many moles of oxygen are required to burn one mole of propanol?

The Laboratory-Based Question: Answer the following essay question. The Section II score weighting is 15%.

5. A student performs a calorimetric experiment to determine the heat of formation for MgO. The student has the following equipment:
    coffee cup calorimeter
    thermometer
    balance
    chemicals: MgO(s), Mg ribbon, HCl(aq)

    The student takes measurements that will enable the calculation of the value of ΔH for the following two reactions:

    (1) $Mg(s) + HCl(aq) \longrightarrow MgCl_2(aq) + H_2(g)$ $\Delta H_1 = ?$

    (2) $MgO(s) + HCl(aq) \longrightarrow MgCl_2(aq) + H_2O(l)$ $\Delta H_2 = ?$

    The student is also given the following:

(3) $H_2(g) + 1/2\ O_2(g) \longrightarrow H_2O(l)$ $\Delta H_3° = -285.8\ kJ/mol$

(a) Write the equation for the heat of formation of MgO. Show how the three equations above can be used to determine the value of ΔH_f (heat of formation) for MgO.

(b) What measurements must be taken to find the ΔH for reaction (1) above?

(c) Without doing any calculations, indicate the equations that must be used to calculate ΔH_1.

(d) The calorimeter constant is the amount of heat that the calorimeter must gain to change temperature by one degree. Why would the use of a calorimeter constant improve the accuracy of the results?

(e) The accepted value for $\Delta H°_f = -602\ kJ/mol$. If the experimentally determined value is $-600.\ kJ/mol$, show how (without doing any calculations) the percent error would be calculated.

Answer the following essay question. The Section II score weighting value is 15%.

6. Using principles of atomic structure or bonding, explain the cause of the following differences:

(a) Water is less dense as a solid than it is as a liquid.

(b) The shape of the meniscus is different for water and for mercury when each liquid has been put into separate burets.

(c) Hydrogen peroxide has both a higher viscosity and a higher boiling temperature than water.

(d) Polyethylene, a polymer, has strength and toughness as a solid, yet it can be melted and formed into a new shape (thermoplastic behavior).

(e) The compound, $C_2H_2Cl_2$, has three isomeric forms due in part to cis–trans isomerism, whereas C_2H_3Cl has only one structure.

ANSWERS TO PRACTICE TEST 1

MULTIPLE-CHOICE ANSWERS

Using the table below, score your test.

Determine how many questions you answered correctly and how many you answered incorrectly. You will find explanations of the answers on the following pages.

1. D	2. C	3. A	4. B	5. B
6. C	7. E	8. E	9. D	10. D
11. D	12. E	13. A	14. D	15. C
16. C	17. E	18. C	19. B	20. A
21. B	22. A	23. C	24. E	25. C
26. E	27. D	28. B	29. C	30. B
31. B	32. E	33. B	34. B	35. D
36. A	37. D	38. E	39. E	40. A
41. C	42. E	43. A	44. E	45. C
46. A	47. D	48. E	49. A	50. E
51. A	52. A	53. A	54. E	55. D
56. D	57. A	58. C	59. D	60. A
61. A	62. C	63. D	64. B	65. D
66. E	67. B	68. A	69. C	70. E
71. E	72. C	73. A	74. D	75. D

CALCULATE YOUR SCORE:

Number answered correctly: _____
Adjust ¼ point for guessing penalty:
    Count the number of questions you
    answered incorrectly, multiply by .25, and subtract: – _____
Determine your adjusted score: _____

WHAT YOUR SCORE MEANS:

Because the test is different each year, the scoring is a little different. But generally, if you scored 25 or more on the multiple-choice questions, you'll most likely get a 3 or better on the test. If you scored 35 or more, you'll probably score a 4 or better. And if you scored a 50 or more, you'll most likely get a 5. Keep in mind that the multiple-choice section is worth 45% of your final grade, and the free-response section is worth 55% of your final grade. To learn more about the scoring for the free-response questions, turn to the last page of this section.

ANSWERS AND EXPLANATIONS

MULTIPLE-CHOICE ANSWERS

1. **ANSWER: D** Acetic acid will produce an acidic solution with a pH below 7 (*Chemistry* 7th ed. pages 131–132, 626–628 / 8th ed. pages 134–135, 642–644).

2. **ANSWER: C** The only ion with color in this group is the Cu^{2+} ion which is blue (*Chemistry* 7th ed. page 955, esp. Table 21.11 / 8th ed. page 964, esp. Table 21.11).

3. **ANSWER: A** Freezing temperature lowering is a colligative property; the more particles in solution the greater is the effect. Because $Ca(NO_3)_2$ provides three particles for each unit and the others all provide fewer, this substance in water would lower the freezing temperature the most (*Chemistry* 7th ed. pages 506–507, 513 / 8th ed. pages 518–519, 525).

4. **ANSWER: B** The solution which is basic (pOH above 7) is the sodium acetate. This is due to a reaction between the acetate ion and water (called hydrolysis) forming OH^- ions: $C_2H_3O_2^- + HOH \rightarrow OH^- + HC_2H_3O_2$ (this is a weak acid, therefore tending to stay mainly in the molecular form) (*Chemistry* 7th ed. pages 655–657 / 8th ed. pages 671–673).

5. **ANSWER: B** This is an ion with 32 valence electrons to consider, $[(4 \times 7) + (5) - 1]$, which form four pairs around the phosphorus and four pairs around each Cl with four bonds between P and the four Cl. This forms a tetrahedron (*Chemistry* 7th ed. pages 369–373, esp. diagram in center left p. 373 / 8th ed. pages 381–384, esp. diagram in center left p. 384).

6. **ANSWER: C** Galvanizing is a process in which steel is dipped into molten zinc, leading a protective coating on the steel. Because zinc is more reactive than steel, this outer coating is sacrificed to corrosion rather than the steel (*Chemistry* 7th ed. pages 815–816 / 8th ed. pages 847–848).

7. **ANSWER: E** Even though oxygen has six electrons, only two are bonding electrons; the nonbonding pairs force the bonding pairs into a bent configuration rather than a linear structure. This is the same general shape seen in the water molecule (*Chemistry* 7th ed. pages 367–373, esp. Figure 8.17 / 8th ed. pages 378–384, esp. Figure 8.17).

8. **ANSWER: E** Ideal gases have particles of zero volume, no forces between the particles, and are described by the relationship $PV = nRT$ (*Chemistry* 7th ed. pages 183, 186–188 / 8th ed. pages 183, 188–190).

9. ANSWER: **D** The two lone pairs of electrons on Xe favor a position of 180° from each other, leaving the four bonded pairs pointing toward the four corners of a square (*Chemistry* 7th ed. pages 373–374 / 8th ed. pages 384–385).

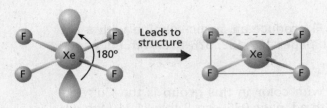

10. ANSWER: **D** The CO_3^{2-} reacts with HOH as: $CO_3^{2-} + HOH \rightarrow HCO_3^- + OH^-$, which gives a basic solution. You can determine that the salts are acidic, basic, or neutral by determining the parent acid and parent base. For salt Na_2CO_3, the parent base NaOH is strong and the parent acid is weak H_2CO_3, so the salt is basic. This is a memory aid and does not really explain the chemistry of why this occurs (*Chemistry* 7th ed. pages 655–660 / 8th ed. pages 671–677).

11. ANSWER: **D** Molar heat capacity refers to the energy required to change the temperature of one mole of the substance, (in this case, steam or water), by 1°C (*Chemistry* 7th ed. page 464 / 7th ed. page 476).

12. ANSWER: **E** The negative slope of the solid/liquid line is due to the density of the solid (ice in this case) which is less than that of the liquid (water). With almost all substances, the solid state is denser than the liquid state; water is nontypical in this property (*Chemistry* 7th ed. pages 467–470 / 8th ed. pages 479–482).

13. ANSWER: **A** The molar volume of any ideal gas @ STP is 22.4 L. From 5.6 L × 1 mol/22.4 L = mol we see that response A is correct (*Chemistry* 7th ed. pages 190–191 / 8th ed. pages 194–195).

14. ANSWER: **D** The K_{sp} is an equilibrium constant with the usual expression, $K_{sp} = [Pb^{2+}][Br^-]^2$. The solid, of course, has a constant concentration and so is not a part of the equilibrium expression (*Chemistry* 7th ed. pages 717–718 / 8th ed. pages 743–744).

15. ANSWER: **C** Increasing the surface area does increase the amount of solid that will dissolve in a given period of time. However, since ions reform the solid on the surface of the solid, the rate of reforming the solid also increases, and at the same rate. Therefore, solubility is not increased by grinding; only the rate of dissolving is increased (*Chemistry* 7th ed. pages 717–718 / 8th ed. pages 743–744).

16. ANSWER: **C** If the Q exceeds the K_{sp} a precipitate forms. Think of the ion product, Q, as a limit which if exceeded causes the equilibrium to shift toward the solid (*Chemistry* 7th ed. pages 724–726 / 8th ed. pages 752–754).

17. **ANSWER:** **E** A buffer resists pH change by reaction with either added H^+ or added OH^-. It may be the weak acid which reacts with OH^-; it may be the anion of the salt of the acid which reacts with the H^+ (*Chemistry* 7th ed. pages 736–737, 687–688 / 8th ed. pages 764–765, 704–705).

18. **ANSWER:** **C** Entropy may be thought of as a measure of the disorder of a system. Solids are more ordered than liquids or gases or solutions. Graphite shows less order than diamond, where all the bonds are identical, because graphite has a 2-D covalent network with delocalized electrons between the layers and diamond has highly localized electrons in a 3-D network of covalent bonds. (Note that you must refer to both substances to get full credit on the AP exam on essay questions on such topics) (*Chemistry* 7th ed. pages 773–779, 457–459).

19. **ANSWER:** **B** Larger, heavier substances are more subject to becoming polar due to interaction with other molecules (*Chemistry* 7th ed. pages 459–462 / 8th ed. pages 471–474).

20. **ANSWER:** **A** $H = (+)$ for bond breaking, which is always an endothermic process. $S = (+)$ for an increase in disorder due here to the increase in the number of particles (*Chemistry* 7th ed. pages 759, 761, 766–768 / 8th ed. pages 783, 785, 790–792).

21. **ANSWER:** **B** Without a salt bridge, one side of the cell would become increasingly positive (on the oxidation side, where electrons leave) and the other side would become increasingly negative. The salt bridge allows for ions to flow between the compartments, to keep the net charge zero as electrons flow through the external circuit (*Chemistry* 7th ed. pages 791–793 / 8th ed. pages 823-825).

22. **ANSWER:** **A** The reaction $Ag \rightarrow Ag^+$ is more likely at the lower concentration of Ag^+; therefore on the left side, the reaction will be $Ag \rightarrow Ag^+ + 1e^-$. Those electrons then flow through the external circuit to the right side (*Chemistry* 7th ed. pages 803–804 / 8th ed. pages 836–837).

23. **ANSWER:** **C** The reaction is $Ag^+ + 1e^- \rightarrow Ag$ (*Chemistry* 7th ed. pages 791–793 / 8th ed. pages 823–825).

24. **ANSWER:** **E** All IA compounds, all nitrates, and $CaCl_2$ are soluble in water (*Chemistry* 7th ed. pages 143–144 / 8th ed. pages 148–149).

25. **ANSWER:** **C** Both 1.0 M HCl and 1.0 M $HC_2H_3O_2$ can deliver the same number of moles of H^+ to neutralize the base. Diprotic H_2SO_4 will neutralize twice as much NaOH (*Chemistry* 7th ed. pages 144–146 / 8th ed. pages 149–151).

26. **ANSWER: E** $Al(NO_3)_3$ contains four ions per formula unit; each mole of $Al(NO_3)_3$ contributes four moles of ions: 1.2 mol/L × 4 = 4.8 mol of ions/L (*Chemistry* 7th ed. pages 134–135 / 8th ed. pages 136–137).

27. **ANSWER: D** Gases with the lowest masses have the greatest effusion rates (*Chemistry* 7th ed. pages 206–208 / 8th ed. pages 213–214).

28. **ANSWER: B** As liquid H_2O changes to solid (ice), heat is lost by the system; this is an exothermic process. Both reactions noted in responses (A) and (D) are common demonstration reactions involving the heating of the solid to form oxygen; they are endothermic. Response (C) is a reaction involving only bond breaking which is always endothermic. Response (E) represents the electron affinity of atomic chlorine gas, another endothermic process (*Chemistry* 7th ed. pages 231–232 / 8th ed. pages 238–239).

29. **ANSWER: C** This law describes the mass ratios for two elements which form two different compounds. These mass ratios can be reduced to small whole numbers (*Chemistry* 7th ed. pages 42–43 / 8th ed. pages 42–43).

30. **ANSWER: B** Note how the compound must be similar to CO_2 and that for purposes of determining shape, a double bond is treated the same as a single bond: $:\ddot{S}=\ddot{S}:$ (*Chemistry* 7th ed. pages 367–376 / 8th ed. pages 378–387).

31. **ANSWER: B** With a critical temperature above room temperature, the gas can be liquefied with increasing pressure due to intermolecular forces, which become more important as the distance between molecules decreases (*Chemistry* 7th ed. pages 468–469, 500 / 8th ed. pages 480–482, 512).

32. **ANSWER: E** The average rate and the instantaneous rate should not be confused. If the concentration is plotted on the *y*-axis vs. time, a curve is generated. For any given point on this curve (i.e., any given time), a tangent may be drawn. The slope of this tangent line is the change in concentration/change in time, which is rate. By convention, this is always expressed as a positive value (*Chemistry* 7th ed. pages 527–533 / 8th ed. pages 540–546).

33. **ANSWER: B** The addition of a catalyst lowers the activation energy by finding a less energy demanding pathway. It does this by forming different intermediate substances (*Chemistry* 7th ed. pages 557–559 / 8th ed. pages 570–572).

34. **ANSWER: B** $Cr_2O_7{}^{2-}$ ions are orange. Other ions and their colors: Cu^{2+} is light blue, $CrO_4{}^{2-}$ is yellow, and K^+ and Cl^- are colorless. Many of the Cr^{6+} complexes have color. Most transition metal compounds are colored because the "d" electrons are easily excited by visible light into other available "d" orbitals (*Chemistry* 7th ed. pages 970–971 / 8th ed. pages 979–980).

35. **ANSWER: D** 0.20 mol OH^- + 0.10 mol H_3PO_4 → 0.40 mol H_2O + 0.10 mol $HPO_4{}^-$ (*Chemistry* 7th ed. pages 149–154 / 8th ed. pages 154–161).

36. **ANSWER: A** The triple bond between the carbons is made of one sigma bond and two pi bonds. No orbitals are hybridized to form the two pi bonds (two unhybridized "p" orbitals). The sigma bond between the two carbons and the sigma bond between the C and H are formed by two "sp" hybrid orbitals on each carbon (*Chemistry* 7th ed. pages 391–401 / 8th ed. pages 404–414).

37. **ANSWER: D** Because $P_{H_2} + P_{H_2O} = P_{total}$, $P_{H_2} = P_{total} - P_{H_2O} = 757 - 21$. (*Chemistry* 7th ed. pages 198–199 / 8th ed. pages 203–205).

38. **ANSWER: E** $Mg(OH)_2(s) \rightleftharpoons Mg^{2+}(aq) + 2OH^-(aq)$. The added H^+ (the pH is 4, therefore acidic) leads to reaction with the OH^- from the $Mg(OH)_2$ to form water. The reaction shifts toward the Mg^{2+} and OH^- ions; some of the $Mg(OH)_2$ dissolves (*Chemistry* 7th ed. page 724 / 8th ed. page 752).

39. **ANSWER: E** The equivalence point will occur when the mol of OH^- from the base and mol of H^+ from the acid are equal (*Chemistry* 7th ed. pages 700–707 / 8th ed. pages 717–725).

40. **ANSWER: A** A salt like AgOH (K_{sp} = 10^{-8}) and a salt like $Ca(OH)_2$ (K_{sp} = 10^{-6}) form a different number of ions in solution. With AgOH, for example, solubility = $\sqrt{K_{sp}}$; whereas with $Ca(OH)_2$, the solubility = $\sqrt[3]{K_{sp}/4}$ (*Chemistry* 7th ed. pages 718–721 / 8th ed. pages 743–746).

41. **ANSWER: C** Due to the hydrolysis of the acetate ion, the solution will be basic (therefore H^+ and OH^- cannot be equal); because some of the acetic acid formed from this hydrolysis reaction will react with water leaving less acetic acid, the acetic acid and OH^- will not be equal. Because some of the acetate ion reacts with water (the hydrolysis reaction again), the $[C_2H_3O_2{}^-]$ will be somewhat less than 0.10 M (*Chemistry* 7th ed. pages 655–657 / 8th ed. pages 671–673).

42. **ANSWER: E** It is important to know that temperature is a measure of <u>average</u> molecular kinetic energy. While the exact relationship, $KE_{ave} = 3/2\ RT$, will most likely not be on the test, you should see that temperature is not simply equal to kinetic energy (*Chemistry* 7th ed. page 204 / 8th ed. page 211).

43. **Answer: A** The additional Ag^+ causes the equilibrium to shift to the left, forming more solid Ag_2CrO_4 (common ion effect). Note that the concentration of a solid is constant; therefore response D is incorrect (*Chemistry* 7th ed. pages 721–722 / 8th ed. pages 746–747).

44. **Answer: E** Because half of the acid has reacted, the [acid] = [A⁻], from $HA \rightleftharpoons H^+ + A^-$, $K_a = [H^+][A^-]/[HA] = [H^+]$. Because the pH = 5, $K_a = 1 \times 10^{-5}$ (*Chemistry* 7th ed. pages 721–722 / 8th ed. pages 746–747).

45. **Answer: C** Entropy or disorder is increasing when a liquid becomes a gas; this is shown with a positive value for entropy (*Chemistry* 7th ed. pages 749–754 / 8th ed. pages 773–779).

46. **Answer: A** Perhaps the quickest way to solve this is to substitute the values given in $\Delta G = \Delta H - T\Delta S$, while being careful with the units. From $\Delta H = -10$ kJ and $\Delta S = -100$ J/K, if the temperature is 100K, then $(-10,000)-(100)(-100) = (-10,000)-(-10,000) = 0$. Any temperature lower than 100K will make ΔG negative, hence a spontaneous reaction (*Chemistry* 7th ed. pages 766–767, 759–760 / 8th ed. pages 790–791. 783-784).

47. **Answer: D** Equation I plus the reverse of equation II/2 = the desired equation, so $(=370)-(-140/2)= -300$ kJ (*Chemistry* 7th ed. pages 768–769 / 8th ed. pages 792–793).

48. **Answer: E** The value for ΔG indicates how far the process is from equilibrium (which direction is favored), not how rapidly it reaches equilibrium (*Chemistry* 7th ed. page 749 / 8th ed. page 773).

49. **Answer: A** From $\log K = nE°/0.0591 = (2)(-0.60)/0.06 = -1.20/0.06 = -20$, you can see (even if you can not calculate the exact value of K) that a negative value means a very small value for K. Because K in this case is K_{sp}, the solubility would be very low (*Chemistry* 7th ed. pages 857–858 / 8th ed. pages 888–889).

50. **Answer: E** The aluminum ion gains 3 electrons and Mg loses 2 electrons. The overall reaction must reflect an equal number of electrons gained and lost (*Chemistry* 7th ed. pages 796–798 / 8th ed. pages 829–831).

51. **Answer: A** If you know the mass of $PbCl_2$, you can determine the moles of Pb^{2+}, then the moles of electrons required, and the number of coulombs of charge needed. This must be multiplied by the current (1A = 1C/s) (*Chemistry* 7th ed. page 818 / 8th ed. pages 849–850).

52. **ANSWER: A** The anode is the site of oxidation. $Zn - 2e^- \rightarrow -Zn^{2+}$ is oxidation (*Chemistry* 7th ed. pages 797–798 / 8th ed. pages 830–831).

53. **ANSWER: A** In this cell the zinc loses electrons ($Zn - 2e^- \rightarrow Zn^{2+}$) and the Ag^+ gains electrons ($Ag^+ + 1e^- \rightarrow Ag$) (*Chemistry* 7th ed. pages 797–798 / 8th ed. pages 830–831).

54. **ANSWER: E** Molarity = mol solute / L of solution (*Chemistry* 7th ed. pages 133–134 / 8th ed. pages 136–137).

55. **ANSWER: D** The 1.0 mol of Pb^{2+} (1.0 mol/L × 1.0 L) is now occupying 2.0 L of the solution, so the $[Pb^{2+}]$ = 1.0 mol/2.0 L = 0.50 M. The $[K^+]$ can be found in like manner. The $[NO_3^-]$ = (1.0 mol/L × 1.0 L + 2.0 mol/L × 1.0 L) / 2.0L = 1.5 M (*Chemistry* 7th ed. pages 134–136 / 8th ed. pages 137–139).

56. **ANSWER: D** Because 30.9/100 atoms are Cu mass 65, and 69.1/100 atoms are Cu mass 63, the average mass of a copper atom is $(0.309 \times 65) + (0.691 \times 63)$ (*Chemistry* 7th ed. pages 77–81 / 8th ed. pages 77–80).

57. **ANSWER: A** Units will help explain this calculation: 5.234 mg/1000 mg/g × 1 mol / 40.18 g × (6.02×10^{23}) atoms/mol (*Chemistry* 7th ed. page 82 / 8th ed. page 81).

58. **ANSWER: C** Arrhenius restricted his acid-base definitions to water solutions. Acids provide H^+ ions (protons) and bases provide OH^- (*Chemistry* 7th ed. pages 168–169 / 8th ed. pages 168–169).

59. **ANSWER: D** Weak electrolytes ionize very slightly, so few ions are formed when they dissolve (*Chemistry* 7th ed. page 131 / 8th ed. page 134).

60. **ANSWER: A** You cannot determine the rate law by just looking at the chemical equation (unlike the equilibrium constant). The rate law and consequently the order are determined by observing how the rate varies with the concentration of reactants (*Chemistry* 7th ed. pages 534–535 / 8th ed. pages 547–548).

61. **ANSWER: A** Because we know that HCl is a strong acid, i.e. it ionizes 100%, there must not be a strong $Cl^- \rightarrow H^+$ affinity (in water). The equilibrium system, $AgCl(s) \rightleftharpoons Ag^+(aq) + Cl^-(aq)$ can be shifted to the left due to the common ion effect (*Chemistry* 7th ed. pages 734–736 / 8th ed. pages 762–764).

62. **ANSWER: C** Note that doubling [A] doubles the rate of this reaction, suggesting that this is a first-order reaction with respect to [A]. Keeping the [A] constant and changing [B] (comparing Exp. 1 to 3), has no effect on the reaction rate, so [B] is not a part of the rate law (*Chemistry* 7th ed. pages 535–537 / 8th ed. pages 548–550).

63. **ANSWER: D** Note from the stoichiometry of the equation, every time a mol of A disappears, two moles of D appear. . . twice the rate of the disappearance of A (*Chemistry* 7th ed. pages 537–538 / 8th ed. pages 550–551).

64. **ANSWER: B** Because this is written in the general form for linear equations ($y = mx + b$), the slope = $-\Delta H°/R$. It is important to note this would allow for the determination of ΔH (and for $\Delta S°$ for $\Delta S°/R$ is the intercept) for a reaction run at several different temperatures and K noted for each (*Chemistry* 7th ed. pages 776–778 / 8th ed. pages 800–802).

65. **ANSWER: D** Reactions are spontaneous if $\Delta G°$ (the Gibb's Free Energy) is negative. This must be the case if ΔH is (–) and ΔS is (+), (from: $\Delta G = \Delta H - T\Delta S$) (*Chemistry* 7th ed. pages 756–762 / 8th ed. pages 780–786).

66. **ANSWER: E** The oxidizing agent is itself reduced. In this case, the $PbSO_4$ is both oxidized and reduced (*Chemistry* 7th ed. pages 158–162 / 8th ed. pages 164–166).

67. **ANSWER: B** The atomic mass for C given in the Periodic Table is the average of the naturally occurring mixture of ^{12}C, ^{13}C, and ^{14}C (*Chemistry* 7th ed. pages 78–79 / 8th ed. pages 78–79).

68. **ANSWER: A** Because the reaction for the collar is $Zn - 2e^- \longrightarrow Zn^{2+}$, it is undergoing oxidation. The site of oxidation is known as the anode (*Chemistry* 7th ed. pages 813–816 / 8th ed. pages 845–847).

69. **ANSWER: C** This is the agreed upon standard for potential (*Chemistry* 7th ed. pages 794–798 / 8th ed. pages 826–831).

70. **ANSWER: E** Using the fact that this is a second-order reaction, rate = $k[X]^2$. Because the concentration of X is tripled (goes from 0.22 M to 0.66 *M*), the rate is $(3)^2$, or 9 times greater (*Chemistry* 7th ed. pages 535–542 / 8th ed. pages 548–555).

71. **ANSWER: E** Though the phrase generally applies to molecules of like polarity, all of the responses do apply (*Chemistry* 7th ed. pages 128–129 / 8th ed. pages 131–132).

72. **ANSWER: C** Because each oxygen has an oxidation state of –2 for a total of $3x - 2 = -6$, and the net charge on this ion is 1–, the nitrogen must have an oxidation state of 5+ (*Chemistry* 7th ed. pages 155–157 / 8th ed. pages 162–163).

73. **ANSWER: A** This weak acid, acetic acid, dissociates (ionizes) only to a slight extent in aqueous solution (*Chemistry* 7th ed. page 131 / 8th ed. page 134).

74. **ANSWER: D** The use of a 1-liter flask constructed with a hairline thin volume marker (calibration mark) at just the 1-liter volume is usually used (*Chemistry* 7th ed. pages 135–136 / 8th ed. pages 140–141).

75. **ANSWER: D** Keep in mind that the pH scale is a log scale based on 10; $pH = -\log[H^+]$. $X/Y = 10^{-7}/10^{-9} = 10^2/1 = 100/1$ (*Chemistry* 7th ed. pages 631–634 / 8th ed. pages 647–650).

SECTION II: FREE-RESPONSE ANSWERS

Question 1: Answers

(a) K_p

Reverse and double equation II:

$$2NH_4Cl(s) \rightleftharpoons 2NH_3(g) + 2HCl(g) \qquad 1/(5.1 \times 10^{15})^2$$

Reverse equation I:

$$2HCl(g) \rightleftharpoons H_2(g) + Cl_2(g) \qquad 1/(2.5 \times 10^{33})$$

Add equation III:

$$N_2(g) + 4H_2(g) + Cl_2(g) \longrightarrow 2NH_4Cl(s) \qquad 3.9 \times 10^{70}$$

$$N_2(g) + 3H_2(g) \rightleftharpoons 2NH_3(g)$$

(b) $K_p = 1/(5.1 \times 10^{15})^2 \times 1/(2.5 \times 10^{33}) \times 3.9 \times 10^{70} = 6.0 \times 10^5$.
Note: Chemical equations are added to give the desired equation. Equilibrium constants are multiplied to give the K_p for the overall process.

(c) Generally a large value for an equilibrium constant indicates that, at equilibrium, there is a higher concentration of products than reactants. Chemists say "products are favored" in this situation.

(d) Each time a mole of ammonia is formed, a mole of hydrogen chloride also forms, and because $K_p = (P_{NH_3})(P_{HCl}) = 5.1 \times 10^{15}$,

$$P_{NH_3} = P_{HCl} = \sqrt{5.1 \times 10^{-15}} = 7.1 \times 10^{-8} \text{ atm.}$$

(e) From $K_c = K_p / (RT)^{\Delta n}$
$(5.1 \times 10^{-15})/(0.0821 \times 298)^{-2} = 3.1 \times 10^{-12}$
(*Chemistry* 7th ed. pages 734–736 / 8th ed. pages 762–764)

Question 2: Answers

(a)

    i. 49.99 mL of KOH has been added:
       0.05000 L × 1.000 mol/L = 0.05000 mol H^+
       0.04999 L × 1.000 mol/L = 0.04999 mol OH^-
       leaving 0.00001 mol of H^+ to survive.
       $[H^+]$ = 0.00001mol/0.09999L = 1×10^{-4}M   pH = 4.0

    ii. 50.00 mL of KOH has been added:
       All the $[H^+]$ reacts with all of the $[OH^-]$ to form water.
       $[H^+] \times [OH^-] = K_w = 1 \times 10^{-14}$ and $[H^+] = [OH^-] = 1 \times 10^{-7}$ M   pH = 7.0

    iii. 50.01 mL of KOH has been added:
       0.05000 L × 1.000 mol/L = 0.05000 mol H^+
       0.05001 Lx 1.000 mol/L = 0.05001 mol OH^-
       leaving 0.00001mol OH^- to survive.
       $[OH^N]$ = 0.00001 mol/0.10001 L = 1×10^{-4} pOH =4.0   pH = 10.0

    iv. There as been a large change in pH (from 4 to 7 to 10) caused by the buffer being overwhelmed by the addition of 0.02 mL of base. This causes the large change seen on the titration curve graph at what is usually called the inflection point

(*Chemistry* 7th ed. pages 697–699, esp. Figure 15.1 / 8th ed. pages 713–719, esp. Figure 15.1).

(b) pH = 2.00, so $[H^+]$ = $10^{-2.00}$ = 0.010 *M*
    0.0100 mol/L × 0.250 L = 0.0025 mol H^+ needed.
    0.0025 mol / 2.00 mol/L = 0.00125 L = 1.3 mL
(*Chemistry* 7th ed. pages 631–634 / 8th ed. pages 647–650)

(c) Concentrations: $[HC_2H_3O_2]$ = (20.5 g/60.0 g/mol)/0.500 L = 0.683 *M*
                      $[NaC_2H_3O_2]$=(17.8g / 82.0 g/mol) / 0.500 L = 0.434 *M*

    $HC_2H_3O_2 \rightleftharpoons H^+ + C_2H_3O_2^-$

    $[HC_2H_3O_2]$ = (0.683 – x) and $[C_2H_3O_2^-]$ = (0.434 + x) where x = $[H^+]$ and is very very small.
    K_a= $[H^+][C_2H_3O_2^-]$/$[HC_2H_3O_2]$ = $[H^+]$ (0.434 + x)/(0.683 – x) = 1.8×10^{-5}
    $[H^+]$ = 2.8×10^{-5}M   pH = 4.55
(*Chemistry* 7th ed. pages 684–691 / 8th ed. pages 701–709)

Question 3: Answers

(a) To break one mole of each H–H and F–F bonds requires (+432) + (+154) = +586 kJ.
    To form 2 mol of H–F bonds releases (–565)2 = –1130 kJ
    The difference (–1130 + 586) gives an estimated ΔH for this reaction of –544 kJ
(*Chemistry* 7th ed. pages 350–353 / 8th ed. pages 361–364).

(b) Bond energy values are average values for species in the gas phase. For individual bond energies, the nearby bonds sometimes affect the strength of the chemical bonds in question. Sometimes the molecular environment is very important (*Chemistry* 7th ed. pages 350–351 / 8th ed. pages 361–362).

(c) In C_2H_6, the C–C bond is a single bond with only one pair of electrons between the carbon nuclei; in C_2H_2, there is a triple bond between the carbons with three pair of shared electrons between the nuclei. More electron pairs means more attractive force between the shared electrons and the nuclei, therefore a shorter bond length (*Chemistry* 7th ed. pages 350–352 / 8th ed. pages 361–363).

(d) The graph shows the change in energy of the system (due to repulsion of electrons by electrons, repulsion between nuclei, and attraction between nuclei and electrons) and how it changes with change in internuclear distance. The atoms move to an optimum distance which achieves the lowest overall energy of the system. In the case shown in this diagram, the lowest energy is achieved between two hydrogen atoms at a distance of 0.074 nm (*Chemistry* 7th ed. pages 330–331 / 8th ed. pages 341–342).

(e) Bonds result when electrons are shared by nuclei. When the sharing is exactly equal between two nuclei, the bond is a "pure" covalent bond. Such a bond is found between H–H, or Cl–Cl. As sharing becomes more uneven, the bond is said to be polar covalent. In the extreme case of uneven sharing, where an electron is completely lost by one nucleus and completely gained by the other, an ionic bond has formed; hence ionic and covalent bonds are extreme types of electron sharing (*Chemistry* 7th ed. page 331 / 8th ed. page 342).

Question 4: Answers

a. (i) $Cr_2O_7^{2-} + 14H^+ + 6I^- \rightarrow 2Cr^{3+} + 3I_2 + 7H_2O$
(ii) The I^- (KI) is the reducing agent.

b. (i) $Pb^{2+} + 2Cl^- \rightarrow PbCl_2$
(ii) The spectator ions are the nitrate ion (NO_3^-) and the H^+ ion.

c. (i) $2C_3H_7OH + 9O_2 \rightarrow 6CO_2 + 8H_2O$
(ii) 4.5 moles of oxygen are required to burn one mole of propanol.
(*Chemistry* 7th ed. pages 145–146, 162–168 / 8th ed. pages 150–151, 166-168)

Question 5: Answers

(a) $Mg(s) + 1/2\ O_2(g) \rightarrow MgO(s)$

Eq. 1	$Mg(s) + 2HCl(aq) \rightarrow H_2(g) + MgCl_2(aq)$	ΔH_1
rev. Eq. 2:	$H_2O(l) + MgCl_2(aq) \rightarrow MgO(s) + 2HCl(aq)$	$-\Delta H_2$
Eq. 3	$H_2(g) + 1/2O_2(g) \rightarrow H_2O(l)$	ΔH_3

Addition of
above: $\qquad$ $Mg(s) + 1/2\ O_2(g) \rightarrow MgO(s)$ $\qquad\qquad$ ΔH_f

So $\Delta H_1 + (-\Delta H_2) + \Delta H_3 = \Delta H_f$.

(b) Data required: mass of Mg ribbon
    initial temperature
    final temperature
    total mass of the solution

(c) Equations required: $q = S \times M \times \Delta T$
        moles Mg = grams Mg / molar mass
        $\Delta H = q$ / mol Mg

(d) Including the calorimeter constant in the calculations would account for any heat gained or lost by the apparatus itself (the calorimeter includes the thermometer as well as the coffee cup calorimeter) and therefore provides a more accurate value for ΔH.

(e) Percentage error
% error = (experimental value – accepted value) / accepted value × 100%
= (–600.) – (–602) / (–602) × 100%
(*Chemistry* 7th ed. pages 235–252 / 8th ed. pages 243–261)

Question 6: Answers

(a) Water forms open crystals as a solid (made of many water molecules) with more space between the H_2O molecules. This arrangement maximizes intermolecular forces in the solid state. As ice melts, the open crystals fall in upon themselves and the liquid state occupies less space, thereby a greater density (*Chemistry* 7th ed. pages 454–456, 435–436 / 8th ed. pages 466–468, 449-451).

(b) Compare the force within the liquid with that between liquid and glass. In the case of water, the attraction for the glass is stronger, making for a concave meniscus. In mercury, the Hg–Hg intermolecular forces are greater than the Hg–glass attraction, making for a convex shape (*Chemistry* 7th ed. pages 426–430 / 8th ed. pages 440–444).

(c) H_2O_2 has more hydrogen-oxygen groups, and therefore more hydrogen "bonding" is possible than in H_2O. That provides groups of H_2O_2 molecules which tangle and resist flow (*Chemistry* 7th ed. pages 429–430 / 8th ed. pages 443–444).

(e) *Cis–trans* isomerism means that when the two Cl groups are on the same side of the double bond, they form a compound different from that when they are on opposite sides of the double bond. There are three structures possible:

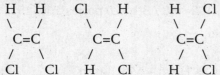

whereas the monochloride has only one possible structure:

```
H     H
 \   /
  C=C
 /   \
H     Cl
```

SCORING THE FREE-RESPONSE QUESTIONS

It is difficult to come up with an exact score for this section of the test. However, if you compare your answers to the answers in this book, remembering that each part of the test you answer correctly is worth points even if other parts of the answer are incorrect (see the section titled "Scoring for the Free-Response Questions" on page 13 of this book), you can get a general idea of the percentage of the questions for which you would get credit. If you believe that you got at least one-third of the possible credit, you would probably receive a 3 on this part of the test. If you believe that you would receive close to half or more of the available credit, your score would more likely be a 4 or a 5.

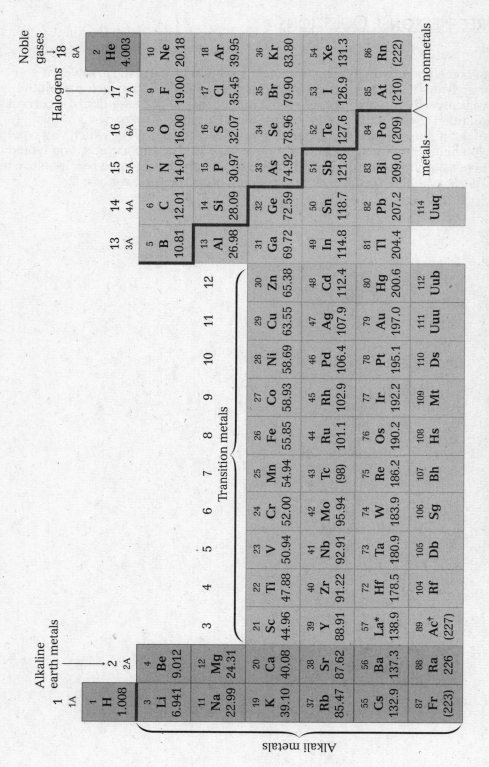

Group numbers 1–18 represent the system recommended by the International Union of Pure and Applied Chemistry.

Practice Test 2

These questions are representative of the AP Chemistry examination, but keep in mind that it is impossible to predict exactly how well you will do on the actual exam. The first section of this test is 50% of your total test grade. Time yourself so that you finish this part in 90 minutes. Remember that on the actual examination there is a guess factor; scoring is the number of questions answered correctly minus one-fourth the number wrong.

AP Chemistry Exam
Section I: Multiple-Choice Questions
Time: 90 minutes
Number of Questions: 75

No calculators are to be used in this section; no tables (except Periodic Table) permitted.

Part A

Directions: The following questions consist of five lettered headings followed by a listing of numbered phrases. For each phrase, you are to select the one heading that is most closely related to it. Headings may be used once, more than once, or not at all.

Questions 1–4 refer to the endings on the names of ions that indicate the formula of the ion.

    (A) -ide
    (B) -ite
    (C) -ium
    (D) -ate
    (E) –ane

Choose the correct ending for each of the following:

1. CO_3^{2-}

2. PO_3^{3-}

3. NH_4^+

4. ClO_2^-

Questions 5–7 refer to the following organic functional groups.

    (A) Carboxylic acids
    (B) Alcohols
    (C) Esters
    (D) Amines
    (E) Ketones

To which functional group family does each compound below belong?

5. $CH_3CH_2CH_2OH$

6. H_3CCCH_3 with $\overset{O}{\underset{\|}{}}$ on the middle carbon

$$6.\quad H_3C\overset{\displaystyle O}{\overset{\|}{C}}CH_3$$

$$7.\quad H_3C\overset{\displaystyle O}{\overset{\|}{C}}OCH_2CH_3$$

GO ON TO NEXT PAGE

Questions 8–10 refer to the following phase diagram.

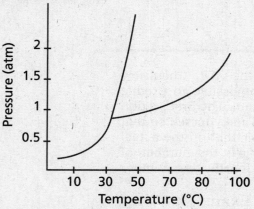

Use the following key to answer the questions.

(A) Sublimation
(B) Condensation
(C) Boiling
(D) Melting
(E) Freezing

8. If the temperature increases from 10 °C to 60 °C at a constant pressure of 0.4 atm, which of the processes occurs?

9. If the temperature decreases from 110 °C to 40 °C at a constant pressure of 1.1 atm, which of the processes occurs?

10. If the pressure increases from 0.5 to 1.5 atm at constant temperature of 50 °C, which of the processes occurs?

Questions 11–14 refer to the following elements.

(A) He
(B) Na
(C) Si
(D) Mg
(E) Ar

11. This element has atoms with the largest atomic radius.

12. This element has the highest first ionization energy.

13. This element is the most metallic in character.

14. This element most easily forms an ion with a 2+ charge.

Questions 15–17 refer to the following elements.

(A) Aluminum
(B) Chlorine
(C) Copper
(D) Oxygen
(E) Tin

15. Forms an amphoteric oxide

16. Found in both brass and bronze

17. Acidifying bleach releases this gas

Part B

<u>Directions:</u> Each of the questions or incomplete statements below is followed by five suggested answers or completions. Select the one that is best in each case.

18. If two atoms have different atomic numbers but the same mass number, what must be true?
    (A) They must be atoms of the same element.
    (B) They must contain the same number of electrons.
    (C) Each must contain the same total number of neutrons and protons.
    (D) The number of neutrons in both must be the same.
    (E) The number of protons in each atom must be the same.

19. Which one of the following is an example of an intensive physical property?
    (A) Aluminum burns in bromine.
    (B) A balloon of hydrogen and oxygen explodes.
    (C) A bottle of rubbing alcohol has a mass of 900 grams.
    (D) The molar mass of sodium is 22.99 g. mol.
    (E) The volume of a helium balloon is 6.5 liters.

20. What is the arrangement of electron pairs around the central atom in the molecule krypton difluoride, KrF_2?
    (A) linear
    (B) trigonal bipyramidal
    (C) trigonal planar
    (D) octahedral
    (E) tetrahedral

21. Which of the following sets of quantum numbers is **not** allowed?

	n	ℓ	m_ℓ	m_s
(A)	3	0	0	$(-)\,\frac{1}{2}$
(B)	2	1	0	$(-)\,\frac{1}{2}$
(C)	4	3	−3	$(-)\,\frac{1}{2}$
(D)	2	0	−1	$(-)\,\frac{1}{2}$
(E)	3	1	−1	$(+)\frac{1}{2}$

22. A substance with strong intermolecular forces of attraction would be expected to have
    I.   a low boiling point.
    II. a low vapor pressure.
    III. a high heat of vaporization.
    IV. a low melting point.

    (A) I only
    (B) II only
    (C) I and II only
    (D) II and III only
    (E) I, II, III, IV

23. Berkelium-243 is prepared by α-particle bombardment of americium-241. How many neutrons are emitted in the process?
    (A) 0
    (B) 1
    (C) 2
    (D) 3
    (E) 4

24. What is the average bond order in the molecular ion NO_2^+?
    (A) 1.0
    (B) 2.0
    (C) 2.5
    (D) 1.5
    (E) 3.0

GO ON TO NEXT PAGE

25. Resonance is required to reconcile the Lewis electron representations with the actual or real structure in which of the following?
    I.   Carbon dioxide, CO_2
    II. Ozone, O_3
    III. Sulfate ion, SO_4^{2-}
    IV. Nitrate ion, NO_3^-
    V.   Azide ion, N_3^-
    (A) I only
    (B) II only
    (C) II and V only
    (D) II and IV only
    (E) I, II, III, IV and V

26. An element has two naturally occurring isotopes. One isotope has an abundance of 80% and a mass of 122.0 amu. The other has a mass of 120.0 amu. What is the atomic mass of the element?
    (A) 120.0
    (B) 121.6
    (C) 122.0
    (D) 121.8
    (E) 120.8

27. If 3.0 moles of pentane (vapor pressure = 88.7 torr) are mixed with 2.0 moles of hexane (vapor pressure = 44.5 torr), what is the total vapor pressure, in torr, above the nearly ideal solution?
    (A) 53.2
    (B) 62.2
    (C) 66.6
    (D) 71.0
    (E) 133

28. 0.24 mole of sodium bicarbonate is dissolved in sufficient water to make 300.0 mL of solution. What is the molarity of the sodium bicarbonate solution?
    (A) 0.24M
    (B) 0.72M
    (C) 0.32M
    (D) 0.80M
    (E) 0.66M

29. The concentration of nitrate ions in a 3.6×10^{-4} M aqueous solution of lead nitrate, $Pb(NO_3)_2$, is
    (A) 3.6×10^{-4} M.
    (B) 1.8×10^{-4} M.
    (C) 1.2×10^{-4} M.
    (D) 4.8×10^{-4} M.
    (E) 7.2×10^{-4} M.

30. The Brønsted-Lowry conjugate base of the bisulfate ion HSO_3^- is
    (A) H_2SO_3.
    (B) HSO_4^-.
    (C) SO_3^{2-}.
    (D) SO_4^{2-}.
    (E) OH^-.

31. Ammonia is converted to nitric oxide at high temperatures by the reaction:

    $$4NH_3(g) + 5O_2(g) \longrightarrow 4NO(g) + 6H_2O(g).$$

    In one particular experiment, equal molar amounts of the reactants were mixed and concentrations of the reactants and products were plotted against time. The graph below was obtained. Identify the components A, B, C, and D.

    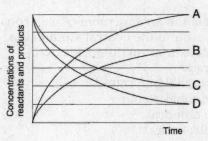

    Choose the line where all the components are identified properly.

	A	B	C	D
(A)	NH_3	O_2	NO	H_2O
(B)	H_2O	NO	O_2	NH_3
(C)	NO	H_2O	NH_3	O_2
(D)	O_2	NO	H_2O	NH_3
(E)	H_2O	NO	NH_3	O_2

32. Hydrogen sulfide, H_2S, in solution (hydrosulfuric acid) has a K_a equal to 1.0×10^{-7} at 25 °C. What is the K_b for its conjugate base, HS^-?
(A) 1.0×10^{-7}
(B) 1.0×10^{-14}
(C) 1.8×10^{-5}
(D) $1.0 \times 10^{+7}$
(E) 2.5×10^{-6}

For questions 33 and 34, consider the following mechanism for the reaction of nitric oxide and hydrogen:

$2NO(g) \rightleftharpoons N_2O_2(g)$ *Fast*

$N_2O_2(g) + H_2(g) \longrightarrow N_2O(g) + H_2O(g)$ *Slow*

$N_2O(g) + H_2(g) \longrightarrow N_2(g) + H_2O(g)$ *Fast*

33. What is (are) reaction intermediate(s) in this reaction mechanism?
(A) NO *and* N_2O
(B) N_2O *and* N_2O_2
(C) N_2O_2
(D) N_2O *and* N_2
(E) N_2O *and* H_2

34. What is the rate law for the reaction according to this mechanism?
(A) Rate = $k[NO]$
(B) Rate = $k[N_2O_2]^2[H_2]$
(C) Rate = $k[NO]^2$
(D) Rate = $k[N_2O_2]^2[H_2]^2$
(E) Rate = $k[NO]^2[H_2]$

35. Which of the following statements concerning the kinetic molecular theory of gases is *false*?
I.   The rate of effusion of a gas is directly proportional to the root mean square speed of its molecules.
II. The molar volume of a gas depends on the molar mass of a gas.
III. The average kinetic energy of the molecules in a sample of gas depends only upon the temperature.
IV. At constant pressure, gases occupy greater volumes at higher temperatures.

(A) I only
(B) II only
(C) III only
(D) IV only
(E) I, II, III, and IV

36.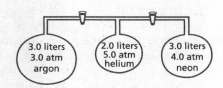

With the stopcocks closed, the three bulbs in the apparatus shown above are filled with the quantities indicated of argon, helium, and neon, respectively. Assume the gases are ideal and that the volumes include the volumes of the connecting tubes. What is the partial pressure of helium inside the apparatus when the taps are opened?
(A) 1.00 atm
(B) 1.25 atm
(C) 1.50 atm
(D) 2.00 atm
(E) 2.50 atm

GO ON TO NEXT PAGE

37. The *principal* reason(s) why the ideal gas law fails as the pressure increases or as the temperature of a system decreases is
    I.   The intermolecular attraction between gas molecules becomes more effective.
    II. The average velocity of the gas molecules increases.
    III. The volume of gas molecules becomes significantly large.
    (A) I only
    (B) II only
    (C) III only
    (D) I and III only
    (E) I, II and III

38. Which one of the following equimolar solutions will act as a buffer solution?
    (A) H_2SO_4 and Na_2SO_4
    (B) Na_2S and KHS
    (C) H_2SO_4 and H_2SO_3
    (D) KOH and KCN
    (E) HI and KI

39. Calculate the equilibrium constant K for the following equilibrium

    $$3\ F_2(g)\ +\ Cl_2(g)\ \rightleftharpoons\ 2ClF_3(g)$$

    with the following equilibrium concentrations: $[F_2] = 2.0M$; $[Cl_2] = 2.5M$; $[ClF_3] = 3.0M$.
    (A) 0.40
    (B) 0.90
    (C) 1.2
    (D) 0.25
    (E) 0.45

40. Technetium is a radioactive element of growing importance in medical imaging, prepared from molybdenum by the following equation:

    $$^{98}_{42}Mo\ +\ ^{1}_{0}n\ \longrightarrow\ ^{99}_{43}Tc\ +\ ?$$

    What particle is necessary to balance the equation?
    (A) α particle
    (B) helium nucleus
    (C) proton
    (D) β particle
    (E) positron

41. In the Lewis dot structure for the molecule S_2O, how many lone pairs of electrons are there around the central atom?
    (A) 0
    (B) 1
    (C) 2
    (D) 3
    (E) 4

42. What is the molar solubility of cadmium sulfide CdS? (K_{sp} of CdS = 3.6×10^{-29})
    (A) 3.6×10^{-16}M
    (B) 6.0×10^{-15}M
    (C) 7.2×10^{-58}M
    (D) 1.8×10^{-16}M
    (E) 6.0×10^{-14}M

43. Below are representations of energy levels (drawn to scale) in the Bohr model of the hydrogen atom. Which electron making a transition, denoted by the arrows, results in the emission of a photon of shortest wavelength?

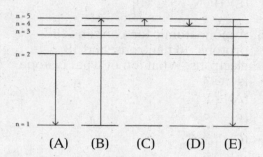

(A)   (B)   (C)   (D)   (E)

44. What is the change in oxidation number of sulfur in the following half-reaction?

$$S_4O_6^{2-} + 10H_2O \longrightarrow SO_4^{2-} + 2OH^- + 14e^-$$

(A) −2 to −8
(B) +4 to +1
(C) +4 to +6
(D) +2 ½ to +6
(E) +3 to +6

45. A U-tube mercury manometer is open to the atmosphere (1 atm = 760 mm Hg) on the right arm and is connected to a glass vessel containing a gas on the left arm as shown.

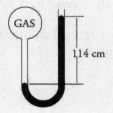

GAS

114 cm

The level of the mercury in the tube is 114 cm(1140 mm) higher in the right arm. What is the pressure of the gas in the glass vessel?
(A) 0.50 atm
(B) 1.0 atm
(C) 1.5 atm
(D) 2.0 atm
(E) 2.5 atm

46. The reaction of ammonia with nitric oxide is described by the following equation:

$$4\ NH_3(g) + 6\ NO(g) \longrightarrow 5\ N_2(g) + 6H_2O(g)$$

At one stage in the reaction, nitrogen is produced at a rate of 30 mol/L/s. At what rate is nitric oxide used up at this stage?
(A) 5.0 mol $L^{-1}s^{-1}$
(B) 6.0 mol $L^{-1}s^{-1}$
(C) 24 mol $L^{-1}s^{-1}$
(D) 30 mol $L^{-1}s^{-1}$
(E) 36 mol $L^{-1}s^{-1}$

47. For which substance does the standard enthalpy of formation ΔH_f° equal zero?
(A) $H_2O(l)$
(B) $Cu(s)$
(C) $N_2(l)$
(D) $O_3(g)$
(E) $Fe(l)$

48. A nonvolatile solute A is dissolved in water. Which of the following affects the vapor pressure of water?
I.   The exposed surface of the water.
II. The density of A.
III. The amount of A dissolved in the water.

(A) I only
(B) II only
(C) III only
(D) II and III only.
(E) I, II and III

49. A 2.0-L sample of an ideal gas at 300 K and 3.0 atm constant pressure is heated until the volume increases to 6.0 L. Then the temperature is held constant and the pressure is increased, restoring the volume to 2.0 L. What final pressure is required?
(A) 4.5 atm
(B) 6.0 atm
(C) 9.0 atm
(D) 10. atm
(E) 12 atm

GO ON TO NEXT PAGE

50. Adding which of the following solutes will increase the solubility of silver carbonate?
    I.   Silver nitrate, $AgNO_3$
    II. Nitric acid, HNO_3
    III. Sodium carbonate
    (A) I only
    (B) II only
    (C) III only
    (D) II and III
    (E) None, they will all decrease solubility

51. Which statement is correct?
    (A) The pH of pure water is always 7.
    (B) At temperatures greater than 25 °C, the pH of a salt solution in water is always less than 7.
    (C) The higher the pH, the more acidic the solution.
    (D) The strongest acid that can exist in aqueous solution is the hydronium ion.
    (E) If the concentration of acid is very high, the acid is said to be strong.

52. When zinc sulfide reacts with 32.0 g of oxygen to yield zinc oxide and sulfur dioxide, how many moles of ZnO are produced?

    $2 ZnS(s) + 3 O_2(g) \longrightarrow 2 ZnO(s) + 2SO_2(g)$

    (A) 0.66
    (B) 1.0
    (C) 1.3
    (D) 2.0
    (E) 2.6

53. Which of the following has the lowest freezing point?
    (A) $0.20\ m\ C_6H_{12}O_6$ (glucose)
    (B) $0.10\ m\ NH_4Br$
    (C) $0.05\ m\ ZnSO_4$
    (D) $0.10\ m\ K_3PO_4$
    (E) $0.20\ m\ Na_2S$

54. Which one of the following choices includes only solutes that are weak electrolytes in aqueous solution.
    (A) HF, NH_3, NH_4F
    (B) $HCl, NaOH, NaCl$
    (C) $HNO_3, H_2SO_4, HClO_4$
    (D) HF, HBr, HI
    (E) HCN, NH_3, HF

55. A monoatomic ion contains 15 protons, 16 neutrons, and 18 electrons. What ion of what isotope is this?
    (A) $^{31}P^{3+}$
    (B) $^{34}Se^{2-}$
    (C) $^{18}Ar^{3+}$
    (D) $^{31}P^{3-}$
    (E) $^{34}As^{3-}$

56. The rate law for a chemical reaction between two substances A and B is: Rate of reaction = $k[A]^2[B]$ where k is the rate constant for the reaction. What happens to the rate of the reaction if the concentration of A is doubled and the concentration of B is halved? (The temperature remains unchanged.)
    (A) The rate halves.
    (B) The rate is reduced to one quarter.
    (C) The rate stays the same.
    (D) The rate doubles.
    (E) The rate increases by four.

57. How many structural isomers can be drawn for butane (C_4H_{10})?
    (A) one
    (B) two
    (C) three
    (D) four
    (E) five

58. Calculate the standard cell potential for the cell

    $Sr(s)|Sr^{2+}(aq)$ $Sn^{2+}(aq)|Sn(s)$ given the half-cell standard potentials:

    $E°$ $Sr^{2+}(aq)|Sr(s) = -2.89$ V

    $E°$ $Sn^{2+}(aq)|Sn(s) = -0.14$ V.
    (A) –3.03V
    (B) +3.03V
    (C) –2.75V
    (D) +2.75V
    (E) +5.05V

59. Which process is always exothermic?
    (A) Evaporation of a liquid
    (B) Dissolving a typical salt in water
    (C) Breaking a hydrogen molecule into atoms
    (D) Sublimation of solid carbon dioxide
    (E) Freezing water

60. Given the standard heats of formation: ΔH_f° CH_3OH $(g) = -239$ kJ/mol; ΔH_f° $CO_2(g) = -394$ kJ/mol; ΔH_f° $H_2O(l) = -286$ kJ/mol,

    what is ΔH_{rxn} in kJ for the following reaction?

    $2CH_3OH(g) + 3 O_2(g) \rightarrow 2CO_2(g) + 4H_2O(l)$

    (A) –727 kJ
    (B) +727 kJ
    (C) –441 kJ
    (D) +1445 kJ
    (E) –1454 kJ

61. In the following reaction $2MnO_4^-$ (aq) $+16H_3O^+(aq) +5$ $Zn(s) \rightarrow 2Mn^{2+}(aq)$ $+24H_2O(l) + 5Zn^{2+}(aq)$, how many electrons are transferred between the reducing agent and the oxidizing agent in the balanced equation?
    (A) 5
    (B) 7
    (C) 10
    (D) 14
    (E) 15

62. A reaction takes place within a system. As a result, the entropy of the system decreases, it becomes more ordered. What *must* be true?
    I.   The reaction is exothermic.
    II. The entropy of the universe increases.
    III. The entropy of the surroundings increases.

    (A) I only
    (B) II only
    (C) I and II only
    (D) I and III only
    (E) I, II and III

63. Balance the following equation for the reaction of dimethylhydrazine with dinitrogen tetroxide:

    $\_\_C_2H_8N_2$ + $\_\_N_2O_4 \longrightarrow$

    $\_\_N_2$ + $\_\_CO_2$ + $\_\_H_2O$

    The sum of all the coefficients in the balanced equation is
    (A) 9
    (B) 11
    (C) 12
    (D) 14
    (E) 15

64. Which of the following is an acid-base reaction?
    I. $Ca(OH)_2(aq)$ + $2HNO_3(aq) \longrightarrow$
         $Ca(NO_3)_2(aq) + 2H_2O(l)$
    II. $2H_2(g) + O_2(g) \longrightarrow 2H_2O(g)$
    III. $CuBr_2(aq)$ + $2NaOH(aq)$
         $\longrightarrow Cu(OH)_2(s)$ + $2NaBr(aq)$

    (A) I only
    (B) II only
    (C) III only
    (D) I and III
    (E) II and III

GO ON TO NEXT PAGE

65. A catalyst
   (A) increases the amount of product at equilibrium.
   (B) changes the route the reaction takes between reactants and products.
   (C) increases the activation energy required for the reaction.
   (D) increases the kinetic energy of the reactant molecules.
   (E) shifts the equilibrium toward the product side.

66. When crystalline ammonium nitrate dissolves in water to make a solution, the solution gets very cold, dropping in temperature about 20 °C. What are the signs of ΔH, ΔS, and ΔG for this process?

	ΔH	ΔS	ΔG
(A)	(–)	(+)	(+)
(B)	(–)	(+)	(–)
(C)	(+)	(+)	(+)
(D)	(+)	(–)	(+)
(E)	(+)	(+)	(–)

67. The graph below shows the speed distributions for three gases, H_2, H_2O, and H_2S at the same temperature.

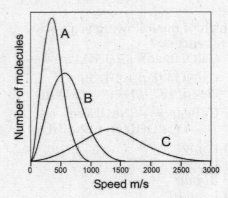

   Match the gas with the distribution.

	A	B	C
(A)	H_2	H_2O	H_2S
(B)	H_2O	H_2	H_2S
(C)	H_2	H_2S	H_2O
(D)	H_2S	H_2O	H_2
(E)	H_2O	H_2S	H_2

68. Ammonia solution is usually diluted for household use. What is the concentration in mole/L of an ammonia solution prepared by taking 100. mL of a 6.50 M solution of ammonia and adding it to enough water to make 2.00 L of solution?
   (A) 0.0650 M
   (B) 0.0875 M
   (C) 0.0975 M
   (D) 0.130 M
   (E) 0.325 M

69. Consider the following disturbances of the equilibrium system shown. In which, if any, direction will the system shift to restore equilibrium in each case? The reaction is endothermic in the forward direction.

   $$2\ NOCl_2(g) \rightleftharpoons 2NO(g) + Cl_2(g)$$

	increase temperature	decrease volume
(A)	Right	Left
(B)	Left	No change
(C)	Left	Left
(D)	No change	Left
(E)	Right	Right

70. Which of the following statements is true?
   (A) Increasing the pressure on an ideal gas at constant temperature slows the molecules down.
   (B) Deviation of a gas from ideal behavior is due only to the finite size of real molecules.
   (C) Argon gas (Ar) diffuses faster than chlorine gas (Cl_2) at the same temperature.
   (D) The smaller the gas molecule, the higher its kinetic energy at the same temperature.
   (E) All molecules in a sample of a gas travel at the same speed at a particular temperature.

71. Which of the following binary compounds would you expect to form a basic aqueous solution?
(A) CO_2
(B) CaO
(C) NO_2
(D) P_4O_{10}
(E) H_2O_2

72. What is the abbreviated electron configuration for the phosphide ion, P^{3-}?
(A) $[Ne]3s^23p^3$
(B) $[Ne]3s^2$
(C) $[Ar]$
(D) $[Ar]4s^2$
(E) $[Ne]3s^23p^5$

73. Which compound contains the metal atom with the highest (most positive) oxidation number?
(A) $KMnO_4$
(B) K_2CrO_4
(C) Cr_2O_3
(D) MnO_2
(E) $K_2Cr_2O_7$

74. A system absorbs 2000 J of heat from its surroundings and as a result, does some work on the surroundings. The internal energy ΔE of the system increases by 500 J. How much work was done?
(A) zero
(B) 500 J
(C) 1000 J
(D) 1500 J
(E) 2000 J

75. Of the colors listed below that are part of the visible portion of the electromagnetic spectrum, which one has the highest frequency?
(A) Red
(B) Orange
(C) Green
(D) Yellow
(E) Blue

GO ON TO NEXT PAGE

A5.5 Standard Reduction Potentials at 25°C (298 K) for Many Common Half-Reactions

Half-Reaction	e° (V)	Half-Reaction	e° (V)
$F_2 + 2e^- \rightarrow 2F^-$	2.87	$Cu^+ + e^- \rightarrow Cu$	0.52
$Ag^{2+} + e^- \rightarrow Ag^+$	1.99	$O_2 + 2H_2O + 4e^- \rightarrow 4OH^-$	0.40
$Co^{3+} + e^- \rightarrow Co^{2+}$	1.82	$Cu^{2+} + 2e^- \rightarrow Cu$	0.34
$H_2O_2 + 2H^+ + 2e^- \rightarrow 2H_2O$	1.78	$Hg_2Cl_2 + 2e^- \rightarrow 2Hg + 2Cl^-$	0.34
$Ce^{4+} + e^- \rightarrow Ce^{3+}$	1.70	$AgCl + e^- \rightarrow Ag + Cl^-$	0.22
$PbO_2 + 4H^+ + SO_4^{2-} + 2e^-$		$SO_4^{2-} + 4H^+ + 2e^- \rightarrow H_2SO_3 + H_2O$	0.20
$\rightarrow PbSO_4 + 2H_2O$	1.69	$Cu^{2+} + e^- \rightarrow Cu^+$	0.16
$MnO_4^- + 4H^+ + 3e^- \rightarrow MnO_2 + 2H_2O$	1.68	$2H^+ + 2e^- \rightarrow H_2$	0.00
$2e^- + 2H^+ + IO_4^- \rightarrow IO_3^- + H_2O$	1.60	$Fe^{3+} + 3e^- \rightarrow Fe$	−0.036
$MnO_4^- + 8H^+ + 5e^- \rightarrow Mn^{2+} + 4H_2O$	1.51	$Pb^{2+} + 2e^- \rightarrow Pb$	−0.13
$Au^{3+} + 3e^- \rightarrow Au$	1.50	$Sn^{2+} + 2e^- \rightarrow Sn$	−0.14
$PbO_2 + 4H^+ + 2e^- \rightarrow Pb^{2+} + 2H_2O$	1.46	$Ni^{2+} + 2e^- \rightarrow Ni$	−0.23
$Cl_2 + 2e^- \rightarrow 2Cl^-$	1.36	$PbSO_4 + 2e^- \rightarrow Pb + SO_4^{2-}$	−0.35
$Cr_2O_7^{2-} + 14H^+ + 6e^- \rightarrow 2Cr^{3+} + 7H_2O$	1.33	$Cd^{2+} + 2e^- \rightarrow Cd$	−0.40
$O_2 + 4H^+ + 4e^- \rightarrow 2H_2O$	1.23	$Fe^{2+} + 2e^- \rightarrow Fe$	−0.44
$MnO_2 + 4H^+ + 2e^- \rightarrow Mn^{2+} + 2H_2O$	1.21	$Cr^{3+} + e^- \rightarrow Cr^{2+}$	−0.50
$IO_3^- + 6H^+ + 5e^- \rightarrow \frac{1}{2}I_2 + 3H_2O$	1.20	$Cr^{3+} + 3e^- \rightarrow Cr$	−0.73
$Br_2 + 2e^- \rightarrow 2Br^-$	1.09	$Zn^{2+} + 2e^- \rightarrow Zn$	−0.76
$VO_2^+ + 2H^+ + e^- \rightarrow VO^{2+} + H_2O$	1.00	$2H_2O + 2e^- \rightarrow H_2 + 2OH^-$	−0.83
$AuCl_4^- + 3e^- \rightarrow Au + 4Cl^-$	0.99	$Mn^{2+} + 2e^- \rightarrow Mn$	−1.18
$NO_3^- + 4H^+ + 3e^- \rightarrow NO + 2H_2O$	0.96	$Al^{3+} + 3e^- \rightarrow Al$	−1.66
$ClO_2 + e^- \rightarrow ClO_2^-$	0.954	$H_2 + 2e^- \rightarrow 2H^-$	−2.23
$2Hg^{2+} + 2e^- \rightarrow Hg_2^{2+}$	0.91	$Mg^{2+} + 2e^- \rightarrow Mg$	−2.37
$Ag^+ + e^- \rightarrow Ag$	0.80	$La^{3+} + 3e^- \rightarrow La$	−2.37
$Hg_2^{2+} + 2e^- \rightarrow 2Hg$	0.80	$Na^+ + e^- \rightarrow Na$	−2.71
$Fe^{3+} + e^- \rightarrow Fe^{2+}$	0.77	$Ca^{2+} + 2e^- \rightarrow Ca$	−2.76
$O_2 + 2H^+ + 2e^- \rightarrow H_2O_2$	0.68	$Ba^{2+} + 2e^- \rightarrow Ba$	−2.90
$MnO_4^- + e^- \rightarrow MnO_4^{2-}$	0.56	$K^+ + e^- \rightarrow K$	−2.92
$I_2 + 2e^- \rightarrow 2I^-$	0.54	$Li^+ + e^- \rightarrow Li$	−3.05

Advanced Placement Chemistry Equations and Constants

ATOMIC STRUCTURE

$$E = h\nu$$

$$\lambda = \frac{h}{mv}$$

$$c = \lambda \nu$$

$$p = mv$$

$$E_n = \frac{-2.178 \times 10^{-18}}{n^2} \text{ joule}$$

EQUILIBRIUM

$$K_a = \frac{[H^+][A^-]}{[HA]}$$

$$E^\Omega =$$

$$K_w = [OH^-][H^+] = 1.0 \times 10^{-14} \text{ at } 25°C$$

$$= K_a \times K_b$$

$$pH = -\log[H^+], \ pOH = -\log[OH^-]$$

$$14 = pH + pOH$$

$$pH = pK_a + \log \frac{[A^-]}{[HA]}$$

$$pOH = pK_b + \log \frac{[HB^+]}{[B]}$$

$$pK_a = -K_a, \ pK_b = -\log K_b$$

$$K_p = K_c (RT)^{\Delta n}$$

where $\Delta n = $ moles product gas – moles reactant gas

THERMOCHEMISTRY/KINETICS

$$\Delta S° = \sum S° \text{ products} - \sum S° \text{ reactants}$$

$$\Delta H° = \sum \Delta H_f° \text{ products} - \sum \Delta H_f° \text{ reactants}$$

$$\Delta G° = \sum \Delta G_f° \text{ products} - \sum \Delta G_f° \text{ reactants}$$

$$\Delta G° = \Delta H° - T\Delta S°$$

$$= -RT \ln K = -2.303 RT \log K$$

$$= -n \Im E°$$

$$\Delta G = \Delta G° + RT \ln Q = \Delta G° + 2.303 RT \log Q$$

$$q = mc\Delta T$$

$$C_p = \frac{\Delta H}{\Delta T}$$

$$\ln[A]_t - \ln[A]_0 = -kt$$

$$\frac{1}{[A]_t} - \frac{1}{[A]_0} = kt$$

$$\ln k = \frac{-E_a}{R}\left(\frac{1}{T}\right) + \ln A$$

E = energy
λ = wavelength
υ = velocity
m = mass
ν = frequency
p = momentum
n = principal quantum number

Speed of light, c = 3.0×10^8 ms^{-1}

Planck's constant, h = 6.63×10^{-34} Js

Boltzmann's constant, k = 1.38×10^{-23} J K^{-1}

Avogadro's number = 6.022×10^{23} mol^{-1}

Electron charge, e = -1.602×10^{-19} coulomb

1 electron volt per atom = 96.5 kJ mol^{-1}

EQUILIBRIUM CONSTANTS

K_a (weak acid)

K_b (weak base)

K_w (water)

K_p (gas pressure)

K_c (molar concentrations)

$S°$ = standard entropy

$H°$ = standard enthalpy

$G°$ = standard free energy

$E°$ = standard reduction potential

T = temperature

n = moles

m = mass

q = heat

c = specific heat capacity

C_p = molar heat capacity at constant pressure

E_a = activation energy

k = rate constant

A = frequency factor

Faraday's constant, $\Im = 96,500$ per mole of electrons

Gas constant, R = 8.31 J mol^{-1} K^{-1}

= 0.0821 L atm mol^{-1} K^{-1}

= 8.31 volt coulomb mol^{-1} K^{-1}

GASES, LIQUIDS, AND SOLUTIONS

$$PV = nRT$$

$$\left(P + \frac{n^2 a}{V^2}\right)(V - nb) = nRT$$

$$P_A = P_{total} \times X_A, \text{ where } X_A = \frac{\text{moles A}}{\text{total moles}}$$

$$P_{total} = P_A + P_B + P_C + \ldots$$

$$n = \frac{m}{M}$$

$$°K = °C + 273$$

$$\frac{P_1 V_1}{T_1} = \frac{P_2 V_2}{T_2}$$

$$D = \frac{m}{V}$$

$$u_{rms} = \sqrt{\frac{3kT}{m}} = \sqrt{\frac{3RT}{M}}$$

KE per molecule $= \frac{1}{2}mv^2$

KE per mole $= \frac{3}{2}RT$

$$\frac{r_1}{r_2} = \sqrt{\frac{M_2}{M_1}}$$

molarity, M = moles solute per liter solution

molality = moles solute per kilogram solvent

$\Delta T_f = iK_f \times$ molality

$\Delta T_b = iK_b \times$ molality

$\pi = MRT$

$A = abc$

Oxidation-Reduction; Electrochemistry

$$Q = \frac{[C]^c[D]^d}{[A]^a[B]^b}, \text{ where } a\,A + b\,B \rightarrow c\,C + d\,D$$

$$I = \frac{q}{t}$$

$$E_{cell} = E^\circ_{cell} - \frac{RT}{n\Im}\ln Q = E^\circ_{cell} - \frac{0.0592}{n}\log Q\,@25°C$$

$$\log K = \frac{nE^\circ}{0.0592}$$

P = pressure
V = volume
T = temperature
n = number of moles
D = density
m = mass
v = velocity

u_{rms} = root-mean-square speed

KE = kinetic energy

r = rate of effusion

M = molar mass

π = osmotic pressure

i = van't Hoff factor

K_f = molal freezing-point depression constant

K_b = molal boiling-point elevation constant

A = absorbance

a = molar absorptivity

b = path length

c = concentration

Q = reaction quotient

I = current (amperes)

q = charge (coulombs)

t = time (seconds)

E° = standard reduction potential

K = equilibrium constant

Gas constant, R = 8.31 J mol⁻¹ K⁻¹

$\qquad$ = 0.0821 L atm mol⁻¹ K⁻¹

$\qquad$ = 8.31 volt coulomb mol⁻¹ K⁻¹

Boltzmann's constant, $k = 1.38 \times 10^{-23}$ J K⁻¹

K_f for H_2O = 1.86 K kg mol⁻¹

K_b for H_2O = 0.512 K kg mol⁻¹

1 atm = 760 mm Hg

$\qquad$ = 760 torr

STP = 0.000°C and 1.000 atm

Faraday's constant, $\Im = 96,500$ coulombs per mole of electrons

Section II: Free-Response Questions
Time: 95 minutes
Number of Questions: 6

Allow yourself no more than 95 minutes to answer these questions. A calculator may be used only on questions 1, 2, and 3. You may refer to the Equations Sheet, the Periodic Table, and the Standard Reduction Table throughout this section.

Part A

There are three mandatory questions in this part, each worth 20% of Section II. You are allowed to use a calculator only in Part A. You are allowed 55 minutes for Part A.

1. For the reaction $PCl_3(g) + Cl_2(g) \rightleftharpoons PCl_5(g)$, $K_p = 0.0870$ at $300\,°C$. A flask is charged with 0.30 atm PCl_3, 0.60 atm Cl_2, and 0.10 atm PCl_5 at this temperature.
    (a) Determine whether the above conditions (the added gases) are at equilibrium. If not, determine to which direction the reaction must proceed to reach equilibrium.
    (b) Calculate the equilibrium partial pressures of the gases.
    (c) Calculate K_c for the above equilibrium.
    (d) Predict and justify what effect increasing the volume of the system will have on the mole fraction of PCl_5 in the mixture.
    (e) The reaction as written is exothermic. Predict and justify what effect that decreasing the temperature of the system will have on the mole fraction of PCl_5 in the mixture.

2. A 0.495 M solution of nitrous acid, HNO_2, has a pH of 1.83.
    (a) Find the $[H^+]$ and the percent ionization of nitrous acid in this solution.
    (b) Write the equilibrium expression and calculate the value of K_a for nitrous acid.
    (c) Calculate the pH of the solution formed by adding 1.00 g of $NaNO_2$ (molar mass = 69.0 $g \cdot mol^{-1}$) to 750. mL of 0.0125 M solution of nitrous acid.
    (d) Determine the pH of a solution formed by adding 1.00 g of $NaNO_2$ to 750 mL of water, H_2O.
    (e) Sketch the pH curve that results when 20.0 mL of a 0.0125 M nitrous acid solution are titrated with 0.0125 M NaOH solution. Label the axes, the equivalence point, and the buffer region clearly.

3. Given the following data:
    $$Fe^{2+} + 2e^- \longrightarrow Fe \qquad E^\circ_{red} = -0.44\ V$$
    $$Ag^+ + e^- \longrightarrow Ag \qquad E^\circ_{red} = 0.80\ V$$
    Answer the following questions with respect to the reaction
    $$Fe^{2+}(aq) + 2Ag(s) \longrightarrow Fe(s) + 2Ag^+(aq).$$

    (a) What is the cell potential, E°, for the reaction?
    (b) Is the reaction spontaneous at standard state conditions?
    (c) What is the value of E° at equilibrium?
    (d) What is the value of the equilibrium constant at $25\,°C$?
    (e) If $[Fe^{2+}] = 0.100$ M and $[Ag^+] = 0.0100$ M, what is the magnitude of E at $25\,°C$?
    (f) For the reaction that is spontaneous, what is the maximum amount of work that can be performed?

GO ON TO NEXT PAGE

NO CALCULATORS ARE TO BE USED FOR THE REMAINDER OF THE TEST.

Part B
You are allowed 40 minutes for this part of the exam.

The Section II scoring weight for Question 4 is 10%.

4. For each of the following three reactions, in part (i) write a BALANCED equation and in part (ii) answer the question about the reaction. In part (i), coefficients should be in terms of lowest whole numbers. Assume that solutions are aqueous unless otherwise indicated. Represent substances in solutions as ions if the substances are extensively ionized. Omit formulas for any ions or molecules that are unchanged by the reaction.
   a. (i) Water is added to a sample of pure calcium hydride.
      (ii) If a phenolphthalein solution is added to the water what color change, if any, would you observe as the reaction proceeds?
   b. (i) The gases boron trifluoride and ammonia are mixed.
      (ii) What is the molecular geometry of both the boron trifluoride and the ammonia molecules?
   c. (i) Calcium metal is heated strongly in nitrogen gas.
      (ii) What is the change in oxidation number of the nitrogen in this reaction?

5. The Laboratory-Based Question: Answer the following essay question. The Section II score weighting is 15% for this question.

   A student weighs out equal amounts of magnesium hydroxide, calcium carbonate, calcium sulfate, and sodium bicarbonate but carelessly forgets to label the containers in which each sample is placed. If the only chemicals available are a bottle of dilute hydrochloric acid and some distilled water, describe a procedure that could be used to identify each solid.

6. The Section II score weighting is 15% for this question.

   Account for the following statements or observations in terms of atomic-, ionic- or molecular-level explanations.
   a. Magnesium exists as 2+ ions rather than 1+ ions in all of its compounds despite the fact that the second ionization energy of a magnesium atom is more than twice as great as the first ionization energy.
   b. Carbon dioxide (CO_2) is a gas at room temperature but silicon dioxide (SiO_2) is a high melting solid.
   c. Nitrogen forms NF_3 but not NF_5 whereas phosphorus forms PF_3 and PF_5. The trifluorides are both trigonal pyramidal and the pentafluoride is trigonal bipyramidal.
   d. Calcium oxide has a much higher melting point (2580 °C) than potassium fluoride (858 °C).

END OF EXAMINATION

ANSWERS TO PRACTICE TEST 2

MULTIPLE-CHOICE ANSWERS

Using the table below, score your test.

Determine how many questions you answered correctly and how many you answered incorrectly. You will find explanations of the answers on the following pages.

1. D	2. B	3. C	4. B	5. B
6. E	7. C	8. A	9. B	10. B
11. B	12. A	13. B	14. D	15. A
16. C	17. B	18. C	19. D	20. B
21. D	22. D	23. C	24. B	25. D
26. B	27. D	28. D	29. E	30. C
31. E	32. A	33. B	34. E	35. B
36. B	37. A	38. B	39. E	40. D
41. B	42. B	43. E	44. D	45. E
46. E	47. B	48. C	49. C	50. B
51. D	52. A	53. E	54. E	55. D
56. D	57. B	58. D	59. E	60. E
61. C	62. E	63. C	64. A	65. B
66. E	67. D	68. E	69. A	70. C
71. B	72. C	73. A	74. D	75. E

CALCULATE YOUR SCORE:

Number answered correctly: _____

Adjust ¼ point for guessing penalty:
Count the number of questions you
answered incorrectly, multiply by .25, and subtract: – _____

Determine your adjusted score: _____

WHAT YOUR SCORE MEANS:

Because the test is different each year, the scoring is a little different. But generally, if you scored 25 or more on the multiple-choice questions, you'll most likely get a 3 or better on the test. If you scored 35 or more, you'll probably score a 4 or better. And if you scored a 50 or more, you'll most likely get a 5. Keep in mind that the multiple-choice section is worth 45% of your final grade, and the free-response section is worth 55% of your final grade. To learn more about the scoring for the free-response questions, turn to the last page of this section.

Answers and Explanations

Multiple-Choice Answers

1.   **Answer:** D carbonate (*Chemistry* 7th ed. pages 57–67 / 8th ed. pages 57–67)

2.   **Answer:** B phosphite (*Chemistry* 7th ed. pages 57–67 / 8th ed. pages 57–67)

3.   **Answer:** C ammonium (*Chemistry* 7th ed. pages 57–67 / 8th ed. pages 57–67)

4.   **Answer:** B chlorite (*Chemistry* 7th ed. pages 57–67 / 8th ed. pages 57–67)

5.   **Answer:** B alcohol (*Chemistry* 7th ed. pages 1010–1016 / 8th ed. pages 1019–1025)

6.   **Answer:** E ketone (*Chemistry* 7th ed. pages 1010–1016 / 8th ed. pages 1019–1025)

7.   **Answer:** C ester (*Chemistry* 7th ed. pages 1010–1016 / 8th ed. pages 1019–1025)

8.   **Answer:** A (*Chemistry* 7th ed. pages 467–470 / 8th ed. pages 479–482)

9.   **Answer:** B (*Chemistry* 7th ed. pages 467–470 / 8th ed. pages 479–482)

10.   **Answer:** B (*Chemistry* 7th ed. pages 467–470 / 8th ed. pages 479–482)

11.   **Answer:** B (*Chemistry* 7th ed. pages 309–314, 876–877 / 8th ed. pages 318–323, 909-910)

12.   **Answer:** A (*Chemistry* 7th ed. pages 309–314 / 8th ed. pages 318–323)

13.   **Answer:** B (*Chemistry* 7th ed. pages 309–314 / 8th ed. pages 318–323)

14.   **Answer:** D (*Chemistry* 7th ed. pages 309–314 / 8th ed. pages 318–323)

15. ANSWER: A (*Chemistry* 7th ed. pages 626–631, 889 / 8th ed. pages 642–647, 919)

16. ANSWER: C (*Chemistry* 7th ed. pages 442–443, 891–892 / 8th ed. pages 455–457, 920-921)

17. ANSWER: B (*Chemistry* 7th ed. page 643 / 8th ed. page 660)

18. ANSWER: C Different atomic numbers, therefore different number of protons. Same mass numbers, therefore same total number of protons and neutrons (*Chemistry* 7th ed. pages 49–52 / 8th ed. pages 49–52).

19. ANSWER: D Molar mass is a physical property; it does not involve chemical change. It is also intensive, as it is independent of the amount present (*Chemistry* 7th ed. pages 237–238 / 8th ed. pages 244–246).

20. ANSWER: B

    Kr: 8 valence electrons

    2F: 14 valence electrons

    Total: 22 val. electrons. Around Kr there are 2 bonding pairs and 3 lone pairs = 5 pairs total

    Therefore: trigonal bipyramidal (*Chemistry* 7th ed. pages 367–377 / 8th ed. pages 378–388)

21. ANSWER: D The absolute value for m_l cannot be greater than l (*Chemistry* 7th ed. pages 293–294 / 8th ed. pages 303–304).

22. ANSWER: D Strong intermolecular forces means that additional energy must be supplied in order to increase the physical distance between the molecules when they move from liquid to vapor state. This results in a greater heat of vaporization. It also means that fewer molecules at a given temperature are able to move into the vapor state; thus the vapor pressure is lower than for other liquids with weaker intermolecular forces (*Chemistry* 7th ed. pages 426–429, 492 / 8th ed. pages 440–443, 504).

23. ANSWER: C

    $$^{241}_{95}\text{Am} + ^{4}_{2}\text{He} \longrightarrow ^{243}_{97}\text{Bk} + 2^{1}_{0}\text{n}$$

    To balance the mass numbers, the number of neutrons must be 2 (*Chemistry* 7th ed. pages 841–846 / 8th ed. pages 873–878).

24. ANSWER: B Both N–O bonds are double bonds; the average bond order is 2.0 (*Chemistry* 7th ed. pages 362–367 / 8th ed. pages 373–378).

25. **Answer: D** O_3 has one single bond and one double bond. The double bond can be drawn in either position. The nitrate ion has two single bonds and one double bond, the double bond is drawn in one of three positions (*Chemistry* 7th ed. pages 362–367 / 8th ed. pages 373–378).

26. **Answer: B** Weighted average = (0.80 × 122) + (0.20 × 120) = 121.6 (*Chemistry* 7th ed. pages 330–333 / 8th ed. pages 341–344)

27. **Answer: D**

Mole fraction of pentane in solution = 3/5

Mole fraction of hexane in solution = 2/5

Partial pressure due to pentane = 3/5 × 88.7 torr = 53.2 torr. This is Raoult's law.

Partial pressure due to hexane = 2/5 × 44.5 torr = 17.8 torr and therefore the total vapor pressure = 53.2 +17.8 =71.0 torr (*Chemistry* 7th ed. pages 497–504 / 8th ed. pages 509–516)

28. **Answer: D** Molarity = number of moles/volume of solution (in liters) = 0.24 moles/0.300 L = 0.80 M (*Chemistry* 7th ed. pages 133–140, 485–488 / 8th ed. pages 136–145, 498-500).

29. **Answer: E** There are two nitrate ions per formula unit, so the concentration is twice 3.6×10^{-4}M (*Chemistry* 7th ed. pages 133–140 / 8th ed. pages 136–145).

30. **Answer: C** The only difference in a conjugate acid-base pair is the removal of a H^+ going from the acid to the base or the addition of a H^+ going from the base to acid (*Chemistry* 7th ed. pages 626–631 / 8th ed. pages 642–647).

31. **Answer: E** The reactants diminish in concentration (C and D). D $[O_2]$ goes down 5/4 faster than C $[NH_3]$. The products are formed (A and B). A $[H_2O]$ is formed 6/4 times faster than B [NO] (*Chemistry* 7th ed. pages 96–106 / 8th ed. pages 97–107).

32. **Answer: A**

$K_w/K_a = 1.0 \times 10^{-14}/1.0 \times 10^{-7} = 1.0 \times 10^{-7}$

(*Chemistry* 7th ed. pages 635–650 / 8th ed. pages 651–666)

33. **Answer: B**

Reactants are NO and H_2.

Products are N_2 and H_2O.

Intermediates are N_2O_2 and N_2O; they are formed and used up (*Chemistry* 7th ed. pages 549–552 / 8th ed. pages 562–565).

34. **ANSWER: E** Slowest step: Rate = $k[N_2O_2][H_2]$.

Assume first step reaches steady state: $K' = [N_2O_2]/[NO]^2$, so $[N_2O_2] = K[NO]^2$ and Rate = $kK[NO]^2[H_2]$ (*Chemistry* 7th ed. pages 549–552 / 8th ed. pages 562–565).

35. **ANSWER: B** The molar volume of an ideal gas is independent of its mass. Review the conditions necessary for the change in volume of a gas (*Chemistry* 7th ed. pages 199–208 / 8th ed. pages 205–214).

36. **ANSWER: B** The partial pressure of the helium = 5.0 atm × (2.0L/8.0L) = 1.25 atm. (*Chemistry* 7th ed. pages 194–206 / 8th ed. pages 199–213)

37. **ANSWER: A** Review the conditions under which a gas becomes less than ideal and what is responsible for this behavior (*Chemistry* 7th ed. pages 208–210 / 8th ed. pages 214–217).

38. **ANSWER: B** The pair must be a weak electrolyte and its conjugate partner (*Chemistry* 7th ed. pages 684–693 / 8th ed. pages 701–710).

39. **ANSWER: E**

$$K = \frac{[ClF_3]^2}{[F_2]^3[Cl_2]} = \frac{(3)^2}{(2)^3(2.5)} = 0.45$$

(*Chemistry* 7th ed. pages 532–534 / 8th ed. pages 545–547)

40. **ANSWER: D** The mass numbers must balance: 99 on both sides.

The atomic numbers must balance: 42 on both sides, so the unknown particle has a mass number of zero and an atomic number of –1. This is a β particle (an electron) (*Chemistry* 7th ed. pages 841–846 / 8th ed. pages 873–878).

41. **ANSWER: B**

Total valence electrons = 2 × 6 for S + 6 for O = 18 total.

Divide 18 by 8, for 2 bonding pairs, leaving 2 electrons or one lone pair (*Chemistry* 7th ed. pages 354–358 / 8th ed. pages 365–369).

42. **ANSWER: B**

$K_{sp} = [Cd^{2+}][S^{2-}] = 3.6 \times 10^{-29}$; $[Cd^{2+}] = \sqrt{3.6 \times 10^{-29}} = \sqrt{36 \times 10^{-30}} = 6.0 \times 10^{-15}$

(*Chemistry* 7th ed. pages 717–724 / 8th ed. pages 744–752)

43. **ANSWER: E** The arrow must point down to show emission and the largest difference corresponds to the greatest difference in energy which means highest frequency and shortest wavelength (*Chemistry* 7th ed. pages 284–290 / 8th ed. pages 294–300).

44. **ANSWER: D**

In $S_4O_6^{2-}$, $4x + 6(-2) = -2$, so $x = +2\frac{1}{2}$

(*Chemistry* 7th ed. pages 154–162 / 8th ed. pages 161–166)

45. **ANSWER: E** The difference in levels is 1.5 atm. The pressure outside is 1.0 atm; therefore, the total pressure of the gas is 2.5 atm (*Chemistry* 7th ed. pages 179–181 / 8th ed. pages 181–183).

46. **ANSWER: E** Rate at which NO is used = rate of production of N_2 × (6/5) = 36 mol/L/s. (*Chemistry* 7th ed. pages 527–532 / 8th ed. pages 540–545).

47. **ANSWER: B** An element in its standard state has a defined standard heat of formation of zero (*Chemistry* 7th ed. pages 246–252 / 8th ed. pages 255–261).

48. **ANSWER: C** Only the mole fraction of the solute and thus the mole fraction of the solvent (Raoult's law) is of consequence (*Chemistry* 7th ed. pages 497–504 / 8th ed. pages 509–516).

49. **ANSWER: C** The volume ultimately remains unchanged. The temperature increases by a factor of 3 and from the given information, one can observe that the volume increases by a factor of three. Thus, to restore the sample to its original volume, the pressure must be increased by a factor of 3 as well (*Chemistry* 7th ed. pages 186–190 / 8th ed. pages 189–194).

50. **ANSWER: B** $Ag_2CO_3 \rightleftharpoons 2Ag^+ + CO_3^{2-}$

Silver nitrate contains Ag^+, a common ion; so this will suppress solubility.

Nitric acid will react with carbonate ion, the conjugate base of a very weak acid, so this will increase solubility because the carbonate ion will be removed from solution. Sodium carbonate contains the carbonate ion, a common ion and so this will also suppress solubility (*Chemistry* 7th ed. pages 591–599, 604–610 / 8th ed. pages 606–613, 620-626).

51. **ANSWER: D** Review what can happen to K_w as temperature is increased and how this would affect values of pH. Review definitions of strong acids (*Chemistry* 7th ed. pages 631–634 / 8th ed. pages 647–650).

52. **ANSWER: A** 32 grams O_2 = 1.0 mol O_2

According to the equation, 1.0 mole O_2 produces 2/3 mole of ZnO or 0.66 mol (*Chemistry* 7th ed. pages 102–106 / 8th ed. pages 102–107).

53. **ANSWER: E** $i= 3$, so the concentration of ions is $3 \times 0.20 = 0.60$. This is the greatest increase in the disorder of the solution (*Chemistry* 7th ed. pages 540–543 / 8th ed. pages 524–525).

54. **ANSWER: E** Review what are considered strong acids and strong bases, remembering that any other acid or base is therefore considered weak, by definition. Also remember that all salts are considered strong electrolytes (*Chemistry* 7th ed. pages 129–133 / 8th ed. pages 132–136).

55. **ANSWER: D**

15 protons means the atomic number is 15, which is P.

16 neutrons means that the mass number is 15+16 = 31.

18 electrons compared to 15 protons means that the charge is 3– (*Chemistry* 7th ed. pages 49–52 / 8th ed. pages 49–52).

56. **ANSWER: D**

Rate = $k[A]^2[B] = (2)^2(1/2) = 4 \times \frac{1}{2} = 2$ (*Chemistry* 7th ed. pages 534–538 / 8th ed. pages 547–551).

57. **ANSWER: B** When drawing structures of organic molecules to determine number of isomers, start by drawing a straight chain hydrocarbon, making certain each carbon atom has four bonds to it. Then remove one methyl (–CH_3) from either end and place it on a carbon that is not on the end. When you do this with C_4H_{10}, you produce only one additional isomer, so the total number of isomers is two (*Chemistry* 7th ed. pages 997–1005 / 8th ed. pages 1006–1014).

58. **ANSWER: D** Change the sign for the oxidation at the anode and add the two half-cell potentials.

+ 2.89 – 0.14V = +2.75 V.

(*Chemistry* 7th ed. pages 794–800 / 8th ed. pages 876–883)

59. **ANSWER: E** The freezing of water involves the release of energy since the water molecules are moving very close to one another to maximize the formation of hydrogen bonds. Bond formation is always exothermic (*Chemistry* 7th ed. pages 774–778 / 8th ed. pages 798–802).

60. **ANSWER: E**

Heat of reaction = heat of formation of products – heat of formation of reactants

$$= [(2 \times -394) + (4 \times -286)] - [(2 \times -239)] = -1454 \text{ kJ}$$

(*Chemistry* 7th ed. pages 242–246 / 8th ed. pages 249–255)

61. ANSWER: C For example: For the oxidation, $5 \text{ Zn}(s) \longrightarrow 5 \text{ Zn}^{2+}(aq)$, 10 electrons are lost. Ten electrons are gained in the reduction, so the equation is balanced (*Chemistry* 7th ed. pages 162–168 / 8th ed. pages 166–169).

62. ANSWER: E II is an expression of the second law of thermodynamics.

If the entropy of the system decreases, the entropy of the surroundings must increase, which is statement III. This can happen only if the reaction is exothermic, which is statement I (*Chemistry* 7th ed. pages 755–759 / 8th ed. pages 779–783).

63. ANSWER: C

$$1C_2H_8N_2 + 2N_2O_4 \longrightarrow 3N_2 + 2CO_2 + 4H_2O$$

(*Chemistry* 7th ed. pages 98–102 / 8th ed. pages 99–103)

64. ANSWER: A (*Chemistry* 7th ed. page 140 / 8th ed. page 145)

65. ANSWER: B (*Chemistry* 7th ed. pages 557–563 / 8th ed. pages 570–575)

66. ANSWER: E The process is endothermic, so $\Delta H = +$.

There in an increase in disorder so $\Delta S = +$.

The process happens, so $\Delta G = -$ (*Chemistry* 7th ed. pages 759–770 / 8th ed. pages 783–794).

67. ANSWER: D Average speeds increase as the molecular masses decreases (*Chemistry* 7th ed. pages 199–208 / 8th ed. pages 205–214).

68. ANSWER: E

$$100 \text{ mL} \times 6.50 \text{ M} = 2000 \text{ mL} \times ??\text{M}.$$

new molarity = 0.325M

(*Chemistry* 7th ed. pages 133–140, 485–488 / 8th ed. pages 136–145, 498-500)

69. ANSWER: A The reaction is endothermic in the forward direction, so increasing temperature by adding heat shifts the reaction to the right. Decreasing volume will force the reaction to the side with

fewer moles of gas, the left (*Chemistry* 7th ed. pages 604–610 / 8th ed. pages 620–626).

70. **ANSWER: C** Review the kinetic molecular theory of gases. Remember that temperature is a measure of average kinetic energy. Also remember that kinetic energy depends on both the mass of the molecules as well as their average root mean square speed. Hence, two gases at the same temperature will be traveling at speeds that are in inverse relationship to their masses (*Chemistry* 7th ed. pages 199–206 / 8th ed. pages 205–213).

71. **ANSWER: B** (*Chemistry* 7th ed. pages 662–663 / 8th ed. pages 678–679)

72. **ANSWER: C** (*Chemistry* 7th ed. pages 302–309 / 8th ed. pages 312–318)

73. **ANSWER: A** In $KMnO_4$, Mn has an oxidation number of +7 whereas the Cr in both K_2CrO_4 and $K_2Cr_2O_7$ has an oxidation number of +6. The Cr on Cr_2O_3 has an oxidation number of +3 (*Chemistry* 7th ed. pages 154–162 / 8th ed. pages 161–166).

74. **ANSWER: D**

$\Delta E = q + w$

500 J = 2000 J + w; $w = -1500$ J or 1500 J of work was done *by* the system (*Chemistry* 7th ed. pages 229–235 / 8th ed. pages 236–243).

75. **ANSWER: E** (*Chemistry* 7th ed. pages 275–277 / 8th ed. pages 285–287)

SECTION II: FREE-RESPONSE ANSWERS

Question 1: Answers

(a)

$$Q = \frac{P_{PCl_5}}{P_{PCl_3} \times P_{Cl_2}} = \frac{0.10}{(0.30)(0.60)} = 0.56$$

Since 0.56 (Q) > 0.0870 (K), the reaction proceeds to the left.

(b)

	$PCl_3(g)$ +	$Cl_2(g)$ ⇌	$PCl_5(g)$
Initial	0.30 atm	0.60 atm	0.10 atm
Change	+x	+x	–x
Equil	(0.30 + x)atm	(0.60 + x) atm	(0.10 – x)

Since the reaction proceeds to the left, PCl_5 must decrease and the reactants increase.

$$K_p = \quad 0.0870 = \frac{(0.10 - x)}{(0.30 + x)(0.60 + x)} = \frac{(0.10 - x)}{0.18 + 0.90x + x^2}$$

Using the quadratic equation and solving for x, $x = 0.078$.
Therefore, at equilibrium, $P_{PCl_3} = (0.10 - 0.078) = 0.022$ atm $= 0.02$ atm

$$P_{PCl_5} = (0.30 + 0.078) = 0.378 \text{ atm} = 0.38 \text{ atm}$$

$$P_{PCl_2} = (0.60 + 0.078) = 0.678 \text{ atm} = 0.68 \text{ atm}$$

(c) Since $K_p = K_c(RT)^{\Delta n}$ where $\Delta n = 1 - 2 = -1$, so $K_c = K_p(RT) = 0.0870(0.08206)(573)$
$$= 4.09$$

(d) Increasing the volume of the container favors the process where more moles of gas are produced, so the reverse reaction is favored; the equilibrium shifts to the left; the mole fraction of PCl_5 decreases.

(e) For an exothermic reaction, decreasing the temperature increases the value of K, favoring the products. The partial pressure of PCl_5 increases.

(*Chemistry* 7th ed. pages 586–588, 591–610 / 8th ed. pages 601–604, 606-626)

Question 2: Answers

(a) pH = 1.83 $[H^+] = 1.5 \times 10^{-2}$M

So percent ionized $= \dfrac{0.015 \times 100}{0.495} = 3.0\%$

(b)

$$K_a = \frac{[H^+][NO_2^-]}{[HNO_2]} = \frac{(1.5 \times 10^{-2})^2}{0.495} = 4.5 \times 10^{-4}$$

The answer will vary slightly if the ionized acid is subtracted from the initial amount of acid present.

$$K_a = \frac{(1.5 \times 10^{-2})^2}{0.480} = 4.7 \times 10^{-4}$$

(c)

$$1.00 \text{ g NaNO}_2 \times \frac{1 \text{ mole}}{69.0 \text{ grams}} = \frac{0.0145 \text{ mol NO}_2^-}{0.750 \ L} = 0.0193 \text{ M NO}_2^-$$

Then using the equilibrium expression, solve for $[H^+]$:

$$4.5 \times 10^{-4} = \frac{[H+][0.0193]}{0.0125} \quad [H^+] = 2.9 \times 10^{-4} \quad pH = 3.54$$

If you use the other value for K_a (4.7×10^{-4}), then pH = 3.52

(d) This is a hydrolysis problem: $NO_2^- + H_2O \rightleftharpoons HNO_2 + OH^-$.

$$K_b = K_w/K_a = (1.0 \times 10^{-14})/4.5 \times 10^{-4} = 2.2 \times 10^{-11}$$

$$2.2 \times 10^{-11} = \frac{[HNO_2][OH^-]}{0.0193}$$

$[OH^-]^2 = 4.2 \times 10^{-13}$; $[OH^-] = 6.5 \times 10^{-7}$; so pOH = 6.19 or pH = 7.81

(e)

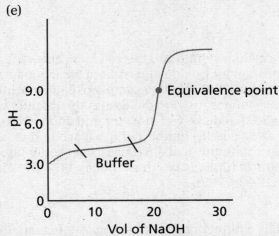

(*Chemistry* 7th ed. pages 696–716 / 8th ed. pages 713–733)

Question 3: Answers

(a)

$$Fe^{2+} + 2e^- \longrightarrow Fe \qquad E^o_{red} = -0.44 \text{ V}$$

$$2(Ag \longrightarrow Ag^+ + e^-) \qquad E^o_{oxid} = -0.80 \text{ V}$$

$$Fe^{2+} + 2Ag \longrightarrow 2Ag^+ + Fe \qquad E^o = -1.24V$$

(b) The reaction is not spontaneous because $E^o < 0$.

(c) $E = 0$ for any reaction at equilibrium.

(d) You can calculate $\log K$ by using the equation

$$\log K = \left(\frac{nE^o}{0.0592 \text{ V}} \right) V$$

$$= \frac{2(-1.24 \text{ V})}{(0.0592 \text{ V})} = -42.0$$

Taking the antilog of both sides yields $K = 1.0 \times 10^{-42}$

(e) E for the reaction is calculated using the Nernst equation:

$$E = E^o - \left(\frac{0.0592 \text{ V}}{2} \right) \log \frac{\left[Ag^+ \right]^2}{\left[Fe^{2+} \right]}$$

$$-1.24 \text{ V} - (0.0296 \text{ V}) \log \left[\frac{(0.0100)^2}{(0.100)} \right] = -1.15 \text{ V}$$

(f) The reaction that is spontaneous is the reverse of the one above, with $E = 1.24$ V.
$W_{max} = -nFE = -(2 \text{ mol e}^-)(96,500\text{C} \bullet \text{mol}^{-1})(1.24 \text{ V})(1\text{J/C} \bullet \text{V}) = 2.39 \times 10^5$ J.
(*Chemistry* 7th ed. pages 800–808 / 8th ed. pages 833–842)

Question 4: Answers

a. (i) $CaH_2 + 2H_2O \rightarrow Ca^{2+} + 2OH^- + 2H_2$
(ii) The phenolphthalein will change from colorless to pink, indicating the presence of the OH^-.

b. (i) $BF_3 + NH_3 \rightarrow F_3B:NH_3$
(ii) BF_3 is trigonal planar and the NH_3 is trigonal pyramidal.

c. (i) $3Ca + N_2 \rightarrow Ca_3N_2$
(ii) The change in oxidation number of the nitrogen is from 0 to -3.

Question 5: Answers

Add water to each solid. The calcium sulfate and the calcium carbonate will be insoluble; however, the magnesium hydroxide and sodium bicarbonate will each be soluble to some degree. Then add the acid to a sample of each insoluble solid; the one which produces bubbles of gas (carbon dioxide) is the calcium carbonate. Then take a sample of each solid that dissolved in water and add acid to each one. The sodium bicarbonate will also produce bubbles of gas (carbon dioxide). The magnesium hydroxide will react with the acid and increase in solubility, but will not produce any gaseous products (*Chemistry* 7th ed. pages 149–154, 212–213, 715 / 8th ed. pages 154–160, 218-219, 732).

Question 6: Answers

(a) The Mg^{2+} ion is smaller and has a higher charge than the Mg^+ ion, so the lattice energy that arises when Mg^{2+} ions form compounds is much greater than what would be observed if Mg^+ ions formed compounds. The increase in lattice energy more than offsets the larger ionization energy of the Mg^{2+} ion (*Chemistry* 7th ed. pages 344–346 / 8th ed. pages 355–357).

(b) Carbon dioxide (O=C=O) molecules are nonpolar and interact with each other only through weak dispersion forces. These weak forces are easily overcome so CO_2 is a gas at room temperature. SiO_2 doesn't have the same molecular structure, because Si does not form double bonds as readily as carbon does. Si–O form single bonds that lead to a network solid held together with strong, covalent bonds, so it is a solid that has a high melting point (*Chemistry* 7th ed. pages 446–448 / 8th ed. pages 459–461).

(c) Nitrogen can form three bonds (NF_3) but not five (NF_5) because it lacks *d* orbitals that are energetically available for the formation of hybrid orbitals (or alternatively, because it is too small to accommodate five atoms). Both NF_3 and PF_3 are trigonal pyramidal because the central atom has three bonding pairs and one lone pair of electrons (leading to sp^3 hybridization). PF_5 is trigonal bipyramidal because it has five bonding pairs (leading to dsp^3 hybridization) (*Chemistry* 7th ed. pages 367–377, 391–393, 397–398, 917–918 / 8th ed. pages 378–388, 404-407, 410-411, 932-933).

(d) Ca^{2+} and O^{2-} ions are attracted about four times more strongly than K^+ and F^- ions. Ions with a 2+ charge are attracted more strongly than ions with a 1+ charge. In addition, the calcium-to-oxygen distance is less than the potassium-to-fluoride distance, leading to an increased force of attraction for the shorter bond (*Chemistry* 7th ed. pages 344–346 / 8th ed. pages 355–357).

SCORING THE DIAGNOSTIC TEST FREE-RESPONSE QUESTIONS

It is difficult to come up with an exact score for this section of test. However, if you compare your answers to the answers in this book, remembering that each part of the test you answer correctly is worth points even if the other parts of the answer are incorrect (see the section titled "Scoring for the Free-Response Questions" on page 13 of this book), you can get a general idea of the percentage of the questions for which you would get credit. If you believe that you got at least one-third of the possible credit, you would probably receive a 3 on this part of the test. If you believe that you would receive close to half or more of the available credit, your score would more likely be a 4 or a 5.